Sustainable System of Farming for Modern Agriculture

NIPA GENX ELECTRONIC RESOURCES & SOLUTIONS P. LTD.
New Delhi-110 034

About the Authors

Prof. (Dr.) Rameshwar Dass Gupta, popularly abbreviated by R.D. Gupta, has the distinction to serve in three Indian Agricultural Universities namely PAU Ludhiana, CSKHPKV, Palampur and SKUAST, J& K Srinagar / Jammu as diligent extension activist, high profile researcher and a commanding teacher in his field of "Soil Microbiology" and "Pedology". He is also unique who has worked under the able guidance of Dr. K.K. Jha, Ex-Vice Chancellor, Rajindra Agricultural University, Samastipur, Bihar, Dr. B.R. Tripathi, Ex-Vice Chancellor, Narendra Dev University of Agricultural and Technology, Faizabad, Uttar Pradesh as well as Dr. N.S. Randhawa and Dr. G. Dev, Ex-DDG, ICAR and Ex-Director Research, CSKHPKV Palampur /Indian Potash Research Institute respectively.

Dr. Gupta is a leading soil scientist and noted environmentalist. Besides having specialized in Soil Microbiology and Pedology, he holds a postgraduate diploma in Ecology and Environment from Indian Institute of Environment, New Delhi with a specialization in "Wildlife of Pir Panjal Himalayan region of Jammu and Kashmir. Dr R.D. Gupta, is one of the pioneer worker whose Ph.D. Agriculture and M.Sc. Agriculture's theses related to the soils of Kangra District, Himachal Pradesh and Jammu and Kashmir which he submitted for earning these degrees from CSKHPKV, Palampur and Ranchi University during 1980 and 1968. He did B.Sc. Agriculture in 1965 from Jamnu and Kashmir University.

Dr. R.D. Gupta started his professional career as an Assistant Extension Specilist (Soil Science) in PAU, Ludhiana during December 1968, promoted as Assistant Scientist in CSKHPKV, Palampur. After completing Ph.D. during 1980 as an in service candidate, he was elevated to Extension Specialist (Soil Science) and posted to Regional Agriculture Research Station Dhaulakuan, Sirmour under the first newly created Krishi Vigyan Kendra (KVK) in May 1983. Thus, Dr. Gupta was the founder of first KVK in Himachal Pradesh.

During February 1985, Dr. Gupta joined as Deputy Director Extension and rose to the position of Head at RARS, R.S. Pura in 1992 Cum Chief scientist of first KVK (Jammu district) of the Jammu region. During 2000 he was appointed as Associate Dean. Infact, he was the founding Associate Dean Faculty of Agriculture of the newly created SKUAST for Jammu who was assigned to look after the work of all research stations of the Jammu region including teaching work of B.Sc. Agriculture as well the Head of the Division of Agricultural Chemistry and Soil Science and Agroforestry.

He has over 150 peered publications like research papers, review in different book chapters and five books to his credit namely "Problems and Management of Soil and Forest Resources of Northwest Himalayas (1991) "Environment and Degradation of Jammu and Kashmir Himalayas and their Control (2005)" "Environment Pollution: Hazards and Control (2006)" "Glimpses of North West

Himalayas, 2009" "Integrated Nutrient Management (INM) in a Sustainable Rice – Wheat Cropping System as Co-author, 2009 with Dr Anil Mahajan" "Wildlife and Wetland Ecosystems of Jammu and Kashmir Himalayas, 2016 and "Agro techniques and Uses of Medicinal Plants" in Collaboration with Dr S.K. Gupta and Dr. S.D. Bhardwaj, 2016.

Dr. R.D. Gupta has guided a number of M.Sc. and Ph.D. students. He has been on the Advisory Board for various scientific bodies and is the resource personnel of a number of organizations like J&K Pollution Control, Board Rural Development Departments, Soil Conservation and Training Schools. He is a life member of more than a dozen of professional societies and has been regularly participating in various programmes on Radio Kashmir Jammu and also on Doordarshan. Above all he also contributes articles relating to Agriculture & Environment, in local newspapers like. The Kashmir Times, The Daily Excelsior, The State Times, The Greater Kashmir.

Prof. (Dr.) R.D. Gupta is the Vice President of Paryavaran (Environment) Sanstha of J&K. He has remained paper setter of Kashmir Administrative Service (KAS) and also that Agricultural Research Service / National Eligibility Test / Examination of Soil Science and Pedology under ICAR, New Delhi, India. It is remarkable that Prof. (Dr) R.D. Gupta, has obtained a number of Professional Awards.

Dr. S.K. Gupta, born in 1959, is currently working as professor and Head/Chief scientist (Oilseeds) in the Division of Plant Breeding and Genetics, SK University of Agricultural Sciences and Technology, FOA, Chatha, Jammu, India, and holds a brilliant academic and service record. For almost two decades he has devoted his research interests to the area of oilseed brassicas.

Dr. Gupta obtained his post-graduate degrees (M.Sc., Ph.D.) from Punjab Agricultural University, Ludhiana, India, in 1984 and 1987, respectively. He is the recipient of a post-doctoral fellowship in plant biotechnology, and has published more than 100 research papers in esteemed national and international journals, mostly on Brassicas. He has already developed eight varieties of rapeseed- mustard. In addition, he has written two books on plant breeding and edited eight books on *Rapeseed Breeding-Advances in Botanical Research,* Vol. 45, Academic Press, Elsevier Publishers. For his excellent scientific endeavors, he has been conferred with the Young Scientist Award: 1993-1994 by the State Department of Science and Technology.

Dr. Anil Mahajan completed his B.Sc. Agriculture (with specialization in Plant Protection) in June 1999, M.Sc. Agriculture (Soil Chemistry and Fertility) in December 2001 and Ph.D. Agriculture (Soil Physics) in January 2006 with distinction from Chaudhary Sarwan Kumar Himachal Pradesh Krishi Vishvavidyalaya a Diploma in Computer Applications from Institute of Computer Education and Sciences Training, Palampur. It is pertinent to mention that he has studied minor subjects such as Agricultural Engineering and Agronomy during his Ph.D.

Agriculture, and Agronomy during his M.Sc. Agriculture. He has worked as Senior Research Fellow (from 27 July 2005 to 31 August 2007) under the Indian Council of Agricultural Research (ICAR), Government of India, in a Geographic Information System-based research project entitled "Developing Mountain Agriculture System Information Files for Planning Niche-based Agriculture Developing in Kangra and Mandi Districts of Himachal Pradesh' at Centre for Geo-informatics Research and Training, CSK, HPKV, Palampur. He has also worked as Research Associate under ICAR-aided research project, 'Modeling Impact and Adaptation for Major Crops in Himachal Pradesh' (from 1 September to 2 December 2007) at Centre for Geo-informatics Research and Training, CSK HPKV, Palampur, and as Research Associate at Horticulture Technology Mission Mode Project, 'Integrated Nutrient Management for Major Fruit Crops of Kandi Region of Jammu' (from 3 December 2007 to 3 January 2008) in the Division of Soil Science and Agricultural Chemistry, Sher-e-Kashmir University of Agricultural Sciences and Technology (SKUAST), Jammu. Presently he is working as Territory Manager, Business Development for rice crop (from 4 January 2008 till date) in a German Multinational Company, Bayer Bioscience Pvt Ltd, (wholly owned subsidiary of Bayer Crop Science Pvt. Ltd., India).

Dr. Anil Mahajan has done remarkable research work in the rice-wheat cropping system during his doctoral programme, and also in rice, maize and vegetables in master's research work. He was a recipient for the University Merit Scholarship during his B.Sc. and Ph.D. programmes, and was awarded Honours Certificate in M.Sc. degree and Best Poster Award at National Workshop on Natural Resource Management for Sustainable Agriculture organized by SKUAST, Jammu, and Soil Conservation Society of Indian in November 2006. He has 30 publications (research review and extension papers) to his credit and has attended several national and international training sessions and workshops. It needs to mention that he twice qualified ICAR National Eligibility Test conducted by Agricultural Scientists Recruitment Board, New Delhi, in the following professional subject: Soil Science: Soil Chemistry/Fertility/Microbiology and Soil Science – Soil Physics, Soil and Water Conservation.

Authors

Sustainable System of Farming for Modern Agriculture

R.D. Gupta
S.K. Gupta
Anil Mahajan

NIPA GENX ELECTRONIC RESOURCES & SOLUTIONS P. LTD.
New Delhi-110 034

NIPA GENX ELECTRONIC RESOURCES & SOLUTIONS P. LTD.

101,103, Vikas Surya Plaza, CU Block
L.S.C.Market, Pitam Pura, New Delhi-110 034
Ph : +91 11 27341616, 27341717, 27341718
E-mail:newindiapublishingagency@gmail.com
www: www.nipabooks.com

For customer assistance, please contact
Phone: + 91-11-27 34 17 17 Fax: + 91-11- 27 34 16 16
E-Mail: feedbacks@nipabooks.com

ISBN: 978-93-90175-09-3

Composed and Designed by NIPA.

Dedicated to

Late Shrimati Mohini Kumari

(Retired Senior Teacher)

W/O Prof. (Dr.) R.D. Gupta

Former Associate, Dean cum Chief Scientist

KVK District Jammu

SKUAST, Jammu, J&K State, India

Foreword

Inspite of attaining rapid growth in industrial and service sectors during the last few decades, agriculture still remains the mainstay for Indian economy accounting for about 14 per cent in Gross Domestic Product. On an average more than 70 per cent of the population of the country depends upon directly or indirectly on agriculture for their livelihood. On an average, India requires 250 million tonnes of food grains, besides other agro-based commodities to feed the current population. By 2030 India will have to produce almost 30 per cent more food grains than the present i.e., more than 300 million tonnes to feed an estimated population of 1500 million (1.5 billion). Indian agricultural production has been exhibited a very healthy trend in the past particularly during Green Revolution which ushered during 1967-68 through the increase in rice and wheat crop yields - the two main crops amenable for bringing Green revolution under irrigated conditions, using their high yielding varieties, responding to more consumption of chemical fertilizers and pesticides. However, later on food production registered a decline during 1984-1993. This declining trend is still going on. Moreover, the declining interest of rural youth is an account of farming in most households not being viable. Over 45 per cent of the farmers interviewed by the National Sample Survey Organization wanted to leave farming. Over half of the farmers are in debt and are committing suicide. The average size of land holding has shrunk to about a half-hectare. The uneconomical farm holding is further exacerbated by deteriorating soil health and dependency on rains. There are more risks from climate change, natural disasters, pest and diseases. Poor access to the rain-fed agricultural technology which still contributes about 42 per cent of the total Indian food grain production, natural disaster, pests and diseases etc.

Diminishing response of crops to applied plant nutrients, flip-flop policies for fertilizer sector, distorting NPK use ratio, availability of adequate feed stocks and its rising prices and interest burden are some major problems or issues of great concern.

It is emerged from the foregoing literature that we must introduce transformative, innovative knowledge vis-a-vis intensive production techniques for both older and younger farmers to address the sustainable food production challenge. Such kind of information stands incorporated in a book having a title, "*Sustainable System of Farming for Modern Agriculture*", written by Dr. R.D. Gupta, Dr. S.K. Gupta and Dr. Anil Mahajan. The book consists of 18 chapters, covering a number of topics concerning with sustainable agriculture, which is the need of the hour. As for examples in Chapter first, the information relating to various agricultural technologies which helped in bringing about Green Revolution in India and signs of Green Revolution plateauing off and poisoning of ecosystem, whereas second chapter deals with "Concept of sustainable agriculture" covering its definition, aims and objectives, and sustainable issues. Chapter three and four tell about soil tillage, its components for sustaining crop productivity and a water key to agriculture sustainability respectively. Chapters five and six deal/ relate to watershed management and organic farming and how they help in sustaining agriculture. Role of integrated nutrient management and bio-fertilizers has been given in chapters seven and eight with regard to sustainable agriculture. Similarly, the chapters nine and ten deal with the sustainability of agriculture through application of vermicompost and organic pesticides usage for checking pests. Chapter eleven comprises of rural women and their role in sustainable agriculture and chapter twelve relates to agroforestry system in sustaining agriculture, respectively.

Chapters thirteen and fourteen speak about cover crops and indigenous technology and their feats in sustainable agriculture. Management of wastelands in respect of agriculture sustainability and some alternatives to sustain productivity of Jhum cultivation stand discussed in chapters fifteen and sixteen. Importance of micro-nutrients and application of extension system approach for sustaining food grain production have been dealt with in chapters seventeen, eighteen respectively.

Thus, the present text book written by Dr. R.D. Gupta, Dr. S.K. Gupta and Dr. Anil Mahajan, having title, "*Sustainable System of Farming for Modern Agriculture*" will prove a boon for the scientists, extension personnels, teachers and students, who are involved in conducting applied and basic research in various aspects of agricultural production on sustainable basis.

As sustainable agriculture improves fertility of the soil and its structure, which in the long run help in increasing productivity, and also prevents soil erosion, one of the major problem in many parts of the country so the present text book written by Dr. R.D. Gupta, Dr. S.K. Gupta and Dr. Anil Mahajan will definitely tackle in solving aforesaid problems.

Pradeep Kumar Sharma
Former, Vice Chancellor
Sher-e- Kashmir University Agricultural Science and Technology
Chatha, Jammu, J&K State, India

Acknowledgement

We the authors of the book entitled "*Sustainable System of Farming for Modern Agriculture*" express our sincere thanks and fully acknowledge various scientists, writers and authors whose work has been referred to in this manuscript from various Journals, Magazines, Newspapers etc.

Our heartfelt thanks are due to worthy Vice Chancellor, Professor Pradeep Kumar Sharma for writing Foreword.

The first author of the book i.e. Prof. (Dr.) R.D. Gupta is deeply indebted to his wife Late Mrs. Mohini Kumari Ex-senior Teacher, retired from Education Department of Jammu, Jammu and Kashmir ….. who always inspired me for writing research papers, extension articles vis-à-vis chapters for the books and text books

Prof (Dr.) R.D. Gupta
Dr. S.K. Gupta
Dr. Anil Mahajan

Preface

Grow "more food campaign" which was later on designated as "Green Revolution". In India, the "Green Revolution", which came to fore during 1967-68, resulted from introduction of high yielding varieties of rice and wheat crops. High yielding varieties of rice and wheat responded to more use of fertilizers, water and pesticides. This Green Revolution gave an impression of grand success for 2-3 decades. Thereafter, it gradually became a costly and unfriendly affair to environment and society. Unable to fulfill the input and output requirements of future agriculture, it gradually affected the productivity of agriculture regenerative capacity of natural resources like soil, sharp decline in population of beneficial flora both macro and micro as well as macro and micro fauna, there was deficiency of macro and micronutrients viz; S, Ca, Mg, Zn, Fe, B etc in soils.

The chemical farming has also been started to affect the life of producers, farmers, by several diseases (cancer, birth of deformed children etc). Since the end of 1990s, increased incidence of farmers suicides in India has been the most dramatic outcome of the hopelessness faced by many farmers. It is attributed to a combination of factors such as high inputs prices, crop failures, due to untimely rains, indebtedness etc. An estimate has shown that 27 per cent of Indian farmers did not like to farming because it was not profitable. In all 40 per cent of the farmers felt that given a choice, they would take up some other careers/professions.

During late 1990s, farmers world over realized the adverse effects of conventional farming and started their own efforts to develop a sustainable system of farming. At the same time, the greatest challenge now being faced by the Indian agriculture is to produce more from shrinking land resources to meet food, fodder, fibre, fuel and raw materials. In view of almost a billion people in Asia including India and Africa who are plagued by hunger, can be solved by adopting a new vision i.e; sustainable agriculture. Thus, in the present publication compilation of various kinds of sustainable agricultural technologies of the country are thoughts of various kinds of sustainable agricultural practices prevailing in different states of the country are discussed in eighteen chapters of the book having title. Towards a "*Sustainable System of Farming for Modern*

Agriculture. The first Chapter relates on Introduction and 2nd Chapter deals with "Concept of Sustainable Agriculture", covering its definitions, aims and objectives, sustainability issues, national response to climate change. Chapter three is related with soil tillage, its components for sustaining crop productivity which is characterized by its definition, classification, soil tilth, objectives of tillage, reasons for adopting conservation tillage and its prospects in sustainable agriculture as well as economics and benefits for certain cropping systems. Chapter four is associated with water-A key to agriculture sustainability and food security states about faulty irrigation, improving irrigation efficiency through using good quality seed, increasing soil fertility, improving on farm water/irrigation management, rain water harvesting technology, ways to tide over scarcity of water, revival of ground water resources. The chapter five is on water shed, its concept and management, problems of watershed and factors affecting its operation, causes of water deterioration, micro watershed its plant, management of watershed development and its objectives and success obtained in certain watershed management of Uttrakhand and Jammu and Kashmir. Chapter six is about organic farming for sustainable agriculture which speaks about its concept including definition, principles and benefits, scope of organic farming, its Indian and world scenario, traditional values of organic farming and its place in Vedic period, options of organic farming and its constraints vis-a-vis its various methods. Chapter seven tells about Integrated nutrient management (INM) system and sustainable agriculture which includes definition of INM and its need and components i.e; organic manures and chemical fertilizers as well as its advantages. Chapter eight consists of bio-fertilizers and their role in sustainable agriculture, all bioferlizers viz; Rhizobium, Azotobacterim and Blue Green Algae cultures Azolla and Mycorrhizal fungi and their application in suitable crops along with advantages stands discussed. Besides, compatibility of certain pesticides with biofertilizers, ecological and physiological considerations benefits of biofertilizers, have also been discussed.

Chapter nine "Vermin-composting/Earthworms: Role in sustainable agriculture," tells a brief about earthworms, their origin and ecology, origin of vermi-compost and its necessity to use, vermin-technology for organic farming through CAPART, Vermi-composting, humification maintenance of soil fertility and improvement in physical and chemical properties, control of soil erosion and reclamation of water logged soil by using Vermi-compost, enhancing of biological nitrogen fixing activities and plant growth by using Vermicompost/vermiwash, uses of vericompost on crops, use of earthwonns as poultry feed, preparation of vermicompost are the other topics which deal with.

Chapter ten "Checking crop pests naturally a novel way to sustain agriculture", mentions about pesticide usage trend, public issues, pesticides residues in human blood and occurrence of disease, contamination of food stuffs, pollution of soil, water and air, pesticide resurgence and poisoning, falling prey of children to pesticides and control measure, use of biopesticides, use of cow's urine and dung, biological control technique, natural pest control measures, judicious use of fertilizers, organic soil amendments, integrated pest management and its aims and objectives. Chapter eleven comprises of "Rural Women and their role in sustainable agriculture", throws light on women and origin of agriculture, myth of green revolution, agriculture models devised by women, farm women and crop diversification and their forest products gathering activities, women and farming system approach, biotechnology and women's empowerment, and other activities, livestock rearing, agro-forestry biogas energy development programme, integrated nutrient management, home and land management as well as environmental conservation. "Role of agroforestry in sustainable agriculture stands mentioned in chapter twelve in terms of the necessity of agroforestry its concept, agroforestry systems, macro and micro nutrient dynamics in leaf letter, decomposition of agro-forest tree. Chapter thirteen, "Covers on cover crops and sustainability of agriculture", devotes on "what are cover crops", advantages of growing cover crops, importance of legumes in cover crops, characteristics of cover crops and their constraints / limitations. Chapter fourteen i.e., "Indigenous technological knowledge: Concept and importance in sustainable agriculture deals with, "what is indigenous technology, its concept and advantages, importance of extension methods of integrated pest management, eco-friendly botanical pesticides, traditional use of organic manures, indigenous fodder tree species, use of organics in ginger cultivation, use of ash to meet the deficiency of potash and cows urine to combat crop pests and diseases. In Chapter fifteen "management of waste lands" has been highlighted in respect of its concept including definition, limitations of wasteland definitions, what is degraded land and classification of wastelands extent of land degradation and its categories, factors responsible for turning productive lands into wastelands, management of wastelands, extent and location of wastelands in Jammu and Kashmir. In Chapter sixteen "Some alternatives to sustain productivity of "Jhum" cultivation" throws light on,"what constitutes shifting cultivation and its classification air pollution and loss of flora and fauna, initiatives and strategies for checking "Jhum" cultivation, task force and constitution, setting up of agro-forestry, terrace cultivation contour cultivation, myths and realities of Jhum cultivation as well as its constraints. In Chapter seventeen "importance of micronutrients in sustainable agriculture" has been given with respect to cause of micronutrients deficiency and extent of micronutrients deficiency in soils, use of zn-amended organic manure, materials for Zinc fertilizers, Boron (B) and its response to various crops effect of B in cropping system, organic manures and B-use efficiency, Fe-deficiency and effect of Fe-carrier and compost on yield, carriers of Fe and rate of Fe-application, Mn and its efficient use, Mn-efficient crop cultivars, use of Cu and Mo in crops, management of micronutrients, role of organics in micronutrient management, chelates and chelating effect of humic substances Eighteen chapter "Agricultural production on sustainable basis through extension system" explains what is meant by system approach, guidelines for trainings and their importance, points to make training programme more effective, use of various training methods utilization of trainings experience training needs, diverstification agriculture.

Authors

Contents

Acronyms

Indian Acronyms Related with Agriculture

AADF : Associated Agricultural Development Foundation, New Delhi
ABMP : Association of Basic Manufacturers of Pesticides, Mumbai
ACPC : Agricultural Costs and Prices Commission, New Delhi
AISLUS : All India Soil and Land-Use Survey (IARI), New Delhi
APEDA : Agricultural and Processed Food Products Export Development Authority New Delhi
ASRB : Agricultural Scientist Recruitment Board, New Delhi
CARI (ANGI) : Central Agricultural Research Institute for Andaman and Nicobar Group of Islands, Port Blair
CAZRI : Central Arid Zone Research Institute, Jodhpur
CCI : Cotton Corporation of India Limited, New Delhi
CCRI : Central Coffee Research Institute, Chikmaglur
CGRT : Centre for Geo-informatics Research and Training, Palampur (Kangra) Himachal Pradesh
CIAF : Central Institute of Agricultural Engineering, Bhopal
CICR : Central institute for Cotton Research, Nagpur
CIRCOT : Central Institute of Research on Cotton Technology (ICAR), Mumbai
CMRS : Central Mango Research Station, Lucknow
CPCRI : Central Plantation Crops Research Institute, Kasaragad
CPRI : Central Potato Research Institute, Shimla
CRIDA : Central Rainfed Institute for Dryland Agriculture, Hyderabad
CRLRRS : Central Rainfed Lowland Rice Research Station, Cuttack
CRRI : Central Rice Research Institute, Cuttack
CRSARD : Centre for Research on Sustainable Agricultural and Rural Development Madras
CRURRS : Central Rainfed Upland Rice Research Station, Hazaribag
CSMRS : Central Salt and Materials Research Station
CSSRI : Central Soil Salinity Research Institute. Karnal
CSWCRTI : Central Soil and Water Conservation Research and Training Institute Dehradun
CTCRI : Central Tuber Crops Research Institute, Thiruvanathapuram
CTRI : Central Tobacco Research Institute, Rajamundry
DARE : Department of Agricultural Research and Education, New Delhi
DPPQ&S : Directorate of Plant Protection, Quarantine and Storage, New Delhi
FAI : Fertilizer Association of India, New Delhi
FCI : Fertilizer Corporation of India, New Delhi
FCI : Food Corporation of India, New Delhi
FRI : Forest Research Institute, Dehradun
IARI : Indian Agricultural Research Institute, New Delhi
IASRI : Indian Agricultural Statistics Research Institute, New Delhi
ICAR : Indian Council of Agricultural Research, New Delhi

ICCC	:	Indian Central Coconut Committee, Kozhikode
ICCC	:	Indian Central Cotton Committee, Mumbai
ICJC	:	Indian Central Jute Committee, Calcutta
ICOC	:	Indian Central Oilseeds Committee, Hyderabad
ICRI	:	Indian Cardamom Research Institute, Myladumpara
ICSCC	:	Indian Central Spices and Cashewnut Committee
ICSC	:	Indian Central Sugarcane Committee, New Delhi
ICTC	:	Indian Central Tobacco Committee, Madras
IFFCO	:	Indian Farmers Fertilizer Co-operative Limited, New Delhi
IGFRI	:	Indian Grassland and Fodder Research institute, Jhansi
IIAER	:	Indian Institute of Agricultural Economics Research, New Delhi
IIHR	:	Indian Institute of Horticultural Research, Bangalore
IISR	:	Indian Institute of Sugarcane Research, Lucknow
IISS	:	Indian Institute of Soil Science, Bhopal
JARI	:	Jute Agricultural Research Institute, Barrackpore
JTRL	:	Jute Technical Research Laboratory, Calcutta.
KRIBHCO	:	Krishak Bharati Co-operative Limited, Surat
MAIDC	:	Maharashtra Agro-Industries Development Corporation Limited, Mumbai
MARKFED	:	The Punjab State Co-operative Supply and Marketing Federation Limited Chandigarh
MSSRF	:	M.S. Swaminathan Research Foundation, Madras
NAARM	:	National Academy of Agricultural Research and Management, Hyderabad
NAAS	:	National Academy of Agricultural Sciences, New Delhi
NABARD	:	National Bank of Agriculture and Rural Development, Mumbai
NAFED	:	National Agricultural Co-operative and Marketing Federation of India Limited, New Delhi
NAPL	:	National Agro Products Limited, Calcutta
NBPGR	:	National Bureau of Plant Genetic Resources. New Delhi
NBSS-LUP	:	National Bureau of Soil Survey and Land-Use Planning, Nagpur
NFL	:	National Fertilizers Limited, New Delhi
NRCA	:	National Research Centre for Agro-forestry, Jhansi
NRCC	:	National Research Centre for Cashew, Puttur
NRCG	:	National Research Centre for Citrus, Nagpur
NRCS	:	National Research Centre for Groundnut, Junagarh
NRCWS	:	National Research Centre for Sorghum, Hyderabad
NRCS	:	National Research Centre for Soybean, Indore
NRCS	:	National Research Centre for Spices, Kozhikode
NRCWS	:	National Research Centre for Weed Science, Jabalpur
PNFC	:	Punjab National Fertilizers and Chemicals Limited, Naya Nangal Chandigarh
RRII	:	Rubber Research Institute of India (Rubber Board), Kottayam
RCNER	:	Research Complex for North-Eastern Hills Region, Shilong
SBI	:	Sugarcane Breeding Institute, Coimbatore
SFAC	:	Small Farmers Agri-business consortium, New Delhi
SWCRDT	:	Soil and water Conservation Research, Demonstration and Training Centres in India
TFRI	:	Tropical Forest Research Institute, Jabalpur
WTC	:	Water Technology Centre (IARI), New Delhi

Glossary

A

Agro-silviculture: A land use system in which agricultural producing crops including herbs and trees both forests and fruits or shrubs are planted together.

Air pollution: Contamination of the air with undesirable solids, liquids and gases or addition of any foreign substance that disturbs the normal composition of the air. The substances that cause air pollution are called pollutants.

Algae: A group of unicellular or multicellular plants containing chlorophyll.

Alley cropping: Growing of annual crops in spaces between the rows of trees or shrubs, often legumes are preferred as they tolerate heavy and regular coppicing. The leafy and woody materials are used as mulch.

Amelioration: Addition of any chemical substance to the soil which can ameliorate or change the soil reaction.

Aminization: The process by which heterotrophs (bacteria, fungi and actinomycetes) break down complex organic nitrogenous compounds or molecules releasing amines and amino acids is known as aminization.

Ammonification: It is a biochemical process by which organic nitrogenous compounds i.e amines and aminoacids are converted into ammonical nitrogen form.

Ammonifying bacteria: Those bacteria which are involved in decomposition of organic nitrogenous compounds into ammonia belong to ammonifying bacteria.

Anaerobic: An organism that survives in the absence of air or molecular oxygen.

Asymbiotic nitrogen fixation: Fixation of molecular nitrogen by free-living organisms like *Clostridium and Azatobacter.*

Available moisture: Moisture range or limit available to plants which lies between field capacity and wilting point of a soil.

Available nitrogen: A small proportion of relatively large amount of nitrogen present in the rooting zone of soil which is directly available to plants. It is mainly in the form of NO_3-N present in soil solution and the dissolved and adsorber: NH_4.

Available nutrients: The proportion of any element or compound in the soil that can be readily absorbed and assimilated by growing plants.

Available phosphoric acid: The part which is soluble in water or in **a** weak dilute acid such as 2 per cent citric acid.

Available water: The amount of water which lies between field capacity and permanent wilting point for plant use. The availability of water to plants is a function of soil-water-plant system rather than soil-water system alone.

Azolla: Azolla, a floating water fern possessing a symbiotic relationship with blue green algae (cynobacteria) is capable a fixing atmospheric nitrogen in the soil to the extent of 150 to 200 Kg ha'[1].

Azospirillum: These are the bacteria which remain in loose association with a number of crops like maize, pearlmillet, sorghum.

Azotobacter: They are aerobic nitrogen fixing bacteria which fix atmospheric nitrogen asymbiotically i.e; without any partnership.

B

Bacteria: Bacteria are unicellular microscopic organisms which are very important for decomposing of soil organic matter which in turn supply nutrients to plants and other organisms.

Bactericide: Any substance or compound which can kill bacteria is known as bactericide

Basic slag: Basic slag is obtained as a bye product in the manufacture of steel from pig iron. Basic slag contains 8 to 10 per cent phosphorus and iron oxide about 15 to 18 per cent and calcium silicate 6 to 8 per cent.

Biodynamic farming: It is a holistic system of agriculture devised by Rudolo Steiner that seeks to connect nature with cosmic creative force. An attempt was made to create a whole farm organism in harmony with its habitat. Compost and special preparations (e.g. plant derived sprays) are preferred to use, whereas use of chemical fertilizers and pesticides is restricted.

Bio-fertilizers: Bio-fertilizers or microbial inoculants, are preparations of living organisms in alone or in combination which may help in increasing crop productivity by way of fixing atmospheric nitrogen in soils, solubilizing of insoluble plant nutrients, stimulating plant growth or decomposing of plant residues.

Biogas slurry: The residual slurry that comes out of the digestion tank of the biogas plant. On an average, it contains 1.8 per cent N, 1.1 per cent P_2O_5 and 1.5 per cent K_2O.

Biological control: Use of natural enemies to control pests including both check with improved natural agents and augmentation and conservation of natural enemies through manipulation of the pest host, the environment and/or enemies themselves.

Biomass: The weight of the material produced by a living organism or group of orgnnisms, plant or animal; here, annual biomass refers to annual plant species and perennial biomass to perennial plant species.

Biotic: It pertains to life, whereas abiotic refers to without life.

Blood meal: A by-product of slaughter house which contains 10 to 12 per cent highly available nitrogen, 1 to 1.5 per cent phosphorous and 1.0 per cent potassium is called blood meal. It is very quick acting manure and is effective on all crops and on all types of soils.

Blue green algae: A heterogeneous group of prokaryotic photosynthetic organisms which contain chlorophyll 'a'. They are unicellular non-symbiotic, free-living nitrogen fixers which grow very well in waterlogged soils like rice fields. Besides fixing atmospheric N, BGA secrete vitamins and growth promoting substances which contribute to better growth of the rice plants. They are also called as cyanobacteria and belong to the class Cyanophyceae or Myxophyceae.

Biological nitrogen fixation: Fixation of atmospheric nitrogen by the living organisms for making its availability to the growing plants. This usually occurs in the soils.

Bone meal: It is an organic fertilizer. Bone meal is used as a manure from time immemorial chiefly as a source of phosphorus and partly also for nitrogen.

Botanical pesticide: Chemicals which behave like synthetic pesticides but owe their origin to plant kingdom are called botanical pesticides.

Bulk density: The ratio of the mass of the soil solid (oven dried soil) to the bulk volume of the soil and determined by Core Sampler method. On an average, the value of the bulk density of general soil is 1.33 Mg m^{-3}. It is always smaller than its particle density, the average value of which 2.65 Mg m^{-3}.

Bulky organic manures: Manures prepared artificially from plant residues and animal waste products are called bulky organic manures. It contains very small quantities of plant nutrients and large quantities of organic matter. Precisely speaking bulky organic manure is one that is the result of some natural product such as farm yard manure, compost, sea birds excreta or guano.

Bund: A ridge of earth placed in a line along the contour of a slope to control water runoff and soil erosion is known as bund.

C

C:N ratio: The ratio of the weight of organic carbon present in soil to the weight of total N of that soil is called C: N ratio. C: N ratio to the extent of 10:1 is found to be most suitable for the availability of nutrients to the mineral soil.

Calcareous soil: A soil containing sufficient amount of free calcium carbonate often mixed with magnesium carbonate and there is visible effervescence when treated with 0.1 N hydrochloric acid (cold) is a calcareous soil.

Carbohydrates: A large group of compounds containing carbon, hydrogen and oxygen only, of the general formula $C_x(H_2O)_y$. Carbohydrates contain CO and H in the same ratio as is found in water. They are produced by plant photosynthesis and are essential to metabolism in all living organisms.

Carrier: A substance which has high organic matter, higher water holding capacity and supports the growth of organism. Generally lignite, charcoal, peat etc. are suitable carrier materials for bio-fertilizers.

Cation exchange capacity: The sum total of exchangeable cations that a soil can adsorb is known by the term cation exchange capacity (CEC).

Chelates: The word chelate is derived from the Greek word, Chela meaning a crabs law. They contain organo-metallic molecules of varying sizes and shapes in which the functional groups of organic matter bind the nutrients in a ring like structure by means of multiple chemical bonds.

Chlorophyll: A light trapping green coloring matter or pigment which is essential for photosynthesis.

Climate: It is an aggregate of weather conditions over a long period of time.

Clod: A natural lump of soil that exists in an isolated entity in the field or formed artificially such as soil disturbances - plowing, digging etc.

Compaction: The increase in bulk density of a soil due to mechanical forces which is caused by the use of heavy agricultural machines.

Compost: Compost is manure similar in properties to farm yard manure, and is made by the decomposition of plant residues mainly, plus urine and dung under action of micro-organisms. On an average, it contains 1.01 per cent N, 0.5 per cent P2O5 and 0.8-0.9 per cent K_2O.

Composting: A biological process by which micro-organisms decompose organic matter during compost or farm yard manure making is known as composting.

Concentrated organic manures: Organic manure which is organic in nature and is made from raw materials of animal or plant origin. It contains higher percentages of primary nutrients as compared to bulky organic manures.

Conventional tillage: It is the traditional tillage system, which combines both primary and secondary tillage operations for growing a given crop in a given geographical area.

Cover crop: It is an annual crop sown to create a favourable soil microclimate, decrease evaporation, and protect soil from erosion, cover crops also produce biomass which can be used for soil fertility management.

Crop production: It is an exploitation of plant morphological (or structural) and plant physiology (or functional) responses within a soil and atmospheric environment to produce a high yield per unit area of land or it is basically conversion of environmental inputs like solar energy, CO_2, water and soil nutrients into economic products in the form of human or animal food or industrial raw materials.

Crop residues: The portion of the crop or plant left in the field after harvesting of the crop, which is not used domestically or sold commercially but can be utilized for compost making is the crop residue. On an average, it contains 0.5 per cent N, 0.6 per cent P_2O_5 and 1.5 per cent K_2O.

Crop rotation: Crop rotation is an order in which the chosen cultivated crops follow one another in a set cycle on the same field over a definite period for their growth.

Crop: Plants sown and harvested by man for economic purposes are known as crops.

Cropping intensity: Cropping intensity is worked out by multiplying the number of crops grown in a rotation with hundred and dividing by the number of years taken to raise the same crops.

Cropping intensity (%) = (Total cropped area/net area)*100 cropping pattern: A sequence in which various crops are grown in a locality or a zone, which is influenced by agrc-dimatic and socio-economic factors prevailing in that zone.

Cropping scheme: A plan according to which crops are grown on individual plots of a farm during a given period of time with the object of obtaining maximum return from each crop without impairing the soil fertility. Thus, it is related to the most profitable use of resources, land, labour, capital and management.

Cropping system: A system in which various crops are grown together on a piece of land like mixed, inter or relay cropping, etc.

Crust: A hard or brittle layer or crust formed on the surface of many soils when become dry.

Cytokines: Large group of proteins, peptides or glycoproteins that are secreted by specific cells of immune system.

D

DDT (Dichloro-diphenyl - trichloroethane): A persistent organochlororine insecticide is the DDT.

Decomposition: The chemical breakdown of minerals and organic compounds into simpler compounds often accomplish with the aid of micro-organisms designates decomposition.

Deep tillage: It is the tillage system in which tillage is done to the deeper depths more than 20 cm with the purpose of conserving moisture in the monsoon season or to break hard and facilitate the plant roots into deeper layers for better crop performance.

Deficiency: It is state of needing something that is absent or unavailable.

Deforestation: Illicit cutting of forest trees either for arable land or to use the timber for construction or industrial purposes is the deforestation.

Denitrification: Denitrification indicates biochemical reduction of nitrate or nitrite to gaseous nitrogen, either as molecular nitrogen or as an oxide of nitrogen or the breakdown of nitrates by soil bacteria resulting into the release of free nitrogen.

Desertification: The degradation of dry lands caused by over cultivation, overgrazing and deforestation and results in soil exhaustion and soil erosion is called desertification. It decreases soil productivity, deprives the land of its vegetative cover, and indirectly affects areas, by causing floods, soil salinization deterioration of water quality, and silting of rivers, streams and reservoirs.

Developmant: Series of changes which an organism goes through during its life cycle, but it may equally be applied to individual organs, to tissues or even to cells. Plant development involves both growth and differentiation.

Disinfection: Killing of harmful and objectionable microorganisms in or on a medium by means of chemicals, heat or ultraviolet light is called disinfection.

Dispersion: The process of movement of fertilizers applied to the soil to short or long distance is called dispersion.

Diversified agriculture: It is a concept wherein more than one type of crops are raised to mitigate the risks of crop failure and price fluctuation of one crop. It is defined as expanding the number of crops in a region with the hope of increasing overall productivity.

Drainage: It is the removal of excess surface or ground water from land by means of surface or subsurface drains.

E

Earthworms: Earthworms are popularly known as annelids (oligochaeta), long, narrow, tubular, brown coloured body divided into a number of segments.

Ecosystem: Ecosystem is an interaction between living (plants and animals including humans) and nonliving (Land, air, water) environments including interactions between them and with their environment.

Ectomycorrhizae: A type of mycorrhizal relationship in which the fungi invade all cells of the root cortex. This association generally exists with evergreen trees and shrubs as well as some deciduous trees.

EDTA: EDTA is Ethylene diamine tetra acetic acid. It is a chelating agent.

Electrical conductivity: The property of a substance to transfer an electric charge (reciprocal of resistance) used for the measurement of the salt content of an extract from a soil when saturated with water. It is measured in milli mhos cm^{-1} or dsm^{-1} at 25°C.

Endomycorrhizae: It is a type of association between mycorrhizal fungi and roots of plant in which fungal hyphae infect the roots up to the cortical region by secreting cellulolytic enzymes. They are more beneficial to the deciduous shrubs and plants.

Endotoxin: A toxin which is generally released by the bacterial cells after their death is known as endotoxin.

Enriched compost: Compost fortified with the addition of fertilizers (urea, SSP, rock-phosphate) during its production to raise its nutrient content and narrow down the C: N ratio.

Environment: The surroundings in which anything exists is called an environment.

Enzyme: It is a catalytic agent, which is biological in nature and speeds up chemical reaction in a living system.

Erosion: The gradual displacement or disappearance of parts of a system under the influence of external factors e.g; erosion of soil by water or wind, erosion of indigenous knowledge, genetic erosion, is called erosion.

Extension: Here, extension refers to agricultural extension activities which disseminate research findings concerning with agriculture sector including horticulture, livestock rearing etc.

External Inputs: All the components which originate from outside the system (Farm, village, region, country) are called inputs. For example, external inputs in case of agriculture are seeds, chemical fertilizers, manures, pesticides, irrigation etc.

F

Farmyard manure: Farmyard manure (FYM) is general manure supplying all constituents of plant food although in low amount. It is a mixture of solid and liquid excreta of farm animals along with litter (bedding material) and left-over materials from roughages or fodder fed to the cattle. On an average, well-rotten FYM contains 0.5 per cent N, 0.2 per cent P2O5 and 0.5 per cent K_2O. It is the oldest and most popular manure used in India.

Farming system: An appropriate combination of farm enterprises viz. cropping systems, livestock, fisheries, forestry, poultry, and other means available to the farmer which are used to raise them for profitability. It interacts adequately with the environment without dislocating the ecological and socio-economic balance on one hand, and attempts to meet the national goals on the other.

Fauna: Animals of a particular region are designated by the term fauna.

Fertilizer use efficiency: The ratio of nutrient removed to nutrient applied and is expressed in per cent. Mathematically, it is computed as:

FUE = [(Nutrient uptake in treated plot - nutrient uptake in control)/ nutrient dose applied] x 100

Agronomically, it is grain yield produced per kg of fertilizer applied. In case of N, the FUE raises from 20 to 30 per cent in lowland rice and as high as 60 per cent in irrigated fields. Flowever in case of P, FUE is 15-20 per cent and that of K is 80 per cent.

Fertilizer: Any chemical material which is added to the soil or applied directly to crop foliage through sprinkling to supply plant nutrients (elements) for plant's growth and development is called fertilizer.

Field capacity: The percentage of water remaining in a soil two or three days after its having been saturated and after free drainage has practically ceased.

Fixation: A process in a soil by which certain chemical elements essential for plant growth are converted from a soluble or exchangeable form to a much less soluble or to a non-exchangeable form, e.g., phosphate fixation.

Flora: These are plant species of any region or these are the plants having life of a geological period.

Fluorosis: It is a disease which is caused by consumption of fluorine compounds.

Fog: It is a collection of tiny water droplets that float in the air. It is similar to clouds except that clouds do not touch the surface of the earth as fog does.

Food additive: Any substance, which is generally added to food in small quantity to ameliorate its flavour, texture and appearance is called food additive.

Food security: Food security may be defined as an access to food both physically and economically which fulfills people's dietary needs vis-a-vis their food preferences.

Fossils: Remains or traces of an organism that have been preserved in earth's crust by natural processes are spoken as fossils.

Free fixation or Azofication: Certain group of heterotrophic bacteria living independently of higher plants, which have the capability of synthesizing their body proteins from the soil atmospheric nitrogen. As they grow and function independently of higher plants, they are called non-symbiotic or free living organisms. The bacteria of the Azotobacter are aerobic and fix atmospheric nitrogen independently. This process is called free fixation of nitrogen or Azofication. Symbiotic Nitrogen Fixation. The most important nitrogen fixing bacteria are those which form nodules on the roots of leguminous plants. These are called Rhizobia and fix atmospheric nitrogen symbiotically i.e. bacteria get energy from host plants and in return for that fix atmospheric nitrogen in nodules. The so fixed nitrogen in the nodules reach the soil via ex-osmosis or after the death and decay of the nodules.

Fungi: These are tiny aerobic, heterotrophic microorganisms having no chlorophyll, many celled.

Fungicides: Any pesticide, which inhibits the growth of fungi (fungus-singular), is known as fungicides. Fungicides are generally used to control those fungi which cause diseases in plants.

G

Genetic resources: Plant and animal stocks with distinct inheritance characteristics use within an agroecosystem are called genetic resources.

Global warming: It is the rise in the average atmospheric temperatures to a level which influences the life forms on the earth surface due to phenomenon of greenhouse effect that was first reported by Swedish Chemist Arrhenius in 1896. The rise in the global temperature is due to greenhouse gases caused by natural and anthropogenic factors and this increase has been between 0.3°C and 0.8°C over the 20th century. It will not only lead to loss of life and property but also cause a major reshuffle in flora, fauna, human population and cropping pattern of the world.

Grain: An indehiscent simple fruit of a grass in which the seed coat is united with the pericarp.

Green manure: Green manure consists of a raising a quick growing leguminous or other crop and burying the same in soil before flowering stage, so as to add organic matter in the soil. Organic matter on decomposition provides all essential plant nutrients to the growing crops.

Green revolution: Growing of high yielding varieties of crops, especially rice and wheat which responded to high doses of fertilizers and pesticides with frequent use of irrigation water which increase the yield of the crops tremendously.

Growth: An irreversible process in which there is an increase in size, shape and volume of an organism.

H

Harrowing: A secondary tillage operation which pulverizes, smoothens and packs the soil in seedbed preparation or control weeds.

Harvesting: The operation of cutting, picking, plucking or digging, or combination of these for removing the useful part or economic part from the plants.

Herbicide: Any chemical substance that is used to destroy weeds or off plants.

High analysis fertilizers: Fertilizers containing a higher percentage of the plant nutrients and have an advantage in the cost of bagging, handling and transportation per unit of plant nutrient. Such fertilizers are urea (46% N), triple super phosphate (45 to 47% P_2O_5) and muriate of potash (60% K_2O).

Humidity: Amount of water vapour present in the atmosphere (air) and measured by instrument known as hygrometers is called humidity.

Humification: It is the process by which organic matter is humified into humus.

Humus: It is complete organic compound which is composed of humic acid, fulvic acid, humin and hymatomelanic acid etc.

Hybrid seed: It is the seed produced by crossing genetically dissimilar plants i.e. of different varieties or species.

Hydraulic conductivity: The ability of soil to conduct water and is expressed as length per unit time or an expression of the readiness with which a liquid such as water flows through a soil in response to a given potential gradient. It generally depends upon both the nature of the soil (the porous matrix) and the fluid (water) properties.

I

Immobilization: It is the conversion of an element from the inorganic to the organic form in microbial tissues or in plant tissues, thus rendering the element into unreadily available form to other organisms or to plants.

Indigenous: Occurring or living naturally in a specific area such as native plants or animals (opposite of exotic) to be differentiated from endogenous, which means having its origin within a specific area is the term which is denoted by indigenous.

Infiltration: The process of downward entry or movement of water into the soil is called infiltration.

Inhibitors: It is an opposite to catalysts as it retards or reduces rate of a chemical reaction.

Inoculation: It refers to use of bacterial culture or any other microbial culture to increase the crop production or to enhance decomposition process.

Inorganic fertilizers: These are fertilizers containing a mixture of chemical compounds which are added to the soil in order to improve its fertility, but can be harmful to the environment if used in large amounts and continuously.

Insecticide: Any pesticide which is used to kill insects is called insecticide.

Integrated nutrient management: Strategy to maintain and partially increase soil fertility for sustaining crop productivity using both organic and inorganic sources of plant nutrients is called by the word integrated nutrient management.

Integrated pest management: A strategy that consists of using both organic and inorganic pesticides to destroy the pests.

Intensive farming: A farming system in which maximum number of crops per year is grown on the same piece of land with an objective of obtaining maximum income per unit area.

Intercropping: It is the practice of growing two or more crops at the same time in the same field. Cropping is intensified in terms of both time and space.

Irrigation: The artificial supply of water applied to the soil for the purpose of supplying moisture essential to plant growth is called irrigation.

L

Landscape: All the features that distinguish one part of the earth's surface from another part is known by the word landscape.

Landslide: It is rapid movement down slope of a mass of soil, rock and debris. In other words landslide is a mass of material that has slipped down the hill.

Leaching: The process of removal of plant nutrients in soil by the passage of water through soil. This is one of the ways in which plant nutrients are lost from the soil and a primary step in the improvement of saline soils. Among the major nutrients, nitrogen is lost in large quantities by leaching through its $N0_3$ form.

Least Significant Difference (LSD): A value used to compare the differences among a set of averages as an aid to detect whether the differences are real or merely due to chance.

Legume: Any plant, tree shrub or herb that belongs to leguminose family and has the ability to fix atmospheric nitrogen, is known as legume.

Ley farming: Growing of food crops and grasses on the same piece of land, which serves as a pasture, is known as ley farming.

Lichen: It is an association of blue green algae or green algae with fungi.

Light soils: Coarse textured soils i.e. those soils which harbor more content of sand and or silt.

Lime: Most of the agricultural lime is calcium carbonate or a mixture of calcium carbonate and magnesium carbonate.

Litter: Uppermost layer of organic material on the surface of soil including leaves, twigs and flowers as well as freshly fallen leaves or slightly decomposed is known by the term litter.

Lodging: Falling of crop plants under the influence of heavy rains or wind.

Low analysis fertilizers: Fertilizers containing lower percentage of plant nutrients in them like Chilean nitrate (16% N) and single super phosphate (16 to 18% P_2O_5). They become very expensive because of higher unit cost and higher storage and handling costs.

Lowland rice: It refers to flooded rice grown on flat land with controlled irrigation. This is also known as irrigated rice or water-logged rice. Lowland rice is commonly flooded when seedlings are 25-30 days old.

M

Macro-nutrients: The nutrients which are required in large quantities such as N, P, K, Ca, Mg and S for the crop growth and development.

Manure: Any organic material, which is mainly prepared from the organic refuse from plants and animals, after its decomposition is called manure.

Manuring: Application of farm yard manure, compost or other organic material which is used to fertilize the soil is called manuring.

Mechanical impedance: Physical resistance offered by the soil matrix to the growing roots and measured by penetrometer is known as mechanical impedance.

Microbial insecticide: A class of insecticides which are produced naturally by certain kinds of microorganisms. For example, *Bacilius thuringiensis* bacterium produces a special type of substance which kills insects.

Microclimate: The temperature, sunlight, humidity and other climatic conditions in a small localized area e.g; in a field, in a stand of trees in the vicinity of a given plant or in the top soil, is the microclimate.

Micro-nutrients: Trace elements which are required in very minute quantities but without their availability the plants canot perform their growth and development functions such as Fe, Zn, Cu, Mn, B, Mo and Cl.

Micro-organisms: A microscopic form of life either plants (bacteria, actinomycetes, fungi) or animals (protozoa, nematodes), is called microorganism.

Mineral: A natural occurring, homogeneous element or inorganic substances found in rocks which have a definite chemical composition and an ordered geometric form is known as mineral.

Mineralization: It is the conversion of an element from an organic form to an inorganic state as a result of microbial decomposition.

Minimum tillage: Soil management practices which seek to minimize labour inputs and soil erosion to maintain soil moisture and to reduce soil disturbance and exposure is called the minimum tillage. Crop stubbles are left to act as mulch. It is also called conservation tillage or reduced tillage. In its most extreme form (Zero or no tillage), seeds are drilled directly into the otherwise undisturbed soil.

Mixed cropping: Growing of two or more crops simultaneously on the same piece of land by using the mixed seeds of various crops, without any definite row arrangement is called mixed cropping.

Mono-cropping: Growing of same sole crop on the same piece of land year after year is called monocropping.

Muck: Highly decomposed organic material in which the original plant parts cannot be recognized or it refers to the soil which is well mixed with high percentage of organic matter.

Mulch: It is a loose covering on the surface of the soil which usually consists of organic residues but may be loose soil produced by cultivation.

Muriate of potash: It is a coarse or a fine salt resembling with ordinary salt and having no bitter taste. It contains 60 to 90 per cent potassium chloride and 60 per cent potash. It is manufactured from potash bearing minerals such as sylvite, carnallite, kainite, langbeinite, sylvinite, etc.

Mycorrhizae: It is a symbiotic association of fungi with roots of vesicular plants and is generally of two types - ectomycorrhizae and endomycorrhizae. It has beneficial effect on plant growth particularly in P deficient soils and effective tool help in the solubilization of the unavailable P in the soil.

N

Natural erosion: It is the Erosion of the natural landscape undisturbed by man or domestic animals.

Natural farming: It is a system of agriculture devised by Masanobu Fukuoka that seeks to follow Nature by minimizing human interference: no mechanical cultivation, no synthetic fertilizers or prepared compost, no weeding by tillage or herbicides, no dependence on chemical fertilizers.

Nematodes: Nematodes or eelworms are found in soils which are microscopic. These are called microfauna (lower animals), many of which are parasitic on plants and animals.

Niche: A space in the ecosystem which, because of specific local ecological, physical and / or social characteristics, is suited to a particular plant or animal species or a particular activity.

Night soil: Human excrement or excreta- solid and liquid which is richer in nutrient content as compared to FYM or compost. On oven-dry basis, it contains 5.5 per cent N, 4.0 per cent P_2O_5 and 2.0 per cent K_2O.

Nitrification and **nitrifying inhibitors:** The chemicals which decrease the activity of nitrifying bacteria and thereby reduce nitrogen losses from wet soils such as oxamide (31%), thiourea (36.8%), urea pyrolyzate (48%), dicyandiamide (42%). Thus, use of these chemicals is very beneficial especially in case of lowland rice fields. Neem cake or neem coated urea and sulphur coated urea also serve as nitrifying inhibitors. In other words, nitrification is the process of biochemical oxidation of ammonium to nitrate ions by certain group of bacteria like Nitrosomonas and Nitrobacter.

Nitrogen fixation: The ability of organisms to convert atmospheric nitrogen into a form which can be used by higher plants is known as nitrogen fixation. These organisms may be free living or live in a symbiotic relationship with the plants.

Nitrosification: Conversion of ammonical nitrogen into nitrite form by the help of nitrite forming bacteria is called as nitrosification.

Non-renewable energy sources: It is an energy sources that is not replaced or is replaced only very slowly by natural processes. Primary examples are the fossil fuels such as oil, natural gas, and coal: They are continually produced by the decay of plant and animal matter, but the rate of their production is extremely slow, very much slower than the rate at which we use them.

Nutrient cycling: The pathway of nutrient substances from their occurrence in the physical environment to their incorporation in the living organisms and their return to the physical environment through the metabolic activity, death and decay of organisms.

Nutrients: The chemical compounds needed for growth and metabolic activities of an organism (plant or animal).

O

Optimum: The best or most favourable condition, degree or amount for a particular situation is called optimum.

Organic farming: A system of farming that encourages having soils healthy and crops grown through such practices as nutrient cycling of organic matter, crop rotations, proper tillage and completely avoidance of synthetic fertilizers and pesticides, is known as organic farming.

Organic fertilizers: Any fertilizer derived from animal products (bone meal, dried blood etc.) or from plant residues (cotton seed-cake, groundnut cake etc.); even bulky manures like FYM.

Organic manures: Manures which are made up of dead plant and animal remains added to the soil specifically for the nutrition of plants and are generally of two types- bulky manures and concentrated organic manures.

Organic matter: Organic matter is partially decayed and partially synthesized plant residues and animal remains. It consists of humic substances (humic acid, fulvic acid and humin) and non- humic substances (polysaccharides and sugars, proteins and amino acids and fats).

Organic phosphorus: Phosphorus that is present as a constituent of an organic compound, e.g. glycerophosphoric acids, inositol, phosphoric acid, phytic acid etc. is called organic phosphorus.

Output: It refers to farm outputs i.e. products or functions which are obtained by means of farming activities and which are consumed, reinvested in farming (used as internal inputs or exchanged or sold).

Oxidation: Energy-yielding process usually resulted from loss of hydrogen or addition of oxygen to a compound. The electron acceptor is called as oxidant.

P

Parasite: Any organism that lives in or on another organism (the host) from which it obtains its food, is referred to as parasite.

Particle density: The weight per unit volume of the solid portion of soil is called particle density. It is determined by Pycnometer method. Generally, the magnitude of particle density of normal soil is 2.65 gm m^{-3}. Soils having higher inorganic constituents like iron oxides and of various heavy minerals increase the value of particle density, whereas, the presence of organic matter and light minerals decreases the value of particle density.

Peat: Unconsolidately soil material consisting of mostly undecomposed or only slightly decomposed organic matter under excessive moisture content is called peat.

Ped: A unit of soil structure such as an aggregate, granule formed by natural forces is called a ped.

Percolation: It is downward movement of water through soil layers due to gravitational pull.

Permanent wilting point: The moisture content of soil on an oven dry basis at which plants begin to wilt and fail to recover their turgidity v/hen placed in dark humid atmosphere.

Permeability: The ease with which gases, liquids, or plant roots penetrate or pass through a soil or soil layer is called permeability.

Pesticide: Any substance which is used for destroying or controlling any pest. It consists of insecticides, herbicides, fungicides etc.

pF: It is the logarithm of height in cm of water column that represents total stress with which water is held by the soil.

pH: It is the negative logarithm of the hydrogen ion activity or concentration of a soil solution. It is a measure of the acidity or alkalinity of soil and expressed in terms of a pH scale from 0-14. A pH of 7 indicates neutrality; higher values indicate alkalinity, and lower values acidity.

Phosphate solubilizing micro-organisms: Micro-organisms which bring about solubilization of insoluble phosphates (tri-calcium phosphate, rock phosphate etc.) into soluble by secreting organic acids such as formic, acetic, propionic lactic, and succinic acids are called as phosphate solubilizing micro-organisms.

Phytohormones: These are plant growth substances produced by the plants.

Plant available water capacity: It is the difference between field capacity (moisture retained at 33.3 kPa suction) and permanent wilting point (moisture retained at 1500 kPa suction).

Plant growth regulators: Organic compounds occurring naturally in plants as well as synthetic other than nutrients, which in small amounts promote, inhibit or modify any physiological process in plants.

Plant nutrient: An element that is essential for growth of the plant.

Pollution: Contamination of the environment with objectionable or offensive matter or accumulation of any elemental, ionic or molecules species at a concentration which has been accidentally raised as a consequence of human activity. It may, pose a problem to human, animal and plant health.

Pore space: The volume of soil mass that is not occupied by soil particles. It is usually occupied by air and water.

Precipitation: Total amount of water falling from the atmosphere in the form of rain, snow, mist etc. on the earth is designated by the term precipitation. It is useful for human beings.

Primary nutrients: Nutrients required by plants in large quantities such as C, H. O, N, P and K are called primary nutrients. N, P and K are not usually available in large amounts in soils and are therefore added through fertilizers.

Puddling: The process in which soil aggregates are broken down into primary particles by repeated ploughings and harrowing near the saturation moisture content. The consequences of puddling are: (i) destruction of soil structure; (ii) decrease in bulk density, soil penetration resistance, total porosity, permeability and gaseous exchange; and increase in moisture retention and soil redox potential.

R

Reduction: Energy-yielding process usually resulted from gain of hydrogen or loss of oxygen to a compound. The electron donator is called as reductant.

Relative humidity: Amount of water vapour actually present in air compared with the maximum amount the air can hold at a given temperature is called relative humidity. It is always expressed in percentage and hence it has no unit. It tells simply relative content and indicates the degree of saturation of air at a given temperature with water vapour. Relative humidity of saturated air is 100 per cent.

Renewable energy sources: Energy capture from existing flows of energy, from on-going natural processes such as sunshine, wind, flowing water, biological processes and geothermal heat flows or a form of energy that is replaced rapidly by a natural process such as power generated from the sun or from the wind. It is subset of sustainable energy.

Rhizobia: The bacteria which are capable of living in symbiotic relation with leguminous plants in the nodules on the roots, association usually being capable of fixing atmospheric nitrogen.

Rhizosphere: It is the soil region in the immediate vicinity of the plant roots in which the abundance of the microbial population is affected by the presence of the roots.

Root growth: It is the lengthening of a root due to elongation of new cells that are formed in the apical region.

Root: The growth of radicle in embryo of a seed.

Rotation: Rotation of crops is an order in which the chosen cultivated crops follow one another in a set cycle on the same field over a definite period for their growth.

Rotational intensity: The ratio of number of crops grown in a rotation to duration of the rotation which is multiplied by 100 and represented in percentage is called rotational intensity.

Run-off: Rainfall water or other water that flows across the soil surface and does not filterate into the soil is called run-off.

Rural or Village compost: It is the compost which is prepared from farm waste products, e.g. straw, crop stubbles, crop residues such as sugarcane trash, groundnut husks and leaves, cotton stalks etc., and weeds, waste fodder urine soaked earth, litter from cowshed and hedge clippings. It contains 0.4-0.8 per cent N, 0.3-0.6 per cent P_2O_5 and 0.7-1.0 per cent K_2O.

S

Secondary nutrients: Ca, Mg and S are next in importance to major nutrients after N, P and K which are required for optimum crop growth. These nutrients are called as secondary nutrients because these are not applied as straight commercial fertilizers, but are applied to the soil, indirectly while adding N, P_2O_5 and K_2O in the form of commercial fertilizers. Thus for manufactures of fertilizers containing major plant nutrients, Ca, Mg and S are present of secondary importance.

Seed: A fertilized ovule consisting of an intact embryo, stored food and seed coat which is viable and has the capacity to germinate.

Seedbed: The soil which is prepared to promote germination of seed and the growth of seedlings.

Seedling: It is the young sprout which primarily depends upon stored food in the seed.

Shifting cultivation: A form of agriculture in which soil fertility is maintained by rotating fields rather than crops. A piece of land is cultivated until soil shows signs or exhaustion or is overrun by weeds, when the land is left to regenerate naturally while the cultivation is done elsewhere. Shifting cultivation is also called Jhum cultivation.

Silvi-horticulture: It is a form of agroforestry where under forest trees are grown along with vegetables and fruit trees.

Slow release fertilizers: Fertilizer whose nutrients are present as a chemical compound or in a physical state such that their availability to plants becomes slowly and uniformly over a period of time such as sulphur coated urea, neem coated urea, urea super granules, prilled urea, thiourea etc.

Sludge: It is the solid portion of sewage system of sanitation. On an average, it contains 4.5-3.5 per cent N, 0.75-4.0 per cent P_2O_5 and 0.3-0.6 per cent K_2O.

Soil alkalinity: The degree or intensity of alkalinity of a soil, expressed by a value greater than 7.0 on the pH scale is known as soil alkalinity. It is due to an excess of hydroxyl ions or having high exchangeable sodium content, or both.

Soil erosion: A process of detachment and transportation of soil by natural agencies such as water or wind is called soil erosion.

Soil fertility: It is the inherent capacity of soil to supply adequate amount of available nutrients in balanced form to plants necessary for its growth.

Soil health: The capacity of a soil to function within ecosystem boundaries to sustain biological productivity, maintain environmental quality and promote plant and animal health is termed as soil health.

Soil management: It is the manner in which soil is managed and used to produce food, fibre and forage etc.

Soil mining: Soil mining is an amount of nutrients mined from the soil. Balanced fertilization leads to soil building while imbalanced fertilization use leads to soil mining. Only soil building leads to a sustainable land use system where most food grain production continues to come from existing agricultural land, while unbalanced use of fertilizers has resulted in the poor crop yields and also deteriorates the physical condition of soil.

Soil moisture regimes: The presence or absence of water in a soil at different times of the year is the soil moisture regime. It is a partial function of climate soil and landform.

Soil pollution: Soil pollution is the build-up in the soils of persistence toxic compounds, chemicals, salts, radioactive materials or diseased causing agents which have adverse effects on plant growth and animal health.

Soil population: All of the living organisms present in the soil including fauna and flora, is designated by the term soil population.

Soil pores: Soil pores are the interstices between soil particles.

Soil porosity: The percentage of the total volume of a soil material which is not occupied by soil particles is called soil porosity. It can be calculated by the formula:

$$\text{Porosity (\%)} = (100\text{- bulk density/particle density}) \times 100$$

Soil productivity: The capacity of soil to produce crop yield per unit area under a specified system of management practices is known by the word soil productivity.

Soil profile: Examination of vertical section of soil which reveals the presence of more or less distinct horizons or layers is called soil profile.

Soil quality: It is an attribute of a soil which cannot be measured directly from the soil, alone but is inferred from soil character and soil behavior under defined conditions. Fertility, productivity and erodibility are some of the examples of soil quality.

Soil structure: The arrangement of soil particles and their aggregate into certain defined patterns. It has main four forms - prism like, plate like, block like and spheroidal structure.

Soil testing: It is a chemical test of the soil that can be made rapidly and with low cost, as compared to conventional methods of soil chemical analysis which are more accurate but more time consuming and expensive. It covers both rapid analysis in the field and laboratory.

Soil texture: The relative percentage of sand, silt and clay in a soil refers to soil texture. A soil is described as a coarse, medium or fine textured depending on the predominant particle size.

Soil tilth: Physical condition of the soil vis-a-vis its fitness for cultivation of crops is known by the term of soil tilth.

Soil: Soil is derived from Latin word 'Solumn' meaning floor. It is defined as a unconsolidated, three dimensional, natural dynamic body on the surface of the earth, composed of complex system made up of organic matter, mineral matter, soil air, soil water and living forms in which plants can grow.

Sole cropping: Growing one crop variety alone in a pure stand at normal density, is called sole cropping.

Starch: It is a polymer of glucose. It consists of D-glucose linked in a linear fraction by 1-4 linkages (amylase) and branched chain fraction by 1-6 linkages (amylopectin).

Straw: It is the dried remnants of crop plants from which the seed has been removed, also called Stover.

Stubble mulching: Stirring the soil with implements that leaves considerable part of the vegetative material or crop residues or vegetative litter on the surface as a protection against erosion and for conserving moisture by favouring infiltration and reducing evaporation is called stubble mulching.

Submergence: Plants that grow partially or completely under water.

Subsistence agriculture: Farming system in which large part of final yield is consumed by the producer himself is called subsistence agriculture. Most of the subsistence systems involve production of some crops for sale, but the ratio of subsistence to each production may vary greatly from year to year.

Sub-soil: The layer or horizon down below the surface soil which is generally compact, light in color and less fertile is called sub-soil.

Sustainable agriculture: Management of resources for agriculture to satisfy changing human needs, while maintaining or enhancing the quality of the environment and conserving the natural resources is designated by the word sustainable agriculture.

Symbiotic nitrogen fixation: Fixation of atmospheric nitrogen by the leguminous plants with the help of nodule bacteria (Rhizobia) is known by the term symbiotic nitrogen fixation.

T

Temperature: It is degree or intensity of hotness and coldness usually measured by a thermometer and expressed in degrees (Celsius or Fahrenheit or Kelvin).

Tillage: It is a mechanical manipulation of the soil to obtain a seedbed of optimum soil tilth for plant growth. It alters the soil environment by reducing the strength of the soil matrix and increasing its porosity thereby improving the permeability of the soil for the flow of both water and gas.

Tilth: A physical condition of soil as related to its ease of tillage, fitness as a seedbed, and its impedance to seedling emergence and root penetration costitute tilth.

Traditional agriculture: Farming systems which are based on indigenous knowledge and practices, and have evolved over many generations are called traditional agriculture.

Transition: The process of changing from one form to another refers here to transition.

Translocation: Transfer of food or other materials such as herbicides from one plant to another is called translocation.

Transplanting: The process of moving the seedlings from seed beds to the field for rice crop is called transplanting.

Trap crops: Plant stands that are grown to attract insects or other organisms so that the target escapes from peak attack are the trap crops.

U

Underground run-off: Water flowing towards stream channels after infiltration into the ground is underground runoff.

Upland rice: Rice which is grown in both flat and sloping fields prepared under dry land conditions and solely depends up on rainfall. This is also called rain-fed rice.

Urban compost: Compost which is prepared from night soil and street/dustbin refuse is the urban compost. It is rich in nutrients (1.0-2.0% N, 1.0% P_2O_5 and 1.5% K_2O) as compared to FYM and rural compost.

Urbanland: Areas so altered or obstructed by urban works or structures, where the identification of soils is not possible is called urbanland.

Urea: It is a white, crystalline, non-protein nitrogen material made synthetically from synthetic ammcnia and carbon-dioxide and contains 44 to 46 per cent N.

V

Vermicompost: Compost which is prepared from refuse and organic wastes by using earth worms *(Eisenia foetida, Eudrillus eugeniae* and *Pheritima elongate),* is called vermicompost.

Vermicomposting: Process of decomposition or organic wastes using earth worms is known as vermicomposting.

Vesicular Arbuscular Mycorrhizae (VAM): A symbiotic association of fungi with the roots of vesicular plants and has beneficial effect on plant growth particularly in P deficient soils.

Viability: The capability of living and developing in a green environment is viability, e.g. in the case of technology of being practised in the long term.

Vitamins: They are organic substances present in minute quantity for human beings and other organisms especially belonging to heterotrophic group to maintain growth and development for normal metabolism. Vitamin A, D, E and K are fat-soluble and are stored in the fat compounds of the body, whereas, vitamins of B group and C are water-soluble and as such cannot be stored.

Volatilization: It is a process where a condensed phase such as liquid or solid, is transformed into vapour by elevation of temperature or reduction of external pressure.

W

Wasteland: Land which is not capable of producing materials or services of much value is the wasteland.

Water harvesting: Collection and storage of water, either runoff or stream flow for securing and improving water availability for crop growth and / or animal and human consumption, is the water harvesting.

Water holding capacity: It is the moisture saturated stage of the soil when all pores (micro and macro) and capillaries are filled with water. The maximum water holding capacity of given soil is determined by the amount of total pore space available in the soil.

Water pollution: It is the loss of any of the actual or potential beneficial uses of water caused by any change in its composition due to human activity or substances like bacteria or viruses or other pollutants which impair the quality of water, rendering it less suitable for use.

Water stable aggregates: A soil aggregate stable to the action of water such as falling drops, or agitation as in wet-sieving analysis is called water stable aggregate.

Water table: The upper surface of ground water, locus of points in soil water at which the hydraulic pressure is equal to atmospheric pressure is called water table.

Water-logged: Any kind of land which gets saturated with water is known as water-logged.

Water-Stable Aggregate: A soil aggregate which is broken down by agitation in water, usually by wet sieving is the water soluble aggregate.

Weather: It is the physical condition of atmosphere with respect to wind, temperature, cloudiness, relative humidity, pressure etc. at a given time. It is dynamic and changes with time.

Weeds: Plants in a place where they are not wanted by humans are called weeds.

Xerophytes: Plants that grow in dry soils (sites) are known as xerophytes.

Y

Yield: It is the harvested produce obtained from a crop grown in a unit area of land, usually expressed as kilogram, quintals or tonnes per hectare. have evolved over many generations are called traditional agriculture.

1

Introduction

Abstract

This chapter deals with the agricultural technology responsible for bringing Green Revolution in India and later on how it become stagnant in agricultural production. Effects of Green Revolution on environment also stand mentioned. Apart from this, why the Green Revolution technologies got fatigued have been explained? All these factors were found to be amenable to bring down unsustainability in Indian Agriculture. Yields of crops, especially of rice and wheat got reduced and as such a number of the peasants of the country became involved in debt. Some of them committed suicides. To prevent this tragic scene, the authors of this book have tried their best to compile all the sporadic and scattered informations responsible to enhance the yield of the crops and eventually looming the farmers income and rendering them free from debts.

Keywords: Green Revolution: Indian Agriculture; agricultural production; environment; debt; yield

India is an agriculture based country which is one of the few developing countries in the world. It successfully achieved Green Revolution during the late 1960s (1967-68). At present, agriculture is a principal means of livelihood for over 58.4 per cent of India's population directly or indirectly (Anonymous, 2011). It is worthwhile to mention that although agriculture continued to be the mainstay of the Indian economy previously (Anonymous, 2010b) yet now its share in the Indian economy is declining year after year (Anonymous 2013d). After Independence, agriculture's sector contribution to GDP swings between 50-60 per cent. Even in 1950-51, it was 55.1 per cent, but fell to mere 17-18 per cent in 2008-09 (Mahajan and Gupta, 2009). During the 2011- 12 share of agriculture in annual GDP remained at 13.9 per cent, which is worry some as the population depends upon the agriculture sector crosses, about 60 percentage mark (Anonymous, 2013d) and still agriculture contributes about 14 per cent of the national GDP and sustains livelihood of about 60 per cent of the population.

1.1 Plateauing of Green Revolution

There is no doubt that Green Revolution in India brought about self-sufficiency in food grain production. This is quite evident from our food grain production which grew from about 50 million tonnes (MT) at the time of our independence i.e. 1947 to about 263.2 MT food grain productions during 2012-13 as per advance estimates. Out of this total production, rice and wheat constituted 106.19 MT and wheat 95.6 MT along with coarse grains and pulses production to the tune of 44.64 and 19.77 MT, respectively (Suryakumar, 2014a). However, after ushering Green Revolution during 1967-68 India's food production registered a 6 per cent decline during 1984-93 according to a World Bank study? Indian Agriculture is in transition now. There are signs of Green Revolution plateauing off and poisoning of ecosystems, especially in the Green Revolution hot spots - Punjab, Haryana and Western Uttar Pradesh. Recent studies indicated that unscientific use of NPK (Nitrogen, phosphorus and potassium) nutrient ratio, through nitrogenous, phosphatic and potassic fertilizers, had a shocking impact on nutrient elements in the soil and reduced the crop yields in the food grains production bowl of the country-Punjab (Sharma, 2015). As a matter of fact, with the use of chemical fertilizers, there is loss of soil fertility. The loss of soil fertility is taken to mean multiple nutrients deficiency and loss of organic matter (Mahajan et al., 2002; Mahajan et al., 2008a). Nutrients mining has increased with intensive cultivation during post Green Revolution period. The situation has now been compounded with low soil fertility status both of major and micro-nutrients not only in Punjab but also in other states of the country where rice and wheat crops are being grown since inception of the Green Revolution. Nearly 89, 80 and 50 per cent of the Indian soils have been found to fall under low to medium categories in case of N, P and K. Now a days S-deficiency has become wide spread. The micro-nutrients deficiency is found to the tune of 49 33, 12, 13, 5 and 3 per cent of the soils for Zn, B, Mo, Fe, Mn and Cu, respectively. The productivity of intensive rice-wheat cropping system of Punjab, Haryana Himachal Pradesh and Jammu and Kashmir has already shown signs of serious decline in their productivity which is associated with soil quality. In Gujarat known as the chemical capital of India, the continuous use of chemical fertilizers has resulted in loss of soil fertility and decrease in farm output to the extent of 60-80 per cent over the past few years.

A parliamentary panel report on "National Project of Organic Farming" tabled recently in Parliament pointed out an alarming level of 39:9:1 NPK ratio in the key agrarian state (Punjab) against an ideal combination of 4:2:1. It has resulted in drastic reduction in crop yields over the years. The report has further indicated that the crop yield per Kg NPK in Punjab has come down to 10 Kg from 50 Kg in the initial years of the Green Revolution (Sharma. 2015). Expansion of agriculture has totally changed India's water flow scenario. High water requirement for high yielding varieties of rice and wheat has made it essential to create more irrigation facilities through surface and to draw ground water by sinking wells/tube-wells. Since the main feature of Green Revolution was to grow rice and wheat in the same field so area under Irrigation had to increase tremendously which has now become almost double. As a result, the water table declined and some of the dug wells dried up. There was lowering of water table by 0.6m a year. Due to this some of the farmers having shallow wells in states like Punjab. Haryana, parts of Himachal Pradesh and Jammu region of Jammu and Kashmir were forced to drill tube wells. In western Uttar Pradesh the Central Ground Water Board has identified 19 blocks, where water table had already been plummeted to a level that has become unviable proposition. According to the National Aeronautics and Space Administration (NASA) agriculture output can collapse in states like Punjab and Haryana, and there can be severe shortage of drinking water in big cities such as New Delhi if efforts were not made to ensure sustainable ground water usage. We require looking afresh for sustainable agriculture, industrial and domestic purposes (Dhaliwal, 2015).

In Punjab, the subsoil water table has been dipping at a fast pace. The state faced a severe drought during 2014 because of the rainfall was below average by 50 per cent. It was another matter that there was no major effect on crop yield because farmers operated their about 14 lakh bore wells (diesel sets and 3 electricity driven ones) overtime to save their crops. The same happened during 2012 and 2009, when the rains were deficient by 47 and 36 per cent respectively, than normal. Earlier in 2007 and 2004, the state faced an equally bad situation in this regard. About 3 crore (30 million) gallons of water is pumped out by industry in Ludhiana (Punjab) alone.

According to a report by the Water Resources and Environment Directorate the water level dipped in 80 per cent of the total area in the state and came up only in about 20 per cent areas. The minimal district wise average fall of 0.38m was noticed in Fazilka from October 2013 to October 2014. The subsoil water level has gone down to more than 58m in Hoshiarpur and Sardulgarh areas of the state over the years due to unregulated rafting (Dhaliwal, 2015).

Falling of water table in some parts of Fatehgarh Sahib District (Punjab) has made tube wells useless. In some parts of Patiala, a few years ago, a centrifugal pump could be installed at 3m, but these days one had to go as deep as 10m. For supply of clean water submersible pump at a depth of 18m is required. The water table has also gone down in some parts of Jammu region like Ramgarh area of Samba district and that of Chatha of Jammu. A number of areas also fall in the same category of Ranbir Singh Pura and Bishnah tehsils of Jammu district. Farmers of Tanda Burj (Samba district) are facing lot of difficulties not only in irrigating the crops but are also getting very less water for drinking. It is because underground water which hitherto could be available at about 4-5 m depth, now one has to go deeper up to 7-5m or even more. Thus, if over exploitation of water mining is not stopped immediately, the aforesaid area may be turned into desert.

A significant proportion of agro-chemicals used in farms runs off into surface or leaches into subsurface water. Excessive amount of nitrate (NO3) in ground water is of the most concern in remote agriculture areas, where its contamination is the most acute and the water so contaminated is mostly supplied by pumping from ground water aquifers. Drinking of this water causes blue baby disease in infants.

The Central Ground Water Board found that water from wells in over 300 sites scattered across Gujarat had high NO3 level. Now wells with high NO3 levels have also been found in Faridabad (Near Delhi), as well as in ground water used in villages, towns and cities of Karnataka including Bangalore and in many parts of Uttar Pradesh.

Heavy metals pollution of mercury (Hg), Cadmium (Cd), Lead (Pb) and Chromium (Cr) has been observed in soils and waters of some areas of the country (Gupta. 2013d). These toxic elements come from geogenic sources or from sewage water, industrial effluents, urban soils wastes and chemical fertilizers. For example, 30 million people inhabiting 8 districts of West Bengal are exposed to Arsenic poisoning through intake of contaminated water and infected food grains. Lead poisoning and chromium poisoning have also been seen in some areas of India, the details of which stands mentioned elsewhere (Gupta, 2013d). About 20 million people in our country (WHO report) are afflicted with flouride (F) poisoning. Worst affected states are Andhra Pradesh, Punjab, Haryana, Rajasthan, UP, Gujarat, Tamil Nadu, Bihar and Maharashtra (Fotedar, 2020).

1.2 Why Green Revolution Technologies Fatigued?

It is summarized from the foregoing literature that Green Revolution technologies (High yielding varieties of rice and wheat, chemical fertilizers, synthetic pesticides, mono-cropping and irrigation etc.) led to rapid agricultural growth, prevented famines and achieved self-sufficiency in food grains production in 1990s (Agarwal, 2009). The growth in agriculture assisted to reduce poverty from 53 per cent in 1977-78 to about 26 per cent in 1990. Green Revolution technologies, however, fatigued with passage of time. The gains so obtained could not be sustained. Slowdown in agricultural growth was observed the greatest in Green Revolt-: states, especially Punjab, Haryana and Western Uttar Pradesh.

There was increase in degradation of natural resources particularly those of soil and water– The problem of water logging, salinity and over extraction of subsurface ground water particularly came to fore in potential irrigated areas. Indiscriminate use of chemical fertilizers, especially on sloppy lands of kandi belt of Jammu and apple growing areas of Sopore (Baramula district) in Kashmir valley has created acidity (Gupta et al., 2011) in soils.

The main soil related constraints in rain-fed uplands rice cultivation may be due to the presence of Al and Mn toxicity, low availability of P and high P. fixation capacity particularly in Latosol soils and Fe deficiency in soils having neutral to slightly alkaline reaction. Apart from this, the red and lateritic soils possess large amount of free iron oxide which often leads to crust formation resulting in poor emergence of rice seedlings after sowing and eventually producing poor crop stand. Moreover presence of Kaolinite as the dominant clay mineral in these soils having low action exchange capacity, there is also low nutrient retention capacity, which leads to a loss of plant nutrients (Mahajan and Gupta, 2009).

Imbalanced or unbalanced use of chemical fertilizers has led to soil nutrients unbalance in those soils where rice and wheat crops had grown continuously. There was deficiency not only of organic carbon but also of N, P, K and S. Micro-nutrients deficiency was also noticed as has been stated earlier. Poor soil organic carbon (SOC) content was the biggest environmental set back of Green Revolution technologies. According to a survey of South-East countries, India was at bottom in SOC content (8.5g per kg) as compared to Malaysia (33.6g per kg) the highest in SOC content. Poor SOC content in India is attributed to extensive mining of soil fertility, low and unbalanced use of fertilizers, removal of crop residue for fodder of household fuel/burning and severe soil degradation. Moreover, the use of high yielding varieties of rice and wheat require great deal of water for optimal yield, had led to drilling of tube wells and large scale depletion of ground water reserves. More than 50 per cent of open wells of many villages of some Indian states have completely dried up which affected agricultural output in the long run.

1.3 Technologies

There is a need of scientific management of agricultural lands through modem technologies developed in agriculture including horticulture by the Central Agricultural Research Institutes, State Agricultural Universities and Departments of Agriculture. Such technologies are

1.3.1 Apply Fertilizers on Soil Test Basis

Amongst the various strategies in sustaining the Indian agricultural productivity the farmers are required to apply fertilizer to various crops on the basis of soil test (Jaggi et al., 2005; Raina et al., 2005). While applying nitrogenous, phosphatic and polassic fertilizers, their ratio should be maintained in the form of 4:2:1. Where, there is deficiency of micronutrients in soils, especially of Zn, Cu, Mn, and Fe, these should be applied as per the recommendation of State Agricultural Universities, Indian Council of Agricultural Research Institutes and State Departments of Agriculture. Whichever amount of organic manure is available with the farmers must be added; rather the farmers are required to adopt organic farming (Mahajan et al., 2007a and b). Organic farming which is also called Rishi Krishi green or organic food, biological farming or diverse farming. Organic farming is Holistic production management system which enhances the health of Agro eco system as it avoids the use of chemical fertilizers and pesticides. Organic farming is very much native to India but was almost forgotten after ushering Green Revolution (Gupta, 2011d). The use of organic manures has proved useful in sustainable agriculture and in saving an environment from being further detrimental (Gupta, 2012c). In Europe recently legislation has been approved banning the use of highly toxic pesticides including those acting as carcinogenic and mutagenic as reported by Thusoo (2014). Use of such pesticides is required to be banned in India also. On the contrary use of organic pesticides is required to be encouraged.

1.3.2 Rainfed Agriculture and Cropping Pattern

On an average, about 70 per cent of the total arable land of India is rainfed, where the farmers cannot take risk to grow rice and wheat. It is because high yielding varieties of rice and wheat crops respond only to assured water supply accompanied with high doses of chemical fertilizers, weedicides, fungicides bactericides and insecticides (Gupta, 2013i). The rainfed area of the country thus, totally remained devoid of Green Revolution.

The dry lands in India are mostly cultivated. The soils of these lands are generally inferior in quality as compared to irrigated lands. They have as uneven and multiple slopes and most often receive low rainfall except those of Siwaliks. Soils are universally poor in nitrogen and phosphorus but high in potassium. The source of water for dry lands is rainfall. *Kharif* and *Rabi* are the main growing seasons.

Upland rice, maize, wheat, barley, sorghum, millet are the main crops grown. Wherever, it is possible green manuring should be practised with dhaincha *(Sesbania aculeata)*. A part of the fodder can also be incorporated as a source of manure. Based upon several research findings, it has been found that about 25+ kg of nitrogen can be added through on farm residues. Possibilities for such a system exist in many regions like Banglore, Bhubaneshwar, Varanasi, Hoshiarpur, Rewa, Agra, Ranchi, Dehradun, Indore and Akola. In these areas, cowpea-finger millet, mungbean-finger millet, mungean-mustard, finger millet gram, maize-gram, soybeam-safflower and mungbean-safflower crop rotations are followed in these areas. Intercropping has also been found beneficial to many regions of the country. Vidarbha region of Maharashtra, Malwa plateau of Madhya Pradesh, Deccan region of Orissa and sub-humid region of Bihar (now Jharkhand) are the versatile examples from this angle (Arora and Gupta 2014)

It is worthwhile to mention that before Green Revolution in India, in a number of sub-mountain areas of Siwaliks lying in Punjab, Haryana, Uttarakhand, Himachal Pradesh and Jammu region of Jammu and Kashmir, farmers usually used to grow pulses. Some area of Jammu Siwaliks locally called Kandi belt of Jammu like Amb Garota, Kote, Chinore, Akhnoor, Samba were once great producer of gram, black and green grams, lentils as well as moth and kulth, but now a days these are no more grown. Taking into consideration on the aforesaid positions the only option to follow timely and effectively space utilization in agriculture by adopting diversified agriculture. Growing of pulses as diversify agriculture will not only help in obtaining higher food grain production but also covers the risk of raising of crop production alone in country's nearly 70 per cent dry land agriculture. Not only this, the pulses so grown will ensure the food nutritional security being rich in protein. Moreover, pulses being leguminous crop will also fix atmospheric nitrogen in soils and, thereby, enhancing their fertility.

1.3.3 Adoption of Agroforestry

Another strategy in sustaining, the agricultural productivity of rain-fed agriculture ecosystems is through adoption of Agroforestry (Arora and Gupta, 2014). Adoption of agroforestry has not only played a significant role in conserving soil and water especially in Kandi belt of Jammu but also helped in planting both of forest and fruit trees as well as setting up of herbal garden (Jamwal and Gupta, 2007; Sharma et al. 2007; Gupta, 2013e). In herbal garden, medicinal plants like harar *(Terminalia chebula),* bahera *(Terminalia belerica),* aonla *(Phyllanthus emblica),* neem *(Azadirachta indica)* and ashwagandha *(Withania somnifera)* have been planted (Gupta et al., 2016).

As many of the areas of these Siwaliks have been eroded/degraded and deforested so these require to be managed through vegetative cover besides water and soil which makes all life possible. This has been done through watershed development, the central theme of which is to have all round development of the area i.e; in agriculture, horticulture, forestry, sheep, and animal husbandry and soil and water conservation (Arora and Gupta, 2014; Gupta et al. 2009).

1.3.4 Status of Small and Marginal Farmers

About 85 per cent of Indian farmers are small and marginal (A land holding of an individual farmer is below 0.6 hectare). The number of such farmers is further growing with fragmentation and division of ancestral land holdings among the sons, grandsons and relatives due to increase in population. An average land holding has been worked out to be 1.4 hectare in size. Merely 15 per cent of the farmers can be called as large and the remaining 85 per cent as small and marginal. The Indian agriculture is now, therefore, heading towards a crisis owing to uneconomical holdings. The agricultural inputs like seeds, fertilizers, machinery, pesticides, diesel etc; have become costly and are going out of the reach of small and marginal farmers (Rana, 2013) These days no bank would finance an individual farmer whose land holding is below 3.2 hectare. Thus, it is difficult for these farmers to purchase agricultural input to use in their fields. Subsidies once a corner stone of Indian economy policy, have now totally dried up (Qazi, 2016).

Nearly half of the farmers of India are in grip of debt. Majority of them had taken loan for capital or current expenditure in farm business. On an average an Indian farmer household had debt of Rs 12,585. The farmers of Punjab top in the list with Rs 41,576 followed by Kerala with Rs 33,907, Haryana with Rs 26,007, Andhra Pradesh with Rs 23,965 and Tamil Nadu Rs 23,963. Their dependence on borrowing has increased but they have little repaying capacity as farming is no longer remunerative. They are caught in virtual debt traps. Since 1995, the farmers have started choosing death over repayment of debt. In the 20 years since the Indian government first started keeping track of farmer suicides, about 3 lakh farmers have ended their lives (Qazi, 2016). According to 2011 census, the suicide rate for farmers is 47 per cent higher than National average. At least 24 cases of suicide have been reported in the past 2 months in Bathinda and Mansa district of Punjab and first such case in Samana. There are many other farmers who are grappling with the rising debt burden due to repeated crop failure. In Punjab 12 per cent of the small families and 5 per cent those of large have left farming with no alternative and gainful employment.

1.3.5 Change of Climate and Necessary Technology for Small and Marginal Farmers

Another serious problem confronting with our agriculture is the disinterest of younger generation in farming (Rana, 2013). Climate resilient agriculture is another area that needs attention. In recent years, global warming and climate change have got huge damage to agriculture. Climate is affecting Indian agriculture in a big way and its impacts are many (Gupta, 2013a). Erratic monsoon, migration of agriculture zones, spread of tropical diseases, sea level rise, changing availability of fresh water, occurrence of storms, drought heat waves, floods and adverse impact on agricultural productivity and human health, are forecast in a report, "climate change and India", released by the Union Ministry of Environment and Forests (Pandit, 2010). In another report, climate change is expected to lead changes in monsoon pattern (Dassgupta, 2015). Most of the climatic models predict that India will be hit by more rainfall events as the world continues to warm in the coming years with more drought and floods (Nazareth, 2014).

1.3.6 Low Cost Technology

To increase the agricultural productivity of small and marginal farmers, it is required to evolve low cost technology for more effective use. The location specific Technologies generated by Krishi Vigyan Kendras are required to be advocated.

1.3.7 Diversification of Agriculture

Crop diversification from staple food (rice, maize, wheat) to vegetables, horticulture and cash crops is desirable for marginal farmers and in those areas where limitation of water has become a major threat. As for example cultivation of vegetable pea in between harvesting of paddy and sowing of wheat has shown the net return increased by 32.5 per cent in Uttarakhand. It is not only a vegetable pea, but cultivation of other vegetables like cauliflower, cabbage, knolkhol in many villages of Solan district, Himachal Pradesh has proved beneficial to the cultivators. Similarly at certain places of mid to high altitude intermedate zone of Jammu region, a number of off season vegetables like tomato, potato, green peas, and beans as well as cabbage, cauliflower, knolkhol and bitter gourd, are grown and sent to plains. Baggarh and Assar areas of Doda district and Chenani of Udhampur are the other glaring examples from this angle. In valley of Kashmir now many of the farmers are reluctant to grow paddy due to lack of water during May to June, and have therefore, shifted to the production of horticultural crops.

There are a number of income generating options like mushroom cultivation, bee-keeping, livestock rearing and sericulture and floriculture which can be adopted by small, marginal and other farmers depending on choice (Gupta, 2019).

1.3.8 Adoption of Integrated Nutrient Management

There is no doubt that in the future earlier years organic farming is not economical. However, in the long run it proves better than inorganic farming as it is eco-friendly both to the human and social health. However, owing to burgeoning human population, organic farming alone will not prove an asset in increasing the agricultural production. Hence, both use of organic manures and inorganic sources of nutrients are needed, which is known as Integrated Nutrient Management (Arora and Gupta, 2014; Gupta 2013f; Mahajan and Sharma, 2005; Mahajan et al., 2003a and b; Mahajan et al., 2008a).

1.3.9 Cropping Pattern

The cropping pattern requires to be changed from monoculture (Growing of Rice-Wheat alone) to multiple cropping systems. More thrust must be given to leguminous crops including pulses. The multiple cropping should be linked with an integrated manner with animal husbandry as an essential component. At the same time such kind of Technology which limits the sustainability by further destroying the natural resources should not be permitted at any cost.

1.3.10 Higher Procurement

Another need is to give highest procurement prices to the farmers and extend their reach of procurement to the crops which are essential for country's food security and nutritional requirement.

1.3.11 Crop Insurance and Provision of More Funds

Enhance crop Insurance and multiple public investments. As now-a-days there is change of climate so time has come for the government to formulate a strategy keeping in view the climate change.

1.3.12 Technology for Preventing Soil Erosion

In the eroded or dergraded / denuded lands all kinds of methods viz; mechanical or engineering, agronomic and biological are required to follow for controlling soil erosion. However, biological methods should be preferred to follow being cheaper. A number of shrubs are used by the farmers for controlling soil erosion in the Kandi belt of north-west Himalayan (Arora and Gupta, 2014). These shrubs mostly consist of Carissa spinarum, Adhatoda vasica, Vitex negundo, *Dodonaea viscosa* (Gupta et al. 2016).

1.3.13 Proper Supply of Electricity

The government should ask the concerned officers/officials to ensure proper supply of electricity so that groundwater could be utilized properly and effectively.

1.3.14 Rain Water Harvesting

As the first and the foremost requirement of any crop is water, so we have to save and use the available water properly. The farmers can fulfill the irrigation requirement for their growing crops after rain water harvesting (Mahajan, 2004). These days, there is an urgent need to propagate the importance of rain water by constructing ponds, reservoirs for storming and the same for the future use. The water so stored can be utilized not only for irrigating the crops but also for other purposes bathing and drinking. Farmers may join together to form groups known as self-help groups (10-15 farmers in each group) construct the ponds, reservoirs and jointly share the rain water harvested in them for irrigating the crops (Gupta, 2013h). By doing this, the farmers, will not only be able to irrigate at right and proper time but can also get loan from the banks.

It is point to mention that harvesting and conserving rain water for irrigation and other households is not a new phenomenon in India, but it is as old as Indus Valley Civilization (Gupta 2013f). To sustain the agriculture we should promote subsidiary farm enterprises like dairy, poultry and piggery as well fishery apart from cultivation of vegetables and mushroom. According to Singh (2016), the government must do something for debit ridden farmer. He said, 'Why cannot the government waive their debt? Alternatively, our Gurudwaras and Temples are flooded with wealth offered by devotees to propitiate the gods. Why cannot we donate this money to the farmers and their families in distress? Out of the several problems to account for bringing forth the low productivity in agriculture in the North West Himalayan States is attributed to soil related problems (Gupta et al. 2011) .Such soil related problems mainly consisted of soil erosion, soil acidity, and salinity and sodicity. The main causes responsible for their development and control measures to increase their productively already stand mentioned separately (Arora and Gupta, 2014).

1.4 Adoption of Family Farming

Family farming is defined as the farming which is usually interchangeable with that of small and marginal farmers. It is however, not about the size but the way of family manages the multilayered characteristics of the land they own, primarily with the help of the family members having limits of the possibilities to make a living, aspire to achieve a decent life by investing the family resources, work with dedication, passion and hard work to develop and improve the farm ensure livelihood of the family and the the farm continuity.

The concept of family farming came to fore in, "The United Nations high level political forum on sustainable development, created during Rio+20 Conference which took place in Rio De Janeiro, Brazil in June 2012 as reported by Gupta (2016a and b). It was emphasized that the practice of sustainability provides the only real bridge from our past to our present and future, and from our planet to our people and prosperity.

To ensure economic sustainability of the farmers and also to prevent further degradation of the natural resources i.e., land and water, various strategies such as conservation tillage, adoption of organic farming, use of fertilizers on the basis of soil testing, educating the farmers about the latest technologies are required to be followed.

1.5 Sustainable Food Grain Production

In the light of the above, it becomes an imperative to compile all the sporadic and scattered information of the research work carried out by various agricultural scientists in the country and abroad for providing continuous technical support in the form of a text book to render the agricultural production on sustainable basis (Anonymous, 2011; Sahai, 2011; Sahai et al., 2014). Thus, it is with this back ground objective that the present text book entitled, "*Sustainable System of Farming for Modern Agriculture*" is proposed to write for providing handsome information to the agricultural scientists, researchers' planners, students and progressive and well educated farmers with regard to sustainable agriculture.

Study Questions

1. How green revolution in India brought self-sufficiency in food grain production?
2. Why the green revolution technologies got fatigued?
3. Explain how agricultural technology was responsible for green revolution?
4. Explain in detail the plateauing of green revolution?
5. Write a short note on 'Adoption of Family Farming".
6. Soil health is deteriorating day by day. How can we overcome this upcoming threat?
7. Explain the effects of green revolution on environment
8. List down the drawbacks of green revolution.
9. Define the following:
    i. Salinity
    ii. Sewage
    iii. Soil degradation
    iv. Green revolution

References

Agarwal, K.P. (2009). Towards a sustainable Green Revolution. Agriculture Today 12 (6):31 -32

Anonymous. (2014a). India should wake up to dangers of reducing land holdings. The Daily Excelsior 40 (144):7 [May 21, Wednesday].

Anonymous. (2010b). Fertilizers fuelling the second Green Revolution. Agriculture Today 13 (12): 16-20

Anonymous. (2011). NREGS and Agriculture: Manifestations and opportunities. Agriculture Today 14 (3): 51-53.

Anonymous. (2013d). Transformative and scientific rationale can boost India's agricultural progress. Agriculture Today 16 (10):51-52.

Arora, S. and Gupta, R.D. (2014). Soil related constraints in crop production and their management options in North Western Himalayas. J. Soil and Water Conservation 13 (4): 305-312.

Dassgupta, C. (2015). India & climate change: Need for updated plan. The Tribune 135 (20): 9 [January 20, Tuesday]

Dhaliwal, S. (2015). Punjab facing a veritable water crises. The Tribune 135 (234): 9 [August 25, Tuesday].

Fotedar, B.L. (2020). Flouride Poisoning. Personnal communication on the address of Prof. (Dr) R.D. Gupta C/o 258-C, Sainik Colony, Jammu, J&K, India

Gill, A. (2011). Save farmers from debt trap. The Tribune 131 (49):13 [February 17, Saturday].

Gupta, R.D, Arora, S. and Sharma, V. (2009). Rainfed agricultural production through water shed management: Success in Jammu region. J. Soil and Water Conservation 8 (1):19-24.

Gupta, R.D. (2011 d). Traditional organic farming. The Daily Excelsior 47 (109):6 [April 11, Monday].

Gupta. R.D. (2012c). Indigenous plant nutrient technology in Siwaliks. The Daily Excelsior 48 (211):8 [August 2, Thursday].

Gupta, R.D. (2013a). Climate change and rendering of refugee. The State Times 18 (186):6 [July 7, Sunday].

Gupta, R.D. (2013d). Heavy metals poisoning-A threat to life. The State Times 18 (273): 6 [October 3, Thursday],

Gupta, R.D. (2013e). Issues concerning agricultural production. The Daily Excelsior 49 (86):8 [March 29, Friday].

Gupta, R.D. (2013f). Rain water harvesting recalls Indus Valley Civilization. The State Times 18 (312):6 [November 12, Tuesday].

Gupta, R.D. (2013h). Strategies for rain-fed agrosystem. The Daily Excelsior 49 (322):8 [December 14, Saturday].

Gupta, R.D. (2013i). Withering wheat green revolution. The State Times 18 (27):6 [March 19, Tuesday].

Gupta, R.D. (2016a). A family farming - A need of the day. The State Times 21 (128): 6 [May 9, Monday].

Gupta, R.D. (2016b). A family farming - A need of the day. The State Times 21 (129): 6 [May 10, Tuesday].

Gupta, R.D., Kumar, A. and Sharma, B.C (2011). Management of problematic soils of North West Himalayas for sustainable hill agriculture (pp. 51-63). In Sustainable Hill Agriculture - An overview (Eds. A. Kumar, B.C. Sharma and V. Sharma). Agrobios (India). Agrohouse Behind Nasrani Cinema Chopasani Road, Jodhpur, Rajasthan (India).

Gupta, R.D., Gupta, S.K. and Bharadwaj, S.D. 2016. Agro techniques and uses medicinal plants. Associated Publishing Company A Division of Astral International Pvt. Ltd. New Delhi - 110002, India

Gupta, R.D. 2019. Diversification for enhancing income. Success story form J&K. Soil and Water Conservation Today. 14(2): 3-4.

Gupta, R.D, Arora, S and Gupta, S.K. (2012). Withering yellow revolution in the Indian context (pp. 285-304). In Technological Innovations Major World Oil Crops (Ed.S.K. Gupta). Springer Science + Business Media (www. Springer.com), Volume 2.

Jaggi, R.C., Mahajan, A. and Sharma, R. (2005). Role of sulphur to onion crop in Himachal Pradesh. Agrobios Newsletter A (3): 17-18.

Jamwal, G.S. and Gupta, R.D. (2007). Watershed management project - A case study of Bari-Badhori, Jammu and Kashmir State (pp. 226- 231). In Natural Resource Management for Sustainable Hill Agriculture (Eds. S. Arora, S.S. Kukal and V. Sharma). Soil Conservation Society of India, Jammu Chapter SKUAST Jammu, J&K State, India.

Mahajan, A. (2004). Watershed - Concept and management. Agrobios Newsletter 3 (1): 28-31.

Mahajan, A. and Gupta, R. D. (2009). Integrated nutrient management (INM) in a sustainable rice-wheat cropping system (pp.1-267). Springer Science + Business Media (www. Springer.com).

Mahajan, A. and Sharma, R. (2005). Integrated nutrient management (INM) system - Concept, need and future strategy. Agrobios Newsletter 4 (3): 29-32.

Mahajan, A. Gupta, R.D. and Sharma, R. (2007a). Organic farming - A need for all times. Gram Vikas Jyoti (July-September Issue): 7-8

Mahajan, A., Bhagat, R.M. and Gupta, R.D. (2008a). Integrated nutrient management (INM) in sustainable rice-wheat cropping system for food security in India. SAARC Journal of Agriculture 6 (2): 149-163.

Mahajan, A., Bhagat, R.M. and Trikha, A. (2003a). Fertilizing bio-fertilizers. Agriculture Today 6 (9): 52-54.

Mahajan, A., Choudhary, A.K. and Bhagat, R.M. (2002). Integrated plant nutrient management (IPNM) system for sustainability in cereal based cropping system. Indian Farmers' Digest 35 (7): 29-32.

Mahajan, A., Choudhary, A.K., Jaggi, R.C. and Dogra, R.K. (2003b). Importance of bio-fertilizers in sustainable agriculture. Farmers' Forum 3(4): 17-19.

Mahajan, A., Gupta, R.D. and Sharma, R. (2008b). Bio-fertilizers - A way to sustainable agriculture. Agrobios Newsletter 6 (9): 36-37.

Mahajan, A., Sharma, R., Gupta, R.D. and Trikha, A. (2007b). Organic farming and sustainable agriculture. Agriculture Today 10 (3): 48-50.

Nazareth, M. (2014). Combination of man's greed and climate change could be behind Kashmir disaster. The Kashmir Times 71 (258):7 [September 19, Friday].

Padder, B.A., Mahajan, A., Sharma, R. and Kumar, S. (2004). Protecting food by radiation - The need of the day. Indian Farmers' Digest 2 (10): 37-38.

Pandit, J. (2010). Climate change in India. The Daily Excelsior 46 (332):8 [December 20, Thursday].

Qazi, M. (2016). No hope of justice for Indian Farmers. The Kashmir Times 73 (7):6 [January 7, Thursday].

Raina, S.K., Suri, V.K., Jagi, R.C. and Mahajan, A. (2005). Get your soil tested for obtaining maximum return from the application of fertilizers. Kheti Dunia 10: 37.

Rana, J.C. (2013). Challenges and Roadmap for Himachal agriculture. The Tribune 133 (32):6 [February 2. Saturday].

Sahai, P., Singh, A.P. and Chander, R. (2014). Bio-fertilizers: A renewable source of plant nutrients. Indian Farmers' Digest 47 (6): 11-14

Sahai, S. (2011). Corporate food saviours. The Tribune 131 (75):11 [March 17, Thursday].

Sharma, B.D; Kumar, R. Mukhopadhyay, S.S and Jassal, H.S. (2007). Productivity potential of soils in Humid Shiwaliks tract for different land uses (pp. 75-86). In Natural Resource Management for Sustainable Hill Agriculture (Eds. S. Arora, S.S. Kukal and V. Sharma). Soil conservation Society of India, Jammu Chapter, SKUAST Jammu, J&K State, India.

Sharma, V. (2015). Fertilizer overuse eating away Punjab soil nutrients. The Tribune 135 (233): 1 & 10 [August 25. Tuesday].

Singh, J. (2016). Help debit ridden farmers. The Tribune 136(18): 8 [January 18, Monday].

Suryakumar, P.V.S. (2014a). Rethinking agricultural extension. Agriculture Today 17 (4):39.

Thusoo, R.K. (2014). Agriculture in crisis. The Daily Excelsior 50 (65):8 [March 7, Friday].

2

Concept of Sustainable Agriculture

Abstract

Sustainable agriculture is the successful management of resources for agricultural production to satisfy the changing needs of today and tomorrow, while safeguard the environment and conserving natural resources. Moreover, it is an emerging concept which aims at sustenance of agricultural productivity, maintenance of environment both physical (land, air, water) and biological (plants and animals) and to satisfy changing human needs according to changed climate. Thus, in this chapter of the book - aspects of sustainable agriculture, concept of sustainability and sustainable agriculture, aims and objectives of sustainable agriculture, measurement of sustainability, sustainable issues, sources for achieving sustainable agriculture, India's Action Plans regarding sustainable agriculture have been presented. Conclusion, future research strategies/priority and study questions have also been dealt with.

Keywords: Sustainable agriculture; sustainability; agriculture production; natural resources; physical environment; biological environment; climate change; India's Action Plan

Agricultural production has made rapid strides in the past four to five decades and has enabled India to achieve self-sufficiency in food grain production. This was only through "Green Revolution" which ushered during 1967-68 onwards, when use of old agricultural practices were mostly done away with. Neglect of old or indigenous agricultural practices such as use of organic manures (farm yard manure - FYM, compost, green manure), cultural control methods like sanitation, tillage/removal of weeds, adjusting sowing/harvesting time crop rotation, trap crops, water and fertilizer management and host plant resistance, and use of plant based insecticides like neem or nim *(Azadirachta indica),* drek *(Azadirachta indica),* nirgudhi or bana *(Vitex negundo)* vis-a-vis biological control measures, has resulted in outbreak of ill-effects in human beings and animals through contaminated food stuffs (Gupta, 2006a). Further, it is added that "Green Revolution" occurred mostly in the irrigated lands where rice and wheat cropping sequence is followed. Growing of high yielding varieties of rice and wheat, responding to more doses in the irrigated areas continuously with indiscriminate use of N, P_2O_5, K_2O fertilizers and inorganic pesticides have not only made the soils sick and deficient in other plant nutrients, but have also either stabilized their yields or have declined. It is not only this, the agriculture scenario in states like Haryana, Punjab and Jammu region of Jammu and Kashmir State, dominated by rice-wheat rotation is causing degradation to soil fertility and fall in underground water table, causing thereby, low productivity (Dhaliwal 2013a). Thus, low productivity in rice and wheat, accompanied with low yields of pulses and oil seed crops, would not project and fulfill food requirement of a growing population. It would be necessary to almost double the rate of growth of agricultural production in the next ten years from the present growth rate of rice (2.3%) and wheat (3%) obtained in the last 30 years (Bhan, 2007). Even at a 2 per cent growth rate, yields for rice and wheat will have to reach 4.7t ha^{-1} and 4.2 t ha^{-1}, respectively in the year 2020. For this purpose, agriculture should receive priority in the 12th Five Year Plan and on wards. The government's policy should therefore, focus on developing sustainable technology that could produce more from less (Patel, 2014)

In the light of the facts as stated above, it becomes imperative to develop such an agricultural or food production system which is able to produce yield of crops regularly in the long run on sustained basis. Infact, sustainable food production systems should be at the heart of global policy. The present chapter of the book, thus, pertains to the concept of sustainability and sustainable agriculture, its aims and objectives, implications etc.

2.1 Aspects of Sustainable Agriculture

The five important aspects of sustainable agriculture as stated by Kanwar (1994) are:

1. Meeting the changing needs of today and tomorrow.
2. Economic viability at enhanced productivity level.
3. Successful management of resources; external or internal, renewable or non-renewable.
4. Maintenance, perfectly enhancement of quality of environment.
5. Conservation of natural resources of soil and water.

2.2 Concept of Sustainability and Sustainable Agriculture

Now-a-days, the word sustainability is of common use in various development sectors. But, what it has its real meaning has been discussed here under:

2.2.1 Sustainability

According to a dictionary, sustainability denotes to hold up or to keep from falling. In other words, it may be defined as an *"attempt or effort to run any developmental system on profitable basis through its regular output, continuously over a long time, without harming the natural resources."* In agriculture sector, sustainability basically refers to the capacity of the crops to remain productive, while maintaining the soil, the water and other resources. The soil that sustains crop plants also requires to be "fed" as much as do the people, who benefit from the food it produced. Similarly, to obtain continuous increased yield of crops, requirement of water becomes an imperative component for the growing crop plants. Genetic resources of various crops are also necessary to maintain their yields.

2.2.2 Sustainable Agriculture

While adding various inputs to soils like seeds, organic manures, fertilizers and pesticides, and water, one must ensure that they have no harming effects on environment/ecology. Thus, according to the Technical Advisory Committee of the Consultative Group on International Agricultural as mentioned by Reijntjes et al. (1992), *"Sustainable agriculture is the successful management of resources for agriculture to satisfy human needs while maintaining or enhancing quality of environment and conserving natural resources".* However, many scientists have propounded a number of other definitions concerning with "Sustainable Agriculture". Some of them are as follows:

2.2.2.1 "Sustainable agriculture, in accordance with Lockeretz (1988), is a loosely defined term that encompasses a range of strategies for addressing many of the problems that afflict agriculture." These problems consist of loss of soil productivity from excessive erosion/intensive cultivation accompanied with loss/depletion of plant nutrients both macro and micro. Occurrence of surface and groundwater pollution from pesticides and fertilizers, and sediments, shortages of nonrenewable resources, and low farm income due to high production costs, are the problems which the farmers are facing these days.

2.2.2.2 American Society of Agronomy (1989) has defined sustainable agriculture as the "System that over the long time, (i) enhances environmental quality and the resource base on which the agriculture depends; (ii) provides food and fibre for human needs; (iii) it is economically viable; and (iv) enhances the quality of life for the farmers and society as whole"

(Takkar, 1996).

2.2.2.3 "Sustainable agriculture is a philosophy and system of farming. It has its roots in a set of values that reflect a state of empowerment, awareness of ecological and social realities and one of ability to take effective"

(Varade, 1992).

It involves design and management producers that work with natural process to conserve all resources, promote agroecosystem resilience and self-regulation and minimize un-favourable environmental impact, while maintaining or improving farm profitability.

Depending upon the institutions or countries involved, there are frequently additional focuses on the sustainability as portrayed in the definitions given below (Eswaran, 1992):

A. *University of Nebraska (Granatstein,* 1988): "A management strategy which helps the procedure to select hybrids and varieties, cultural practices, soil fertility programmes, and pest management approaches which reduce costs of purchased inputs, minimize the impact of the system on the immediate and off-farm environment, and provide a sustained level of production and profit from farming."

B. *Food and Agricultural Organization of the United Nations, Rome (1989):* "Sustainable agriculture should involve successful management of resources for agriculture to satisfy changing human needs while maintaining or enhancing the quality of environment and conserving natural resources."

C. *Agriculture Canada (Baier, 1990):* "Sustainable agriculture systems are those that are economically viable, and meet society's needs for safe and nutritious food, while conserving or enhancing Canada's natural resources and the quality of the environment for generations."

D. *Technical Advisory Committee (1988):* The Technical Advisory Committee of Consultative Group on International Agricultural Research (1988), as quoted by Kanwar (1994), states that "Sustainable agriculture is the successful management of resources for agriculture to satisfy the changing human needs, while maintaining or enhancing the quality of the environment and conserving natural resources."

Thus, sustainable agriculture can be defined as "a management system for ameliorating the natural resources including environment and that provides food, livelihood and income on sustained basis for the present generations and the posterity. And the sustained produce so obtained must be economically viable and socially acceptable."

2.3 Aims and Objectives of Sustainable Agriculture

The main aim/goal of sustainable agriculture includes:

- To develop such farming systems which are more productive and profitable?
- To enhance health and safety. It refers to that various food stuffs produced by agriculture, horticulture, olericulture and animal husbandry must be free from any kind of contamination and pollution.
- To conserve the natural base - soil. Soil is the key component of food production system as it forms the pivot and the most important natural resource. It must be managed scientifically in determining the sustainable productivity of an agro-ecosystem. While managing the soil as resource, we must take into account the present and future needs of the people, its productivity and quality needs to be maintained continuously.

- To protect or safeguard the environment, it represents that life supporting systems viz., land/soil, air and water should not be degraded, disturbed and polluted.
- To have enough yields of the various crops sown by the farmers, enabling them to make self-sufficient for food requirements and/or enough income, and gain sufficient returns to warrant the labour and costs involved.
- To have social justice which refers to that resources and power are distributed in such a way those basic requirements of all the members of the society are met with. Moreover, their rights to land use, adequate capital, technique and assistance and marketing opportunities are ensured.
- To have adaptability, which denotes that rural communities are capable of adjusting to the continuous change for farming population growth, new strategies/policies, market demands etc?

2.4 Measurement of Sustainability Quantitatively

For measuring the sustainability quantitatively, we must accept the criteria proposed by Lynam and Herdt (1988), which states that a sustainable system is the one with non-negative trend in measured output. In other words, a food production system can be sustainable over a defined period if the outputs do not decrease when inputs are not increased (Kanwar, 1994).

2.5 Parameters for Measuring Soil Quality

There are many parameters for measurement of soil quality. Kanwar (1994) stressed upon to establish standards or threshold values of these parameters to quantify the changes therein. For all practical purposes, the following changes are required to be measured as indicators of the changes in the quality of soil under the influence of external or local inputs over a defined time:

- Changes that occur in soil depth and water holding capacity.
- Changes which take place in soil reaction, salinity and sodicity, acidity vis-a-vis in nutrients availability.
- Changes which happen to occur in hydrological regime, water-logging and water depletion.
- Changes produced in erodibility and soil loss especially of surface layer.
- Changes in the physical environment which are caused due to cultivation and soil erosion or any other factor.
- Changes that are resulted in soil productivity as an integrated index of all the above said factors.

Thus, any system of agricultural or food production, which over a period causes adverse effect on these important features determining the changes in soil quality would be considered contributing to unsustainable agriculture. Eswaran (1992) has described the following ten soil qualities which are very important for sustainable agriculture:

1. Available water holding capacity
2. Nutrient retention capacity - cation exchange capacity
3. Nutrient availability - pH and base saturation
4. Nutrient fixation
5. Chemical constraints - acidity, sodicity
6. Physical constraints - low hydraulic conductivity, permeability, high bulk density, and crusting
7. Effective soil volume-depth to root restricting layer, stoniness
8. Surface tilth
9. Erodibility
10. Water-logging

According to Lai et al. (1990), as reported by Varade (1992), the sustainability can be expressed in terms of output-input as a function of time. It can also be expressed as output per unit consumptive use of the most limiting manageable resources. The following equation has been suggested:

$$S = F(P, E, D, C, Q)\, T$$

Where, S is sustainability index, P is agronomic productivity, E is total energy input, D is a measure of degradative process, C is carbon flux from soil and biomass into the atmosphere, Q is a measure of water quality and T is time.

Parameters D and Q can be defined in relation to the specific process. Soil degradative process may be due to physical, chemical or biological process. Consequently, the parameter D may be expressed in terms of rate of soil erosion, reduction in soil organic matter or cation exchange capacity, changes in soil porosity or decrease in infiltration rate. The parameter Q may be expressed in terms of total sediment or dissolved salt or concentration of pollutant - NO3, pesticides etc.

2.6 Sustainability Issues

In spite of significant agriculture growth, there is failure of several new farming systems owing to requiring high inputs. These farming systems are now showing of fatigue (Virmani, 1994). Out of many of the current farming systems, the following two are quoted as examples:

A. Cotton based cropping system.

B. Rice-wheat cropping system.

The crop yields of both the aforesaid cropping systems have either plateaued or, are showing a declining trend in some eco-regions like Indo-Gangetic plains of India, Pakistan, Bangladesh and Nepal (Mahajan and Gupta, 2009). The hazards of chemical residues have tremendously increased due to unbalanced or excessive application of nitrogenous fertilizer and indiscriminate use of the pesticides.

The problems of pesticide use have assumed serious proportions because of the non-uniform use of the plant protection chemicals in the above said cropping systems and indiscriminate use and misuse, and abuse of these chemicals. For example, in the rainfed areas of India, which constitutes about 60 per cent of its total cropped area, the pesticide use is minimal. However, in the irrigated areas, the pesticide consumption has been high for the cultivation of some crops. Crops like cotton, rice, fruits and vegetables which accounted for less than 35 per cent of the cropped area, consumed 84 per cent of the cropped area (Subba Rao, 1999; Gupta and Singh, 2008). Contrary to this, wheat, coarse cereals, millets and pulses contributing to 54 per cent of the cropped area consumed only 3 per cent of the used pesticides.

2.7 Reckless Use of Pesticides

Reckless use of pesticides in Punjab and Haryana has not only polluted the soil, but has also given rise to brown hopper (a very common pest of rice crop) which was hitherto unknown. It is further added that only a small amount of the applied pesticides in fields reaches the organisms that are supposed to be controlled. Major portion reaches the soil, water and air where it has a damaging effect on living organisms. Aquatic organisms are particularly sensitive to pesticides. Pesticides that do not break down easily are absorbed in the food chain and cause considerable damage to insects, insect consuming animals and prey birds and eventually to the human beings (Gupta, 2005c; Gupta, 2006a). Pesticide residues in various food stuffs have become a matter of threat to man. Even small quantities of pesticide residues ingested daily along with food, can build-up high levels in body fat. The long-term effects of pesticide residues in human beings include a number of infirmities and diseases. Details of various infirmities and diseases have been appeared in the literature (Gupta and Singh, 2008). It has been observed that exposure of pesticides besides leading to cancers - prostrate, breast, brain tumors, leukemia-enhances losses of female foetuses, endocrine disruption and related syndromes also. Asthma and allergies, cerebral damage and dysfunction, male sterility, premature aging, anaemia, etc. are the other diseases/disorders which have been produced by the indiscriminate use of pesticides.

The above findings are based on the study conducted by Agriculture Health Study (1993), National Cancer Institute, USA and Pesticide Resource Data - US Environment Protection Agency and Pesticide Literature Review, April 2004, Ontario College of Family Physicians, Canada.

2.8 Indiscriminate Use of Fertilizers

Like indiscriminate use of inorganic pesticides, injudicious application of N, P2O5, K2O fertilizers to various soils has created un-sustainability in agricultural production. Indiscriminate use of nitrogenous fertilizer without phosphate and potash has damaged the productivity of crops especially of rice and wheat. This could be attributed to scrutinizing the ratio of N, P_2O_5, K_2O fertilizers which has been worked out to be 8.9 to 5.7 : 2.2 to 3.0 : 1 for 1998-99 to 2004-05 as against 4:2:1 (Gupta and Singh, 2006). Indiscriminate use of fertilizers has not only decreased the agricultural productivity of the land but has also created a number of health hazards both to the soils and human beings.

Soils showed the deficiency of secondary nutrients (S, Mg, Ca) and micro-nutrients (Zn, Fe, B). There was loss of beneficial macro and micro-organisms. Use of nitrogenous fertilizers alone has produced negative effects in acidic soils in respect of maize and wheat, and other crops. Palampur in Himachal Pradesh and Ranchi in Jharkhand, India provide good evidence, where plots receiving nitrogenous fertilizer alone produced low crop yields in maize and wheat even when compared with control (Gupta and Sharma, 2006). Contamination of ground water with NO_3 and heavy metals (Cd, Cr, Cu, Fe, Hg, Ni, Pb etc.), incidence of F and eutrophication are other effects which have occurred due to use of indiscriminate fertilizers.

Application of nitrogenous fertilizers alone without using phosphate and potassic fertilizers, render to form some specific types of compounds in the plant cell during N-metabolism, Such compounds named as nitrosamines, are suspected to act as carcinogenic agents (Gupta and Singh, 2008). Loss of biodiversity damage to water quality and land degradation has also increased due to following of intensive agriculture (Virmani, 1994).

Recent studies have shown that indiscriminate and injudicious use of chemical fertilizers and pesticides in agriculture has resulted in number of associated adverse effects like environmental pollution, ecological imbalances, pesticide residues in food grains, fruits, vegetables, fodder, soil and water, pest resurgence human and animal health hazards, development of resistance to pests etc. (Bhadauria and Singh, 2013). Therefore Government of India has adopted Integrated Pest Management (IPM) as cardinal principle and main plank of plant protection in the overall Crop Production Programme. IPM has not only found an eco-friendly approach but also proved a boon to sustain agriculture production.

Green manuring practice has been proved a cheap monetary input for sustainable agriculture which is attributed to enhance soil fertility and supplement for nutrients (Ara et al., 2009) as well as to prevent soil erosion and to control weeds. It also improves soil structure besides serving a very good way of increasing the fertility of the soil and eventually its productivity (Kumar et al. 2011).

2.9 Means or Sources for Achieving Sustainable Agriculture

The means of achieving sustainable agriculture are application of "low input methods" and skilled management. These means seek to optimize the management and use of internal production inputs i.e., various on farm resources, which provide sustained crop yields as well as livestock production at deserted or acceptable levels and result in economically profitable returns. This approach stresses upon different cultural and management practices. These practices mainly consist of to follow right kind of crop rotation, cropping system, recycling of animal manures, use of well decomposed crop residues as source of plant nutrients and conservation tillage to maintain the soil productivity (Parr et al. 1989; Parr and Homick, 1990; Mahajan and Gupta, 2009).

Balanced use of chemical fertilizers viz., nitrogenous, phosphatic and potassic fertilizers by maintaining their ratio in the form of 4 : 2 : 1 to various crops, use of zinc sulphate ($ZnSO_4$), gypsum ($CaSO_4$), ferrous sulphate ($FeSO_4$) and borax to control deficiency of Zn, S, Fe and B in soils, respectively. Adoption of all soil and water conservation measures i.e. agronomic, engineering or mechanical and biological, are the other sources which can assist in bringing forth the sustainability in agriculture. Apart from following agronomic, mechanical or engineering and adoption of diversified agriculture has proved beneficial, especially for the peasants dwelling in *Kandi* belts of India covering parts of Punjab, Haryana, Uttarakhand and Jammu Region of Jammu and Kashmir State (Gupta, 2009b). The first and the foremost strategy for crop diversification of *Kandi* belt is afforestation of the eroded/denuded areas (Anonymous, 2013a).

Afforestation of local tree species like kikar (*Acacia nilotica),* phulai (A. *modesta),* khair (A *catechu),* shisham *(Dalbergia sissoo),* sarin (*Albizia lebbeck),* should be preferred to grow. Among the local fruits which have been found suitable for growing in the degraded or eroded areas are: phalsa (*Grewia asiatica*), amla (*Phyllanthus emblica),* jamun *(Syzygium cumini),* ber (*Zyzyphus mauritiana),* Shatoot (*Morus alba*) and rattanjot *(Jatropha curcas)* have been found to grow very well in soils of *Kandi* belt of Jammu, Kathua, Rajouri and Udhampur (Sharma et al., 2009). Medicinal plants like neem (*Azadirachta indica),* ashwaganda (*Withania somnifera),* kaal megh *(Andrographis panniculta Nees),* kuargandal (*Aloevera* spp.), neer brahmi (*Bacopa moniera Linn),* shatawar *(Asparagus racemosus)* and safed musli *(Chlorophytum* borivilianum) have shown very good growth in *Kandi* belt of Akhnoor and Jammu tehsils of Jammu district, Jammu region of Jammu and Kashmir state (Gupta et al., 2016).

Apart from the above means, use of recent concepts such as zero tillage, integrated nutrient management, bio-fertilizers including vermicompost, indigenous technology have played a big role in sustaining the productivity in various crops, particularly in rice and wheat cultivation vis-a-vis cotton (Mahajan and Gupta, 2009; Gupta, 2015a). As such all these sources or means are required to be utilized to maintain or sustain the agricultural productivity. Details with regard to use of bio-fertilizers, integrated nutrient management, vermicompost and zero tillage have been described in separate chapters of this book.

As most of the hilly areas of India, Pakistan and Nepal are deforested, denuded and overgrazed, thereby, resulting in unsustainable production in various crops, so adoption of watershed development or watershed management is one of the important means in this context (Mahajan, 2004). Detailed account of watershed management stands discussed separately.

There is no doubt that like other parts of the country, the North Western Himalayan states like Uttarakhand, Himachal Pradesh and Jammu & Kashmir including sub-mountainous areas of Punjab and Haryana have gone through a number of revolutions like Green Revolution, followed by Golden Revolution White Revolution and Yellow Revolution. However, productivity level of food grain crops like rice and wheat has either become constant or declined. Out of the several reasons to account for bringing forth the low productivity in food grains, "problematic soils and their management" is one of them which stand elaborated elsewhere (Gupta et al., 2011). Soil erosion and its control measures viz. land capability classification, conjoint use of agronomic and engineering practices and biological methods be followed. Soil acidity, salinity and sodicity are the other main problems of the soils which can be controlled using gypsum in soil salinity, alkalinity and liming in acidity.

2.10 India's National Action Plans to Promote the Objectives of Sustainable Development

India's National Action Plan stresses to:

- Maintain high population growth rate
- Increase living standards of vast majority of the people
- Reduce people's vulnerability to impact of climate change
- Promote the objective of sustainable development

2.10.1 National Response to Climate Change

India's National Action Plan on Climate Change was released by the Prime Minister on 30th June, 2008, which outlines its strategy to meet the challenge of climate change. The strategy promotes firstly, the adaptation to climate change and secondly, to enhance the ecological sustainability of India's development (Meena, 2008).

This Action Plan recognizes that climate change is a global challenge and it should be tackled through globally collaborative and cooperative efforts on the basis of equity principle. India is willing to play its role as a responsible member of the International Community to make its contribution. However, this not only requires sustainable production processes but is also responsible for sustainable lifestyle across the globe. In this respect, every citizen of the planet must have an equal share of the planetary atmospheric space. Long term convergence of per capita green house gas (GHS) emission is the only equitable base for a global agreement to overcome climate change. The Action Plan ensures that India's per capita GHS emissions would never exceed the per capita GHC emissions of developed countries, despite India's developmental imperatives.

India's National Action Plan stresses upon to maintain a high growth rate which becomes imperative for increasing living standards of the vast majority of the people of India and reducing their vulnerability owing to the impact of climate change. Accordingly, the Action Plan has identified Eight National Missions, representing multipronged, long-term and integrated strategies for achieving key goals in the sustainable development of India in general and in the context of climate change in particular. The focus is on promoting understanding of climate change adaptation and mitigation, energy efficiency and natural resource conservation. While a number of these programmes are already going on as a part of the current action, the Action Plan seeks to enhance them in scope and effectiveness and implemented them in an accelerated through time-bound plans. These Eight National Missions are:

2.10.1.1 The National Solar Mission

The main aim of the National Solar Mission is to increase the share of solar energy to the total energy mix through development of new solar technologies while attempting to expand the scope of other renewable and non-fossil options such as nuclear energy, wind energy and biomass.

2.10.1.2 The National Mission on Enhanced Energy Efficiency

This mission consists of four new initiatives namely, a market based mechanism for trading in certified energy savings in energy intensive large industries and facilities, accelerating the shift to energy efficient appliances in designated sectors, demand side management programmes in all sectors by capturing future energy savings and developing fiscal instruments to promote energy efficiency.

2.10.1.3 The National Mission on Sustainable Habitat

It attempts to promote energy efficiency in buildings, management of solid waste and modal shift to public transport including transport options based on bio-diesel and hydrogen.

2.10.1.4 The National Water Mission for Green India

Its main objective is to conserve the water by minimizing wastage and ensuring more equitable distribution both across and within states.

2.10.1.5 The National Mission for Sustaining and Safeguarding the Himalayan Glaciers and Mountain Ecosystem

The National Mission for sustaining the Himalayan Ecosystem is aimed at evolving management measures for sustaining and protecting the glaciers of Indian Himalayas and mountain ecosystem as a whole.

2.10.1.6 The National Mission for a Green India

It focuses on enhancing ecosystem services and carbon sinks through afforestation on degraded forest land in line with the national policy of expanding the forest and tree cover to 33 per cent of the total land area of the country.

2.10.1.7 The National Mission for Sustainable Agriculture

It would develop strategies to make Indian agriculture resilient to climate change through development of new varieties of thermal resistant crops, mechanisms for improving productivity of rain-fed agriculture and new credit/insurance.

2.10.1.8 The National Mission on Strategic Knowledge for Climate Change

It is intended to identify the challenges of, and the responses to, climate change through research and technology development and ensure findings of high quality and focused research into various aspects of climate change.

2.11 Conclusion

Contrasting edaphic requirements of crops, especially of rice and wheat or any other non-rice crop grown in the sequence on the same piece of land then go for special management practices. Sustaining the productivity of these crops is a great challenge to the scientists. The general issues on sustainability are: (i) deterioration of soil physical properties accompanied with primary (NPK), secondary (Ca, Mg, S) and micronutrients (Fe, Cu, Mn, Zn, B, Mo) problems, leading to imbalance of the farming system and resource depletion; (ii) fatigue stress in the production system is discernible in the form of decreasing factor productivity; and (iii) human health and animal problems. These issues have taken place due to indiscriminate use of chemical fertilizers, and pesticides. This, therefore, requires to fully understanding with regard to the concept of sustainability and sustainable agriculture. The same has been highlighted in this chapter of the book.

2.12 Future Research Strategies/Priorities

Over-exploitation of natural resources such as exhaustion of soil fertility with intensive chemicalised agriculture and depletion of underground water has not only created un-sustainability in agricultural productivity in various agricultural crops like rice and wheat, but also has degraded the environment. Soil pollution, water pollution and air pollution are on the increase. Other ecological disasters are like depletion of ozone layer, global warming, acid rains and extinction of several animal and plant species annually world over leading to genetic erosion. This calls for the need to conserve soil, water and other agricultural production related resource to sustain the yield of the crops. For this purpose, following strategies are essential to follow:

- Need of the hour is to switch over from highly mechanized, chemicalised and pesticide driven agriculture to low input/organic agriculture (Gupta 2019 a & b). Although in organic agriculture, the yield of the crops is low in the initial years (3-4 years after its commencement) yet later on it gets stabilized. Organic farming saves energy up to ten-fold. Now the question arises, how this energy is saved?

The current agricultural/food production is linear in design, treating inputs like energy and raw materials as infinitely available (which actually they are not) and the environment as infinitely capable of absorbing waste (which it is not). Hence, this system of food production is not sustainable. To change this, we require a tax system that factors in the full cost of all these finite items and uses the proceeds to subsidize organic, low input and localized agricultural systems. In contrast, organic production systems are an example of sustainable circular methods of food production in harmony with the natural ecosystem. In many countries, growing of organic food production is increasing.

- Developing a sustainable food production system should become a major government policy based on setting targets for:
- Sustainable production
- Import substitution
- Fair trade
- Local sourcing

These targets are required to be achieved within specific time scales.

- The very large external/environmental costs of transportation must be internalized. It means that after agricultural production, the costs of water purification and pesticides runoff or of damage to the environment from long distance transportation that exacerbates climate change must be taken into account.
- All food products need to be labeled to indicate the environmental impact of distribution. And organic and other assurance schemes should take the lead by introducing the proximity principle into certification.
- A sustainable food production system should promote human health and certainly not harm him. For example, there is increasingly convincing evidence that chemicalised or industrialized farming production systems do the reverse. As for example, with continuous massive increase in the use of pesticides have shown their contamination in various food stuffs causing, thereby, brain diseases in human beings like Parkinson's and Alzheimer's.
- Globally, what is making so much of the world's food systems unsustainable is climate change. The drying out of croplands, the growth of continental and Indonesian fire and rising frequency and ferocity of storms, cyclones, flooding and rising sea level, increasingly are putting a risk to the prospect of feeding up to 9 billion people on this planet by 2050.

Hence, agricultural production systems require to be thoroughly drifted in accordance with climate change.

- Attaining sustainable agricultural and allied sectoral productivity in tune with climate change/environmental degradation is more challenging in developing and underdeveloped countries than those in developed. As green revolution would not have been possible without the development of dry land agriculture, and without active involvement of the farmers/ peasants, we must involve the entire rural population in our massive efforts with aimed at increasing production.
- As the productivity of wheat crop has stabilized for the last 10 to 12 years, so adoption of conservation agriculture, precision farming, balanced use of N, P, K fertilizers and improved varieties of wheat would accelerate the growth, and thereby, the sustainability of the wheat production (Paroda, 2006).
- Integration or diversification of various farm enterprises Gupta (2019) viz., cropping, agroforestry systems like agro-horticulture, agro-silviculture, agro-olericulture, agro-animal husbandry and rearing of livestock, offer great opportunities not only to sustain agricultural productivity but also to raise farm income especially of small and marginal farmers.
- Like wheat, the yield of rice has also stagnated so need of the hour is to evolve a new strategy by which area especially under rainfed rice (having low productivity) could be reduced with simultaneous increase in yield during hybrid rice, IPM and conservation agricultural technologies. Main focus should be now on rainfed lowland rice producing states in the eastern region, including eastern Uttar Pradesh, Bihar, Orissa, West Bengal and Assam. Hybrid rice technology has shown good potential and must be exploited in this region. Also increased use of conservation agriculture (zero tilled rice), rice transplanter, super-granulated urea, IPM (like in Indonesia) would help in sustainable rice production (Paroda, 2006).

Study Questions

1. What is meant by sustainable agriculture?
2. How we can define sustainability?
3. Write in brief about the main objectives of sustainable agriculture.
4. What are the various means or sources for achieving sustainable agriculture?

5. Write a paragraph on India's National Action Plan with regard to sustainable development.
6. What are the various sustainability issues? Write one of them in details.
7. How the following define sustainable agriculture?
   i. American Society of Agronomy
   ii. FAO
   iii. Agriculture Canada
   iv. Technical Advisory Committee

References

Anonymous. (2013a). Crop diversification: An ideal way to deal with food insecurity. Agriculture Today 19 (4): 50-51.

Ara, T., Gul, A. and Paul, S. (2009). Green manuring: Eco-friendly to the environment. Indian Farmers' Digest 42 (9): 20-22.

Baier, W. (1990). Characterization of the environment for sustainable agriculture in semi-arid tropics (pp. 90-128). In: Sustainable Agriculture: issues, perspectives and prospects in semi-arid tropics. Proc. Int. Symp. on Natural Resource Management for Sustainable Agriculture (New Delhi), Indian Soc. Agron.

Bhadauria, B.P. and Singh, Y. (2013). Exploration of bio-fertilizers and bio-pesticides in IPM for sustainable agriculture. Indian Farmers' Digest 46 (7): 14-17.

Bhan, S. (2007). Conservation agriculture as strategy for enhancing the integrity of natural resources and productivity in South Asia. J. Soil and Water Conservation 6 (4): 153-161.

Dhaliwal, S. (2013a). Centre accepts diversification proposals, sanctions, Rs 224 crore. The Tribune 133 (181): 5 [July 11, Thursday].

Eswaran, H. (1992). Role of soil information in meeting the challenges of sustainable land management (18^ Dr. R.V. Tamhane memorial lecture). J. Indian Soc. Soil Sci. 40 (1): 6-24.

FAO. (1989). Sustainable agriculture production: For international agricultural research. Report of the Technical Advisory Committee, Consultative Groups on International Agricultural Research (CGIAR), Pergamon, New York.

Granatstein, D. (1988). Reshaping the bottom line: On-farm strategies for sustainable agriculture. Land Stewardship Project (unpublished mimeo).

Gupta, R.D. (2005c). Environmental Degradation of Jammu and Kashmir Himalayas and Their Control. Associated Publishing Company, Karol Bagh, New Delhi (India).

Gupta, R.D. (2006a). Environment Pollution: Hazards and Control. Concept Publishing Company, New Delhi (India).

Gupta, R.D., Gupta, S.K. and Bhardwaj, S.D. 2016. Associated Publishing Company, A Division of Astral, International Pvt. Ltd. New Delhi-110002, India

Gupta, R.D. 2019a. Organic farming need of the day. The State Times. 24(5):6.

Gupta R.D. 2019b. Organic farming need of the day. The State Times. 24(6):6

Gupta, R.D. (2009b). Diversified agriculture in Kandi belts. The Daily Excelsior 45 (320): 6 [November 18, Wednesday].

Gupta, R.D. (2015a). Indigenous and traditional agricultural knowledge. The Daily Excelsior 45 (320): 6 [July 4, Saturday].

Gupta, R.D. and Sharma, S.P. (2006). Soil pollution and its solution through integrated nutrient management (pp. 32-38). In Balanced Fertilization for Sustainable crop Production (Eds. S.P. Sharma, S.K. Sharma, R.P. Sharma and T.S. Verma). Proceedings National Seminar on INM for Balanced Fertilization held on November 17-18, 2006.

Gupta, R.D. and Singh, H. (2006). Indiscriminate use of fertilizer poses health hazards. Farmers' Forum 6 (9): 20-24.

Gupta, R.D. and Singh, H. (2008). Indiscriminate use of pesticides in agriculture: Public health issues and their control. Indian Farmers' Digest 41 (1): 8-13.

Gupta, R.D., Kumar, A. and Sharma, B.C (2011). Management of problematic soils of North West Himalayas for sustainable hill agriculture (pp. 51-63). In Sustainable Hill Agriculture - An overview (Eds. A. Kumar, B.C. Sharma and V. Sharma). Agrobios (India). Agrohouse Behind Nasrani Cinema Chopasani Road, Jodhpur, Rajasthan (India).

Kamal, M.Y. (1998). Sustainability and rural masses. Newsletter (Chennai Agro Development and Research Institute) 2 (9): 1-3.

Kanwar, J.S. (1994). Relevance of soil management in sustainable agriculture (pp. 1-11). In Soil Management for Sustainable Agriculture in Dry land Areas (Eds. T.D. Biswas, G. Narayanasamy, J.S.P. Yadav, G. Dev, J.C. Katyal and P.S. Sidhu). Indian Soc. of Soil Sci., New Delhi.

Kumar, R., Bohra, J.S., Kumawat, N., Shukla, U.N. and Singh, S. (2011). Green manuring: Boon for rice farmers. Agrobios Newsletter 11 (12): 30-31.

Lai, R., Ghuman, B. and Shearer, W. (1990). Sustainability of different agricultural production systems for a rainforest zone of southern Nigeria (pp. 186-191). Trans. 14th Int. Congr. Soil Sci. VI, August 1990, Kyoto Japan.

Lockeretz, W. (1988). Open questions in sustainable agriculture. American J. Alternative Agriculture 3 (4): 174-181.

Lynam, J.K. and Herdt, R.W. (1988). Sense and Sustainability: Sustainability as an objective in International Agriculture Research. Paper presented to CIP - Rockefeller Conference on "Farmers and Food Systems", Lima, Peru (September 26-30, 1988).

Mahajan, A. (2004). Watershed - Concept and management. Agrobios Newsletter 3 (1): 28-31.

Mahajan, A. and Gupta, R. D. (2009). Integrated nutrient management (INM) in a sustainable rice-wheat cropping system (pp.1-267). Springer Science + Business Media (www. Springer.com).

Meacher, M. (2006). Policies for sustainable food system, national and global. Sustainable food systems should be at the heart of global policy, not (as now) another device for the exercise of imperial power by the strongest nations. The Kashmir Times 63 (132): 13 [May 13, 2006],

Meena, N.N. (2008). National response to climate change. The Daily Excelsior 44 (276): 6 [October 3, Friday].

Paroda, R.S. (2006). Strategy for increasing productivity growth rate in agriculture. Farmer's Forum 6 (8): 1-6.

Parr, J.F. and Homick, S.B. (1990). Research developments in alternative agriculture in United States. In Proc. Int. Conf. On Kyusal Nature Farming (Oct. 17-21, 1990). Khon Kaen, Univ. Khon Koen. Thailand.

Parr, J.F., Papendick, R.L., Homick, S.B., Youngberg I.G. and Meyer, R.E. (1989). Sustainable agriculture in the United States - Chapter 4 (pp. 50-67). In Proc. Int. Symposium on Sustainable Agricultural Systems. Soil and Water Conservation Society, Ankeny, Iowa,

Parr, J.F., Stewart, B.B., Hornick, S.B. and Singh, R.P. (1990). Role of alternative farming methods in modern production agriculture. Adv. Soil Sci. 13:1-8.

Patel, A. (2014). Achieving food security sustainability. Agriculture Today 17 (1): 40-41.

Reijntjes, C., Haverkort, B. and Bayer, A.W. (1992). Farming for the future - An introduction to low external input and sustainable agriculture. The Macmillan Press Limited London and Bastingstoke, pp. 211.

Sharma, A.K., Gupta, R.D. and Rai, P.K. (2009). Crop diversification is the need of the hour for conserving soil and water in Kandi belt of Siwaliks (pp. 69-71). In Soil and Water Conservation Training Manual (Eds. B.S. Hansra, P. Vijaykumar, R. Nanda and P.K. Rai) organized by School of Agriculture, IGNOU, New Delhi and KVK, SKUAST of Jammu, R.S. Pura Jammu (Jammu and Kashmir State), India.

Subba Rao, I.V. (1999). Soil and environment pollution - A threat to sustainable agriculture. J. Indian Soc. Soil Sci. 47 (4): 611-633.

Takkar, P.N. (1996). Micro-nutrient research and sustainable agricultural productivity in India. J. Indian Soc. Soil Sci. 44 (4): 562-581.

Varade, S.B. (1992). Soil management or sustainable productivity. J. Indian Soc. Soil Sci. 40 (1): 25-38.

Virmani, S.M. (1994). UNCTAD Agenda-21: The new challenges for soils research. J. Indian Soc. Soil Sci. 42 (4): 516-524.

3

Soil Tillage
Essential Component for Sustaining Crop Productivity

Abstract

Tillage operations in various forms have been practised from the very beginning of growing plants. Primitive man used tools and implements to disturb the soils for placing seeds. Jethro Tull, who is considered as father of tillage suggested that thorough ploughing is necessary so as to make the soil into fine particles. The tillage systems are sequence of the operations that manipulate the soil and influence the soils physical, chemical and biological properties which in turn affect the soil productivity. This chapter of the book relates to soil tillage, in which its definition and classification i.e., various kinds of tillage viz., conventional tillage, reduced tillage, zero tillage, conservation tillage, mulch tillage, ridge tillage and deep tillage have been detailed. Besides, physical condition of soil known as soil tilth, its qualities, types of tilth, methods of improving soil tilth, ideal characteristics for soil tilth, have also been described. Some of the objectives of tillage, reasons for conservation tillage, prospects of conservation tillage in sustainable agriculture in respect of improvement in soil texture/aggregation, porosity and bulk density, soil conservation, control of wind and water erosion, water conservation, run-off and infiltration, evaporation, soil temperature, soil organic matter, weed control, plant constitute the other aspects of this chapter. In the end, conclusion, future res references have been given.

Keywords: Soil tillage; sustainable agriculture; conventional tillage; reduced tillage; zero tillage; conservation tillage; mulch tillage; ridge tillage; deep tillage; soil tilth; soil texture; porosity; bulk density; soil conservation; water conservation; organic matter; run-off; infiltration, evaporation; soil temperature; weed control; plant diseases

In modern times, soil tillage has emerged as an integral part of a crop system. Soil tillage or soil surface management has been found important in alleviating soil related constraints encountered in crop production. An important effect of soil tillage on crop productivity is through its influence on soil processes, soil properties and crop growth. At the same time, soil tillage has its impact on the environment e.g. soil degradation, water quality, emission of green house gases from soil related processes etc. The need to attain self-sufficiency in agricultural production has therefore arisen particularly in several tropical eco-regions and soil ecosystems having low nutrients carrying capacity especially in the tropics. Appropriate tillage systems are soil and crop specific and their adoption is governed by both bio-physical and socio-economic factors (Mahajan, 2006). In addition to increasing crop yields, tillage methods must also facilitate soil and water conservation, improve root system development and maintain a favourable level of soil organic matter content.

To achieve self-sufficient in food, fibre and edible oils for efficient use of natural resources, emphasis has been laid on intensive cultivation of high yielding varieties in a multiple cropping system. Through the intensive use of resources, Indian agriculture in general and its food grain segment in particular have attended an impressive increase from 51 million tonnes in 1950-51 to 257 million tonnes in 2014-15 (Ghose, 2005; Saha, 2009; Sandhu, 2015). In India, out of total geographical area of 329 m ha, about 175 m ha is subject to varying degree and forms of soil degradation. Frequent stirring of soil enhances organic matter decomposition, destruction of soil aggregates, soil erosion and continuous decline of soil productivity. So, there is need of practice that will minimize the degradation of land. The concept of conservation tillage has been emerging as a solution to this problem. Conservation tillage is not only essential for sustenance of soil productivity but is also equally essential to minimize the energy consumption in the form of fossil fuel (Mahajan et al., 2004).

3.1 Definition of Tillage

The word tillage is derived from the Anglo-Saxon words tilian and teolian, meaning to plough and prepare soil for seed to sow and cultivate, and to raise crops. It is the oldest art associated with development of agriculture. Tillage is thus defined "as ***the mechanical manipulation of the soil to obtain a seedbed of optimum soil tilth for plant growth".*** Jethro Tull, who is considered as father of tillage pointed out that the purpose of tillage is to obtain a seedbed having optimum soil tilth to break weeds, insects and diseases cycles as well as to bury plant residues, incorporate fertilizers and amendments, break surface crust and fracture plough pans or hard subsoil horizons. Presently a number of implements such as mould board plough, disk plough and harrow etc. are designed for tillage operations.

3.2 Classification of Tillage

According to Abrol et al. (2008), the tillage systems are sequence of the operations that manipulate the soil and influence the soils physical, chemical and biological properties which in turn affect the soil productivity. It is an integral part of cropping system aimed at optimizing crop production by solving specific soil related ecological constraints to crop production. Tillage alters the soil environment by reducing the strength of the soil matrix and increasing its porosity, thereby, improving the permeability of the soil for the flow of both water and gas. A granular soil structure is desirable to allow rapid infiltration and good retention of rainfall water to provide adequate moisture and air exchange capacity within the soil and to minimize resistance to root penetration. A good seedbed generally implies finer particles and greater firmness in the vicinity of the seeds. The classification of tillage system (Fig. 3.1) is discussed as

3.2.1 Conventional Tillage

It is the traditional tillage system, which combines both primary and secondary tillage operations for growing a given crop in a given geographical area.

3.2.1.1 Primary Tillage

The operation performed to open up any cultivable land with a view to prepare seedbed for growing crops is termed as primary tillage. It is mainly done with heavy implements like plough, disc harrows etc. Arable soils on an average are ploughed twice each time in diagonally directions. The aim is to cut, lift, twist, and turn the soil while incorporating organic wastes into the ploughed layer.

3.2.1.2 Secondary Tillage

Lighter and finer operations performed on the soil after primary tillage, but before and after seed placement are termed as secondary tillage. It is mainly done with light implement like rages etc.

The secondary tillage includes operations for obtaining seedbed for sowing. Since the soil is in loose form due to primary tillage, hence there may not be any necessity for ploughing the soil. Blade harrows are worked once or twice to kill the weeds, break up clods to obtain smooth/leveled surface for subsequent sowing. After the crop is sown, the soil may require further tillage to control weeds, break up crushing or create soil mulch. All such tillage operations are known as inter-tillage or intercultural operations. The reasons for less adoption of conventional tillage are attributed to increasing erosion risk, accelerating organic matter decomposition and deteriorating soil structure.

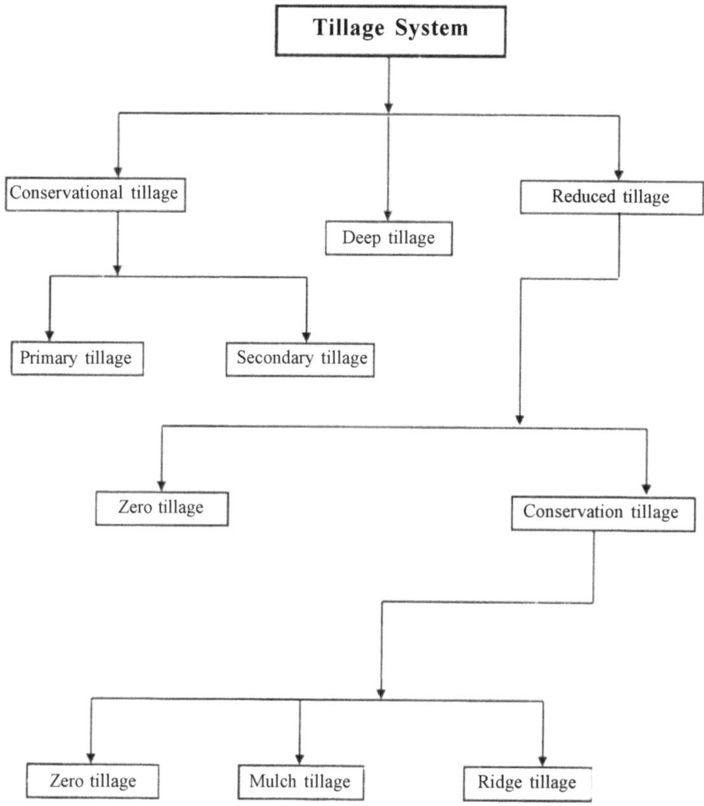

Fig. 3.1: Flow chart showing the classification of tillage system

3.2.2 Reduced Tillage

Reduced tillage or minimum tillage or limited tillage may be defined as the system in which the primary tillage operation is performed in conjunction with special planting procedure in order to reduce or eliminate secondary tillage operations or the minimum soil manipulation, which is necessary to meet tillage requirements for crop production, is called minimum tillage. In other words, minimum tillage means elimination of one or more tillage operations from a conventional tillage. Land preparation and seeding may be completed in one operation. The main disadvantages of reduced tillage are as under:

- Sowing operations are difficult with ordinary implements.
- Lower seed germination.
- Modulation is affected in some leguminous crops such as pea, beans etc.
- More nitrogen has to be added as the rate of decomposition of organic matter is slow.

3.2.2.1 Zero Tillage

Zero tillage may be defined as the "practice of not adopting any tillage operations or to adopt minimum tillage operations in soil so that it is least disturbed'. It is a method of planting crops that involves no seedbed preparation other than the opening of soil for placing seed at the intended depth. It usually involves opening a narrow slit or punching a hole into the soil, where chemical weed control is normally used. In other words, the soil is left undisturbed from harvest to planting except for nutrient injection. Through this method reduction in soil erosion up to 70 per cent has been obtained. Zero tillage is possible where the soil has coarse textured surface, good internal drainage and adequate quantity of residue as mulch. The other advantages of zero tillage are as under:

- Less time is required for sowing.
- Less labour requirement involves in this kind of tillage.
- Less diesel requirement involves in the system.
- Less energy is required.
- Low costs are involved.
- There is an avoidance of delay in wheat sowing crop after rice crop harvesting.
- It improves physical and chemical properties of the soils.
- Increased soil moisture conservation takes place. It is due to increased surface crop residues, resulting in the increased conservation.
- Zero tillage leads to nutrient build up in soil as a result of increased microbial activities.
- Higher yield and income are possible.
- There is reduction in the population of *Phalaris minor.* As the soil is not tilled so the seeds of *Phalaris minor* present at depth are not able to germinate.
- Zero tillage helps in building up organic matter in soil and it resembles the original soil of native grassland as the soil is least disturbed (Anshu and Bhardwaj, 2004).
- It helps to ameliorate the environmental conditions as it assists to lessen the use of fossil fuel because of minimum tractor use. As a result there is reduction in emission of greenhouse gases.
- The quality of seedbed improves and there is reduction in moisture loss from the soil.

- Fertilizer placement is along with seeds, which increases the fertilizer use efficiency.

3.2.2.2 Conservation Tillage

Initially conservation tillage was known as the minimum tillage, but in 1977, Soil Conservation Service of USA changed the term to conservation tillage. According to the Mannering and Fenester (1983), "any tillage operation, which reduces loss of soil and water relative to conventional tillage, is termed as conservation tillage", it has two characteristics:

- It is non-inversion tillage that retains some residue mulch on the surface.
- It involves tillage operations less intensive than conventional tillage.

According to Conservation Technology Information Center (1992) conservation tillage defined as "any tillage and planting system in which at least 30 per cent of the soil surface is covered by residues after planting in order to reduce soil erosion by wind'. No tillage, ridge tillage and minimum tillage practices are different forms of conservation tillage as long as adequate crop residue levels (30% of ground covered by crop residue) are maintained on the soil surface. The essence of conservation tillage lies in minimum disturbance of soil and maintenance of some residue at the surface.

3.2.2.2.1 Mulch Tillage

In mulch tillage, the soil is disturbed before planting. Tillage tools such as field cultivars, disks, chisels, sweeps or blades are used. Weed control is accomplished with herbicides or by cultivation. Mulch tillage is also used to manage residues of various crops that use sub-surface tillage for loosening the soil and at the same time leaving the residue on the surface to control the erosion hazards (Saxena and Mishra, 2004).

3.2.2.2.2 Ridge Tillage

In ridge tillage, the soil is left undisturbed from harvest to planting except for nutrient injection. Planting is completed in a seedbed prepared on ridges with disk openers, sweeps or row cleaners. Residue is left on the surface between the ridges. Weed control is accomplished with herbicides or cultivation.

3.2.3 Deep Tillage

It is the tillage system in which tillage is done to the deeper depths more than 20 cm. It is resorted in order to break hard pans and facilitate the plant roots into deeper layers for better crop growth and performance (Mahajan and Gupta, 2009). According to National Resources Conservation Service Conservation

Practice Standard (NRCS, Georgia 2008), deep tillage is defined as *"Performing tillage operations below the normal tillage depth to modify the physical or chemical properties of a soil"*. Deep tillage operations should be preferred when soil moisture is less than 30 per cent of field capacity, according to "feel test". Tillage equipments such as chisels, sub-soilers or rippers, to reach the required depths, are used. Its purpose is to:

- Fracture restrictive soil layers
- Mix soil deposits from wind or water erosion or flood over wash
- Reduce concentration of soil contaminants which inhibit plant

Deep tillage ameliorates puddling induced traffic pan and improved wheat yield is shown in Table 3.1. In both the study years, the grain yield of wheat was significantly higher in plots where deep tillage (T_4) treatment was adopted which were followed by conventional tillage (T_2 and T_3). This could be associated with improvement in soil physical properties and plant growth parameters. During the second year of the crop study, the grain yield in T_7 (where deep tillage-no tillage sequence was followed with a puddled rice crop in between) was significantly higher than T_1, T_2 and T_5. Since the puddling effects are mostly confined to the upper layers and the lower layers are left as such which may be the reason of more pronounced effect of deep tillage in lower layers. This also clearly indicated the positive effect of deep tillage even after inclusion of a puddled lowland crop. However, significantly lower grain yields were recorded in no tillage during both the years of the study due to crop-weed competition. This practice therefore, requires to be done after every two to four years.

Table 3.1: Effect of deep tillage on yield of wheat after rice in Himachal Pradesh (Adapted from Mahajan and Bhagat [2006])

T treatments	Grain yield (Mg ha'^{1})	
	2002-03	2003-04
T_1	2.10	2.16
T_2	2.39	2.66
T_3	2.86	3.00
T_4	3.35	3.53
T_5	2.36	2.55
T_6	2.90	2.76
T_7	3.39	2.99
LSD (0.05)	0.18	0.15

*Ti - No tillage, T_2 - Tillage up to 10 cm depth, T_3 - Tillage up to 20 cm depth, T_4 - Tillage up to 40 cm depth, T_5 - Tillage up to 10 cm depth, T_6 - Tillage up to 20 cm depth, T_7 - Tillage up to 40 cm depth during first wheat growing season (2002-03) and T, - No tillage, T_2 - Tillage up to 10 cm depth, T_3 - Tillage up to 20 cm depth, T_4 - Tillage up to 40 cm, T_5 - No tillage, T_6 - No tillage, T_7 - No tillage during second wheat growing season (2003-04)

Adapted from Mahajan and Bhagat (2006)

Tillage operations can also be broadly classified into two group's viz., on season tillage and off season tillage, which are based upon the time during which they are carried out. Tillage for a crop from the start of a crop season to the crop harvest is known as "on season tillage". It consists of preparatory tillage and inter-tillage. Preparatory tillage refers to tillage operations done for raising a crop up to sowing or planting as the case may be. It is of two types i.e., primary tillage and secondary tillage. After the crop is sown, the soil may require further tillage to control weeds, break up crushing or create soil mulch. All such tillage operations are known as inter-tillage or intercultural operations. It includes inter cultivating, harrowing, hoeing, weeding, earthling-up, forming ridges and furrows etc. Inter tillage helps to incorporate top dressed manures and fertilizers, to earth up and to prune roots. Secondly, off season tillage refers to the tillage operation is done for conditioning the soil during uncropped season with the main objective of water conservation, leveling to the desirable grade, leaching to remove salts for soil reclamation reducing the population of pest and diseases in the soils. The different tillage implements and their works are as under:

- Chisel plough - Subsoil breaking.
- Disc plough - Deep ploughing in grass fields.
- Rotary plough - Cut and pulverizes the light soil.
- Ridge plough - Earthing-up and form ridges and furrows.
- Tractor drawn cultivator - Destroy weeds and helps breaking clods.
- Harrows - Preparation of seedbed, destroy weeds.
- Seed dressing drum - used to treat the seed with chemicals.

3.3 Soil Tilth

The physical condition of a soil described by its bulk density, porosity, structure, roughness and aggregate characteristics is related to water, nutrient, heat and air transport; stimulation of microbial and micro faunal population and microbial processes; and impedance to seedling emergence and root penetration (Mahajan et al., 2007). In other words, it is a physical condition of soil as related to its ease of tillage, fitness as a seedbed, and its impedance to seedling emergence and root penetration.

Soil tilth varies not only on granule formation but also depends on such factors as bulk density, soil moisture content, degree of operation, rate of water infiltration, drainage pattern and capillary water capacity (Saxena and Mishra, 2004). As might be expected, tilth often changes rapidly and markedly as for example, the working properties of fine textured soils may be altered abruptly by a change in consistence caused by a slight change in moisture.

3.3.1 Qualities of a Good Soil Tilth

Good soil tilth refers to the favourable physical conditions for germination and growth of crops. Soils with good tilth are generally loose, friable and well granulated. They tend to have higher organic matter contents than soils with poor tilth because organic matter (including polysaccharides and other decomposed products of bacteria, fungi and actinomycetes) promotes the granulation process and increase water and nutrient retention capacities. Soils with good tilth permit air entry while retaining of soil water stimulates microbial processes. Soils with good tilth permit drain rapidly and can accommodate tillage and wheel traffic than soils having poor tilth.

3.3.2 Poor Soil Tilth

It is often characterized by and associated with mean pore sizes that are either too large for optimum water retention or too small for optimum root growth. Soils with large pores are usually coarse and are difficult to manage for proper soil seed contact.

Soils with poor tilth have the tendency to form surface crusts that impair seedling emergence, decrease infiltration, increase surface run-off and enhance soil erosion. Natural or traffic induced compaction is another major problem in some soils with poor tilth.

3.3.3 Types of Tilth

The different types of tilth are described as below:

I. ***Fine Tilth:*** It refers to the powdery condition of the soil.

II. ***Coarse Tilth:*** It refers to the rough cloddy condition of the soil.

III. ***Fine seedbed:*** It is required for small seeded crops like ragi, onion, berseem, tobacco.

IV. ***Coarse seedbed:*** It is needed for bold seeded crops like sorghum, cotton, chickpea etc.

3.3.4 Methods of Improving Soil Tilth

- Use of conservation tillage.
- Management of crop residues.

  Use of environmentally acceptable drainage programmes
- Stimulation of soil microbial and faunal populations.
- Implementation of nutrient management practices that maintain favourable soil fertility and soil pH.

- Use of manures, green manures, cover crops, or following of rotation involving grasses, forages or other crops with fibrous root system.

3.3.5 Ideal Characteristics for Good Soil Tilth

- Soils are generally loose, friable, well granulated and offer minimum resistance to soil penetration.
- Soils tend to have higher organic matter content than soils with poor soil tilth which promote granulation processes and increase water and nutrition retention capacity.
- Soils facilitate the placement and coverage of green manures and organic residues and provide maximum resistance to erosion.
- Soils permit free intake and moderate retention of rainfall and thus drain rapidly.
- Soils permit air entry while holding the competition between air and water for occupancy of the pore space volume at minimum and promote microbial activity.
- Good soil tilth can accommodate tillage and wheel traffic than soils with poor soil tilth.

3.4 Soil Tillage and Its Objectives

The mechanical manipulation of the soil with an aim and objective of promoting fine tilth in relation to higher crop production is called soil tillage. Soil tillage systems such as conventional tillage, reduced and deep tillage are considered the most important soil management practices. These practices alter the soil physical environment and affect the plant and root growth, thereby, water and nutrient uptake and crop yields. The major objectives of soil tillage for crop production are:

- To obtain a seedbed of optimum soil tilth favourable for plant growth.
- To loosen the soil for easy root penetration and proliferation.
- To remove other sprouting materials in the soil.
- To control weeds and remove surface trash.
- To improve soil physical conditions;
- To remove the stubbles of the previous crops while preparing land for the next crop.

- To conserve soil moisture by creating dust mulch by repeated tillage.
- To minimize soil erosion (wind and water) by practices such as contour tillage and proper placement of trash, and to manage plant residues.
- To harvest efficiently the effective rain water.
- To incorporate and mix fertilizers, pesticides or soil amendments into the soil by tillage
- To break the surface crust for increasing infiltration and soil aeration.
- To maintain or improve organic matter content of the soil.
- To reduce the insect and disease incidence by ploughing under of certain crop residues.
- To reduce compaction or firming of soil to decrease soil permeability or percolation losses in rice fields.
- To enhance the metabolic processes of living plants and micro-organisms.
- To modify thermal regime of the soil
- To improve soil productivity.
- To level the field for efficient water management.

3.5 Reasons for Adopting Conservation Tillage

- Wind and water erosion is reduced, and water in the soil profile is increased.
- Maintains mulch on the soil surface retains water and insulates the soil surface, thereby reducing soil temperature.
- It also reduces the number of field operations, which reduces input cost for labour, fuel, tractors and other equipments.
- Conservation tillage generally decreases the compaction of soil; decreases run-off contains fertilizers, herbicides and other pesticides due to increased infiltration rates and increased soil water retention.
- It improves timing of planting and harvesting of a crop.
- Water use efficiency is also enhanced.
- It minimizes the breaking up of the surface area.
- It improves crop yield and has high economic returns.

3.6 Prospects of Conservation Tillage in Sustainable Agriculture

Soil tillage particularly conservation tillage is an important component of sustainable agriculture and a better approach for soil management by improvement in soil structure, soil and water conservation, favourable soil temperature, improvement in soil fertility, control of weeds and plant diseases, energy conservation, saving time and labour, improved crop yield and high economic returns. The essence of conservation tillage lies in minimum disturbance of soil and maintenance of some residue at the surface (Mahajan et al., 2004). Moreover, conservation tillage is more suited for use on sloping lands. These systems are also well suitable for well-drained and moderately well-drained soils. Conser vation tillage stores more available moisture (Acharya, 2013).

3.6.1 Improvement in Soil Structure

Properly implemented, no till and conservation tillage systems improve soil structure, increase the relative proportion of bio-channels and macro-pores and increase water infiltration and decrease susceptibility to crusting. In some soils no till also increases homogeneity of soil structure and decreases bulk density. Improvement in soil structure by conservation tillage is attributed to high soil organic matter content and high biotic activity, especially earthworms. Earthworm's activity and intensive tillage are highly incompatible. Hence, there are few earthworms in most cultivated soils.

3.6.2 Soil Aggregation

Soil aggregation refers to the cementing or binding together of several soil particles into secondary units. In general, as tillage intensity increased the soil aggregation decreased. Water stable aggregates are of special importance for high water infiltration, good soil structure and good plant growth. Large stable aggregates at the soil surface are important for controlling wind and water erosion. While feeding on organic material and burrowing in soils, earthworms secrete gelatinous substances that coat and stabilize soil aggregates. Water stable aggregates are formed also with water insoluble gum like substances secreted by bacteria, fungi and actinomycetes, the populations of which are more in reduced tilled soils.

3.6.3 Porosity and Bulk Density

Soil porosity and bulk density are inversely related; therefore any practice that affects one also affects the other. Bulk density is a widely used soil parameter to know the structure status of the soils. The magnitude of change in bulk density of soil depends upon its antecedent properties, time of measurement, depth and intensity of tillage operations. Soil loosening decreases and compaction increases the bulk density. Thus, the bulk density of the tilled layer is normally lower in plowed soil than in unplowed soil. Conservation tillage decrease bulk density and mechanical impedance of the soil due to the presence of surface crop residues and also results in the increase in soil porosity. In an investigation conducted in rice-whaat cropping system under two tillage systems such as puddling and direct sowing along with plant nutrients manage ment practies signiticantly affected the soil physical quality (Sinha at al., 2019). Application of organic manure or practising green manusing improved the tilth (Gupta, 2019).

3.6.4 Soil Conservation

Soil conservation is itself a land management system involving the application of the best practice to result in greatest profitable production without damaging the land. Soil erosion is the process of detachment of soil particles from the parent body and transportation of the detached soil particles by natural agencies such as wind and water. Accordingly, it is called as water erosion or wind erosion. Soil erosion is effectively controlled by conservation tillage (Table 3.2).

Table 3.2: Estimated soil losses with three different tillage systems

Sr.No.	Tillage systems	Estimated soil loss	
		tonnes/acre/year	tonnes/hectare/year
1.	Conventional tillage	9.0	22.5
2.	Conservation tillage	2.0	5.0
3.	Zero tillage	0.8	2.0

3.6.4.1 Control of Wind Erosion

The urgent need to control soil erosion by wind effectively in the late 1930's and early 1940's stimulated the development of the conservation tillage. The influence of wind erosion is extremely complex and includes the processes of soil particle movement, transport, sorting, abrasion, avalanching and deposition. The principle function of surface residue is to decrease the force of wind on the soil itself. The force that moves erodible particles is reduced when the force of the wind is transferred to the residue kept on the surface. Wind erosion is eliminated when the force at the soil air interface are reduced to a threshold value below which erosion will not occur. Conservation tillage provides greater protection against wind erosion as compared with conventional tillage because the surface can be kept rougher due to few tillage operations and more residues on the surface.

3.6.4.2 Control of Water Erosion

Water erosion is the dominant problem on about 148.9 m ha of the total geographical area. Conservation tillage controls water erosion effectively by decreasing raindrop impact on the soil, increases water infiltration, decreases run-off velocity and decreases soil detachability. The principle mechanisms involved in erosion control are:

- Reduced soil erosion due to improved aggregation and a high proportion of water stable aggregates. Aggregation is improved by a high soil organic matter content, high biomass carbon and high biotic activity of soil fauna.
- A high proportion of roots concentrated in the top soil horizons also reduce soil detachability.
- Reduced inter-rill erosion and soil splash due to the prevention of raindrop impact on the soil surface,
- Reduced rill erosion due to decrease in run-off rate, amount, and velocity. The high infiltration rate in conservation tillage is due to the lack of a surface crust and the presence of stable and continuous bio-channels. Crop residue mulch reduces the velocity of water run-off.

3.6.5 Water Conservation

A major dry land production goal in sub-humid and semi-arid regions is to store enough water in the soil between crops so that the subsequent crop will not be severally stressed for water and therefore will produce a favourable yield. Even in more humid regions, storage of additional water in soil is beneficial for alleviating the adverse affects of short-term droughts. To conserve water, the water that would normally be lost by runoff must infiltrate into the soil and it must then be protected against loss by evaporation or use by weeds. Conservation tillage systems will help to a considerable extent in this regard.

3.6.5.1 Run-Off and Infiltration

Infiltration is one of the most important processes in the soil phase of the hydrological cycle because its rate often determines the amount of run-off over the soil surface during rainstorms. Conservation tillage can modify soil surface conditions to retain water for longer time to promote greater infiltration in soils that are slowly permeable.

The process of run-off and infiltration are closely related to that water which infiltrates a soil and is prevented from leaving it as run-off. However, run-off reduction or even prevention doesn't necessarily mean that the water will infiltrate into the soil because the water may be lost by evaporation. The large decrease in run-off with conservation tillage results in from surface residues, which dissipates the energy of falling rain drops, decreased dispersion and surface sealing, and increase infiltration. So, due to increase in infiltration of water in soil, the water content of soil increases which eventually can be used by crops later on.

3.6.5.2 Evaporation

In many areas, evaporation accounts for the major loss of water from agricultural soils. Soil water evaporation occurs in three stages- initial constant rate, intermediate falling stage and residual slow rate stage. The greatest potential for decreasing soil water evaporation lies within the first and second stages. Potential methods include:

- Decreasing turbulent transfer of water vapour to the atmosphere;
- Decreasing capillary continuity and
- Decreasing capillary flow and water holding capacity of surface soil layers.

The crop residues left on the soil surface have been studied as potential mulch for decreasing the rate of evaporation under conservation tillage relative to conventional tillage. Mulches shade the soil from solar radiation, insulate the soil from heat in the air, and impede movement of water vapour from soil to air, thus developing high humidity and permitting condensation within the residue mulch. Mulches greatly affect the first stage of evaporation, but the long term effect of mulches on evaporation is difficult to establish because of their interacting influences on water infiltration, distribution in soil and subsequent evaporation.

3.6.6 Soil Temperature

High soil temperature is a typical problem in dry land tropics and semi-tropics but conservation tillage has been found better than conventional tillage in reducing and stabilizing soil temperature under such conditions. This results from the insulating effect of residual cover on the surface in the form of organic mulch which reduces soil temperature during early part of growing season. In addition to this, residue delays soil drying, thereby, maintaining warmth in soil at optimum level.

3.6.7 Soil Organic Matter

The organic matter content of the soil usually increases with conservational tillage, but remains fairly constant, or may decrease with conventional tillage. Because there is less soil disturbance with conservational tillage as compared with conventional tillage. Moreover plant residues, fertilizers and other soil amendments donot get mixed into the soil as much and plant roots tend to proliferate in the top few centimeters. These conditions tend to cause the organic matter content to increase or to decrease at a slower rate compared to under conventional tillage.

3.6.8 Weed Control

Use of crop residues as mulch on the surface contributes significantly towards controlling weeds besides reducing soil erosion and preventing direct impact of rain drops on the soil surface. Those weeds not controlled by residue mulch can be killed by direct application of contact herbicides like paraquat or translocated herbicides like glyphosate. These herbicides give effective control of pre-plant vegetation

3.6.9 Plant Diseases

Plant diseases occurrence in humid regions is a major impediment to the successfully adoption of conservation tillage due to pathogen interactions with soil physical parameters, continuous moist soil conditions and changes in crop residue management. However, from the considerable amount of literature on the subject, it becomes apparent that conservation tillage practices can variously increase, decrease or have no effect on plant diseases (Table 3.3).

Table 3.3: Examples of decreased disease with conservation tillage as compared with conventional tillage

Sr.No.	Crops	Disease	Causal organisms
1.	Potato	Black scurf	*Rhizoctonia solani*
2.	Corn	Stalk rot	*Fusarium* species
3.	Sorghum	Stalk rot	*Fusarium moniliforme*
4.	Peas	Root rot	*Aphanomyces euteiches*

3.6.10 Energy Conservation

Tractor fuel is the largest single use of fossil fuel in agriculture. Conservation tillage conserves fuels simply by reducing the number of trips of tractor over the field. Ploughing with wooden plough consumes 70 per cent of the total energy expended in carrying out the field operations. Only 10 per cent of the ploughs in India are iron ploughs, but they are about four times effective as wooden ploughs.

3.6.11 Environmental Aspect

The practice of conservation tillage is an eco-friendly farming operation as compared to intensive tillage system by decreasing soil erosion; improving water quality, particularly surface water; enhancing soil quality that may result in increasing soil productivity. It also helps in reducing the use of forest fuels because the minimum tractor use that leads to reduction in emission of green house gases.

3.6.12 Saving Time and Labour

Eliminating tillage obviously means of eliminating the time required for each tillage operation. For example, zero tillage reduces time and labour by 50 to 60 per cent of that required for conventional tillage per acre.

3.6.13 Economics and Benefits for Certain Cropping Systems

When production expenses are decreased and yields are increased, remain constant, or even slightly decreased, the conservation tillage systems are more economical than intensive tillage systems. In irrigated areas, additional benefits from conservation tillage systems are derived from greater water conservation, which results in higher yields with reduced expenses for energy, equipment and labour for pumping irrigation water.

The effect of tillage and residue management on hydrothermal regime, nutrient uptake and yield of wheat was studied in Entisol under Himachal Pradesh soil conditions, one of the hilly northern states of India (Bhagat, 1990). Incorporation of organic residues as mulch improved hydrothermal regime of soil for the growth of wheat. The increase in grain and straw yields of wheat under FYM incorporation and organic residue mulch treatment over conventional tillage were 49-70 and 29-44 per cent, respectively. Main benefits that are accrued from mulches are detailed elsewhere (Gupta and Banerjee, 1991). Through various experiments, it has been concluded that incorporation of zero tillage in certain cropping systems has led to yield increment (Anshu and Bhardwaj, 2004).

- Wheat sown by zero tillage after harvesting of rice crop.
- Wheat taken after soybean/maize/sorghum.
- Pulses crop can be planted with zero tillage without any problem.

Green gram can be planted by this technology immediately after harvesting of wheat which can be used for grain and green manure as well.

The different technologies and improved techniques for reduced tillage to address agricultural sustainability in cool and humid climates are described in Table 3.4 as under:

Table 3.4: Technologies and improved management techniques for reduced tillage

Sr.No.	Sustainability concept	Technologies integrated into reduced tillage	Practical impact
1.	Energy efficiency	Precision or custom tillage Use of legumes Crop rotation/diversity Biological control	Reduce use of fossil fuel Reduce N fertilizer use Reduce agriculture chemicals
2.	Maintain resources	Mulches and residues Cover crops Drainage	Reduce soil erosion Increase soil organic carbon Protect soil surface Protect non-crop period Reduce soil compaction
3.	Biological structuring	Crop rotation/diversity	Pest and weed control Nutrient cycling
4.	Reduce system "leaks"	Mulches and residues Nutrient management Farming by soil Crop diversity	Immobilize nutrients Conserve nutrients Reduce inputs Range of root depths and uptake efficiency

3.7 Why Producer Acceptance of Conservation Tillage is Slow?

- Lack of knowledge of the economic as well as other benefits of conservation tillage.
- Conservation tillage greatly increases dependence on herbicides which may lead to possible health and environmental hazards.
- Higher herbicide costs for weed control and more difficulty in controlling certain weed infestation such as Johnson grass.
- Crop residue mulches left at the soil surface cause slower warming of soils during the spring season and have been identified as a serious problem.
- The high level of crop residues on the soil surface presents the following main constraints to the adoption of conservation tillage:
    a) Mechanical interference with the seedling operations.
    b) Decreased plant productivity due to allelopathic or residue derived toxins. The main phytotoxins are water soluble organic acids, aromatic and short chain fatty acid, oxalic acid, acetic acid and patulin.
    c) Increased incidence of rodents, crop diseases and insect-pests attack.
- It may not work in all types of soils, e.g. heavy and poorly drained soils.
- Too difficult for average farmers; requires too much management practices especially for weed control, use of chemicals etc.

- The main constraint of using conservation tillage on wet soils is that it creates an anaerobic soil environment which results in the inability to remove excess water early enough to allow the soil to warm up in time enough to plant the crop. Anaerobic conditions also affect the microbial activities considerably.
- Crop yields may be less than conventional tillage yields if adopted under non-conducive environments.
- It requires higher level of skill and management ability as compared to conventional tillage.

3.8 Limitations of Soil Tillage

- Exposing of loose soil to wind and water erosion hazards.
- Decrease in soil organic matter content.
- Decrease in water stability of aggregates.
- Creation of hard pans at plow depth that restrict water movement and root development.
- Very expensive input.
- Main problem of zero tillage are low seed germination, low mineralization and build of volunteer plants.

3.9 Conclusion

The technology to grow crops successfully using a variety of conservation tillage systems is available to our farmers. The alliance of farmers, scientists and agribusiness men has transferred crop residue management strategies and tillage methods from an idea to a system that effectively reduces erosion, soil degradation, which is cost-effective and environmentally acceptable. Soil properties and their ecological environment determine the limitations and suitability for using conservation tillage methods. Conservation tillage is an approach that emphasizes the use and improvement of the natural resource rather than its exploitation and mining its productivity for quick economics gains.

3.10 Future Research Strategies/Priorities

- Any successful approach to conservation tillage must be profitable and sustainable, protect our soil and water resources which become environmentally sound and acceptable.

- There is an urgent need to determine the amount of mulch required for different soils and climates in order that we may select conservation tillage systems that are most appropriate for a given site-specific environments.
- The development of new and more effective biological herbicides allows flexibility in choice of crops, crop rotations, and tillage methods.
- Long term and large-scale ecosystem studies are necessary ensuring the effectiveness of conservation tillage systems in reducing the transport of sediments and nutrients.
- Minimizing energy related inputs of tillage, fertilizers and bio-pesticides are major objectives that deserve to be a high priority.
- Greater efforts must be made by researcher in government and industry to provide the most up-to date information in suitable form for extension workers.

Study Questions

1. Define the following:
   i. Soil tillage
   ii. Soil tilth
   iii. Mulch tillage
   iv. Soil aggregation
   v. Soil erosion
   vi. Ridge tillage
   vii. Soil Conservation
2. Differentiate the following:
   i. Soil tillage and soil tilth
   ii. On season tillage and off season tillage
   iii. Primary tillage and secondary tillage
   iv. Preparatory tillage and inter-tillage
   v. Conventional tillage and conservation tillage
   vi. Conventional tillage and deep tillage
3. Draw up a flow chart diagram showing various types of soil tillage classification?

4. Write down briefly the various objectives of soil tillage?
5. Enlist the disadvantages of reduced tillage.
6. Define tilth and explain their different types.
7. Describe in details about the various qualities of soil tilth?
8. List down the different tillage implements and their works.
9. Write a short note on deep tillage.
10. What are the reasons for adopting the conservation tillage?

References

Abrol, V., Sharma, P., Kumar, R. and Sharma, V. (2008). No Tillage: A boon for sustainable agricultural production. Indian Farmers' Digest 41 (12): 34-35.

Acharya, C.L. 2013. Soilquality improvement and water conservation: Must to usher seeond green revolution. Indian Society Soil Science. 61(4): 132-149.

Anonymous. (2004-05). Fertilizer Statistics. The Fertilizer Association of India, New Delhi.

Anshu, K. and Bhardwaj, A.K. (2004). Zero tillage: A way to sustainable agriculture. Indian Farmers' Digest 37 (9): 35-36.

Bhagat, R.M. (1990). Effect of tillage and residue management on hydrothermal regime, nutrient uptake and yield of wheat in a river deposit. Soil Tillage Res. 17 (3&4): 315-326.

Conservation Technology Information Center (1992). National Survey of Conservation Tillage Practices. CTIC, West Lafayette, In.

Ghose, A. (2005). Second green revolution should aim at raising crop yield per unit area. The Kashmir Times 62 (293): 13 [October 21, Friday].

Gupta, R.D. and Banerjee, S.K. (1991). Problems and management of soil and forest resources of North West Himalayas. Mahajan Book Centre, Last Morh Gandhi Nagar, Jammu (Jammu and Kashmir State), India.

Gupta, R.D. 2019. Organic Farming need of the day. The State Time. 24(5): 6.

Mahajan, A. (2006). Soil water dynamics, root growth and wheat yield in a sequentially deep tilled post-rice soil. Ph.D. Thesis, College of Agriculture, Department of Soil Science, CSK Himachal Pradesh Krishi Vishvavidyalaya, Palampur (India).

Mahajan, A. and Bhagat, R.M. (2006). Effect of depth of tillage on root growth parameters of wheat vis-a-vis wheat yield in post-rice soils in Palam Valley of Himachal Pradesh (pp. 35-41). In Natural Resource Management for Sustainable Hill Agriculture (Eds. S. Arora and V. Sharma). Soil Conservation Society of India, Jammu Chapter, SKAUST-J, (Jammu and Kashmir State) India.

Mahajan, A. and Gupta, R. D. (2009). Integrated nutrient management (INM) in a sustainable rice-wheat cropping system (pp.1-267). Springer Science + Business Media (www. Springer.com).

Mahajan, A., Bhagat, R.M. and Sharma, S. (2004). Conservation tillage - An ecological approach for better soil management to improve crop productivity. Indian Farmers' Digest 37 (3): 16-20.

Mannering, J.V. and Fenester, C.R. (1983). What is Conservation tillage? J. Soil Water Conservation 38 (3): 35-54.

NRCS, Georgi (2008). Deep tillage (Ac.), code 324, September 2008.

Saha, B.K. (2009). Stagnation in agricultural output: An introspection. Agriculture Today 12 (2): 46.

Sandhu, J.S. (2015). Nutrient rich varieties of rice in the making. Agriculture Today 18 (5): 16.

Saxena, R. and Mishra, H.S. (2004). Soil tilth and tillage in relation to climatic conditions for profitable crop Production. Indian Farmers' Digest 37 (9): 27-29.

Sinha, N., Chopra, U., Kumar, S.A., Kumar, N. Mohanty M. and Somasundaram, J. 2019. Soil physical quality as quantified by S. index under different management system in rice-wheat system. Soil and Water Conservation. 18(1): 27-34.

4

Water - A Key to Agriculture Sustainability and Food Security

Abstract

Water is essential for life on the earth. It has been a decisive factor in growth and development of human civilization throughout the history. However, owing to increased deforestation and consequent loss of top soil, much of the rain water is lost as run-off. Hence, in India even the places like Cherrapungi which gets the highest rainfall faces drinking water and food production problems. Even Kerala with 42 tiny rivers, several lakes, lagoons and heavy rainfall during the south-west and north-east monsoon season poses water crises if there is failure of monsoon in a year. Infact, Indian agriculture is a gamut of monsoon rainfall as bulk of the cultivable land is under rainfed farming. Hence, for economic reconstruction and regeneration of agricultural development has to be accorded to the highest priority not only to sustain the food production but has to be substantially increased to keep pace with the expanding human population. It is too essential to build up the grass roots economy of the country whose more than 60 per cent of the people depend upon agricultural based activities for their livelihood and occupation. The present chapter of the book, therefore, deals with water - a key to agriculture sustainability and food security. Consequences of faulty irrigation, improving irrigation efficiency and agricultural productivity as well as improving on farm water management, are detailed. Soil and crop management practices, participatory irrigation management, rain water harvesting technique, construction of check dams and other ways to tide over the scarcity of water like revival of ground water resources *i.e.* wells, revival of ponds, are the other topics which have been fully narrated. Digging of trenches and digging of ponds around agricultural fields have been described. Change of crops to cropping system under eastern, northern and north-eastern hill states of India in relation to rainfall vis-a-vis food security and nutritional security also stand discussed.

Keywords: Indian agriculture; human civilization; food production; agricultural productivity; water crises; deforestation; rain water; run-off; rainfall; monsoon; rain-fed farming; faulty irrigation; irrigation efficiency; participatory irrigation management; rain water harvesting technique; check dams; ground water resources

Water is precious, but it is a finite natural resource. Although it accounts for about three-fourth of the earth's surface yet very small fraction of it i.e., only 2.7 per cent of the global water is available as fresh water. Around 2.8 billion people live in areas of high water scarcity (Anonymous, 2014d). Out of the total amount of the fresh available water merely 30 per cent is withdrawn to meet the water requirements of the human and livestock population while the remaining 70 per cent is required to produce food crops that fuel human activities. These days, however, there is limited access to water, which is very seriously becoming a constraint to food production. This highlighted the importance of water during the year 2002, on the eve of World's Food Day; sixteenth of October having theme, "Water Source of Food Security". Food and Agriculture

Organization (FAO) has infact, recognized the essential role, which the fresh water is playing in feeding the global population and ensuring food security. Currently, 800 million people don't have enough to eat. Farmers are, therefore, exerted to use water judiciously (Jhunjhunwala, 2011). They should grow those crops which require less water like mustard, groundnuts and pearl millet instead of chilies, grapes, rice which are water intensive crops. Moreover, farmers can be given incentive to establish water recharging structures.

The theme serves as a call to governments, civil societies and international communities to identify the importance of water in achieving food security. It aims at creating awareness about problems of hunger and to work out a long term solution to guarantee food for all. India supports more than 16 per cent of the world's population with only 2.3 per cent of the world's land and 4 per cent of the world's fresh water resources. However, the per capita water availability is decreasing day by day. In 1951, it was more than 5300 m^3 $year^{-1}$ but reduced to 1905 m^3 $year^{-1}$ in 1999 and is likely to be less than 1500 m^3 $year^{-1}$ by 2025 (Singh, 2002). It is worth mentioning that per capita availability of water less than 1700 m^3 $year^{-1}$ is considered to be as a "stress level" (Rao, 2000; Munshi, 2005). This level of availability of water is classified as "scarce" which is thought to be a severe constraint on socio-economic development and environmental quality. Several regions or areas of India are already facing a severe water crisis. Rain water is the primary source of water, whether it is rain-fed agriculture or irrigated. Actually, it is the precipitation in the catchment areas that finally fills up the dams and reservoirs which are very vital for irrigation; again it is the rainfall that recharges ground water. Both these aspects were amply highlighted by the repercussions faced due to the widespread drought occurred during 2002. Hence, the need of the hour is to harvest the rain water more effectively both for drinking and irrigation purposes through adopting various techniques/ methods. The same stand discussed here under:

4.1 Faulty Irrigation

We have invested over Rs. 48,000 crores in surface irrigation projects since 1951. However, at least 50 per cent of the potential created through these projects is still unutilized and the productivity of the lands served by them continues to be very poor. While premature situation is eating into the storage capacity of reservoirs, the fertility of substantial areas in the Command Area Development Projects is becoming less and less due to water-logging and salinization of the soils.

4.2 Improving Irrigation Efficiency

Improving irrigation efficiency is a vital factor influencing agricultural sustainability and food security. The common forms of irrigation - flood and sprinkler can result in more water being applied than actually be taken up by crops. Efficient and localized methods such as drip irrigation which put water only where it is needed can significantly reduce water requirements for agriculture. Rainfed agriculture can also be benefited from such agricultural practices adapted to uncertain water availability. By conserving water harvesting structures - small furrows to check dams - can conserve rain water and directed it to crops. This practice can boost the yield of all crops to the extent of 2-3 times than conventional methods of irrigation.

Drip irrigation method has also been found useful in water management of citrus, especially in central India where world famous Nagpur mandarin is cultivated (Shirgure et al., 1999), water scarcity in most part of the year is the major stumbling block in order to meet water requirements. Drip irrigation has also proved beneficial for close row crops (Musaida, 2011).

The FAO has started a number of attempts to manage water use efficiency in many countries. The communities in the arid province of Gold mine in Western Morocco have traditionally relied up on food irrigation to divert water to their crops. But years of drought and wind erosion took its toll on canals and dams. Since water resources fell on the province, so they too did agriculture and livestock production.

Water resoures are keys to the sustainable agricultural production (Gupta, 2016; Makwana, 2019). They are also very helpful in sustainable economic and social development. However, due to human interventions, in dustrialization change in land use, water pollution, and climate change, floods and droughts have become more frequent and thus, possess more concern for sustainable development of agriculture as well as sustainable development of human beings. Hence, to under stand fully the physical processes in volved in rainfall, runoff loss modellings and adoption of desired water shed management practices have been fully found suitable in planning and water shed management (Makwana et al., 2019).

4.3 How to Improve Agricultural Productivity

The FAO has fully convinced that the solution to meet future food and water requirements is to ameliorate the agricultural productivity and water use efficiency. By using better and high quality seeds (Suryakumar, 2014b), boosting soil fertility, and improving "On Farm Water Management", the farmers can enhance the yield of various crops and obtain optimum gains from precious water supplies.

4.3.1 Better and High Quality Seeds

Better and high quality seeds i.e., which are not shriveled and are free from an assault of insects and diseases, can increase yield up to 10 per cent.

4.3.2 Increasing Soil Fertility

Soil fertility refers to the inherent capacity of soil to supply adequate amount of a vailable plant nutrients in balanced form to plants necessary for their growth. There are 16 essential nutrients meant for the growth of plants. Out of these sixteen nutrients (C, H, O, N, P, K, Ca, Mg, S, Fe, Mn, Zn, Cu, Mo, B and Cl) except C, H, O all are taken up by the plants from the soils (Jaggi et al., 2001b). Their amount in the soils can be increased through use of organic manures (FYM, Compost, green manure etc.) as they contain all these nutrients though in small amounts.

4.3.3 Improving "On Farm Water Management"

There is an immense scope for improving conservation, distribution and on farm utilization of water. The water available for irrigation must be allocated equitably and optimally to the competing regions and crops. One way to improve access to water may be to treat it as an economic good as well as social right (Mahajan, 2004). A pricing policy that renders wasting water expensive is one of the best incentives to save water. The small and marginal farmers with limited resources must be treated fairly. They should neither have to pay the full cost of water nor should they get free of cost.

The other means for efficient "On Farm Water Management" consist of optimum scheduling irrigations based on suitable methods, conjunctive use of water from different sources, improved agronomic practices. Continuing competition from domestic and industrial sectors will enforce us to develop more and more water use efficient technologies and cropping patterns on a continuous basis. Irrigation scheduling having limited irrigation resources is great challenge to agricultural scientists.

4.4 Soil and Crop Management Practices

To attain high water use efficiency, it becomes imperative to maximize yield of crops with a given amount of water. For this purpose, it becomes necessary to adopt improved soil and water management practices to overcome constraints for getting high yields.

Tillage operations at the proper time, following good crop rotations, use of mulches and organics in conjunctions with small amount of in-organics, and combinations thereof have substantially increased crop yield and, thus increased water use efficiency by reducing evaporation and deep percolation components of soil water balance (Abrol et al., 2013). As already stated, modern irrigation techniques such as drip irrigation and sprinkler irrigations must be promoted, wherever it is possible. Such irrigation techniques have an importance in areas where water is scarce, and the topographic and soil conditions do not allow conventional methods.

Mulching as the barrier to loss of water vapor has been recognized as an effective means of soil moisture conservations not only in food grain crops but horticultural crops also (Mahajan et al., 2004). Use of various mulch materials such as straw, grass, dry leaves and polyethylene, has proved to save moisture by 10 per cent and increase in citrus yield by 30 per cent in clayey soil (Shirgure et al., 1999).

As much as 50 per cent of water released at the project head works is lost in transmission to the canal outlet. Additional loss occurs in water courses which is directly proportional to their lengths and duration of water flow. This happens more so in tail end areas as has been occurred in Ranbir Singh Pura, Bishnah and Jammu tehsils of Jammu district (Jammu region, Jammu & Kashmir State) where very little water remains left for irrigation. Lining of water courses reduces these losses. Lining of main canals may also be considered if losses are excessive or there is a threat of water-logging and soil salinization.

4.5 Participatory Irrigation Management

Participatory irrigation management can take place in different forms. Which particular kind will suit, that depends on various physical, physiographical economic and social factors. Broadly speaking, participatory irrigation management can be categorized into three classes which are described below:

- Organizations of the beneficiaries from the water development scheme such as the irrigation deliveries by the canal systems (e.g. Irrigation Associations).
- Voluntary organizations interested in undertaking the responsibility for socially beneficial work (e.g. bund protection).
- The non-government organizations of enlightened persons interested in the spread of knowledge and information (e.g. Science or Nature Clubs).

4.5.1 What is Irrigation Management?

Irrigation management is defined as the optimum utilization of available water for the sustained production of food grains and fibre without deteriorating the inherent natural health of the soil. Proper irrigation is the integrating but participatory approach to mobilize the organizational set up, involving farmer's participation in supplying water to the fields. According to Singh et al. (2010), irrigation water management is the key to obtain full benefit from the available water resources.

4.5.2 Participatory Approach

The productivity of Indian agriculture even in irrigated areas is very low as compared to other countries. This low productivity is attributed to lack of effective participation of the farmers or the beneficiaries in irrigation management. Wherever the beneficiaries have constituted certain groups for using irrigation water to their fields, commonly known as "Water Users Associations (WUAs)", these have substantially raised the area under irrigation and production. Productivity of the various crops has also increased largely and the income as well. Such associations have given an impetus to self-reliance and joint action for all operations at the right and proper time.

In foreign countries, especially in Mexico and Turkey, the participatory irrigation management system has proved very successful. In Mexico, out of 80 per cent irrigation districts, 65 have been transferred to the WUAs and in Turkey 1.6 m ha of land has been transferred to 407 legally registered associations. There is also a shift in agency management in Latin America, Morocco, Egypt, China, Indonesia, Philippines and Pakistan.

In India, only one per cent of the irrigated area has been transferred to the registered WUAs. Some of the states, which have taken a lead in this aspect, are: Maharashtra, Gujarat, Kerala, Karnataka, Tamil Nadu and Andhra Pradesh. In Jammu and Kashmir State also, participatory irrigation management system has been initiated. In Jammu command area development two such distributaries have been taken up by two NGOs separately. The progress on D-2 of Tawi Canal is going on very well and WUA has since been registered. The estimate for rehabilitation of D-2 by Ravi Tawi Canal Department has been prepared The progress of D-1 of Ravi Canal is very slow and NGO has not been able to register WUA so far. However, Ravi Tawi Canal Department has now drawn up the rehabilitation estimate of this distributaries and Memorandum of Understanding has been submitted by the Chief engineer Ravi Tawi Irrigation Complex, Jammu to the Government and awaiting nod for implementation on trial basis.

The leading research think tank said the crisis is due to rapid population growth and urbanization in developing countries. Water use for households industries and agriculture will increase by at least 50 per cent in the next 20 years. Increased competition for water will severally limit the availability of water for irrigation which in turn will seriously constrain the world's production of food, it added further. As the immediate fall out of the current trends in water policy and investment, we will soon face threats to the global food supply, environment degradation and ongoing health hazards for hundreds of millions of people lacking for access to clean water (Gupta, 2005c; Gupta, 2006a). Using sophisticated computer modeling, it has been found that by the end of 2025 water scarcity will create annual global losses of 350 million metric tonnes of food production - slightly more than the entire current United States grain crop production

Unless we change the policies and fix the priorities in the next twenty years, there would not be enough water for cities, households, the environment or the growing food. In light of these facts, experts and scientists, therefore, opine that rain water harvesting is the only way to tackle recurring drinking water problem vis-a-vis to fortify the food production (Kumar and Kher, 2012). They also stress up on the need for restoring the traditional water bodies like ponds, baulies, tanks etc., which hitherto had been playing a pivotal role in meeting the water requirements of the populace across the country from time immemorial both for drinking purposes, agricultural use and other uses. For this purpose, a conference between October 3-5, 1998 was organized by the Centre for Science and Environment (CSE) pertaining to the potential of "Water Harvesting Technologies, Policies and Social Mobilization" (Vaidyanathan, 1999), The conference brought together cross section of people from all over India and abroad. The participants presented a number of papers in the conference, covering a wide range of issues on "Rain Water Harvesting" for irrigation and domestic use. The conference recommended the creation of a "National Water Harvesting Network" to pursue various ideas and water harvesting techniques, proposed by the various scientists.

4.6 Rain Water Harvesting Technique

Harvesting and conserving rain water for irrigation and other households use is not a new phenomenon on especially in India. The diggings in Indus Valley have unearthed a number of startling facts about the water managers of Gujarat. The Dholaviran exploration is one of the versatile examples like opening a full look on Indus which has imparted answers to many riddles (Gupta, 2013f). The most important is the giant reservoirs (the largest measuring 80.4 m by 12 m and 7.5 m depth) holding an amazing 2,50,000 m^3 of water. This shows that Dholaviran's knew the very good art of rain water harvesting and conserving it nicely. If Harappa's with simple hydraulic engineering skills could conserve the harvested rain water about 3,000- 4,000 years ago, why we can't do it?

Roof water harvesting is common in areas having high rainfall intensity well distributed in the year e.g. Himalayan areas, north eastern states, Andaman Nicobar, Lakshadweep Islands etc. Rain water harvesting is now being followed increasingly for meeting the drinking water needs of rural areas particularly during the period of drought as well as for irrigating the crops. There are many methods or techniques available for harvesting of rain water. One of the techniques in this regard is discussed below:

The rain water harvesting technique consists of harvesting the rain water by permitting it to percolate into the ground for groundwater recharge. In this regard, the grooved structure at the roof tops is necessary to channelize the rain water for storing in an underground tank for its use directly. A number of bore wells are required to construct around the building for recharging the rain water. Nine meter deep recharge bore wells have given very good results in many cases. Apart from these structures, an underground tank and check bunds are needed to harvest the rain water. Check bunds made of bricks have proved better to prevent water from flowing rapidly to promote recharge of water through the recharge bore wells, so building up of such structure is of absolutely necessary.

Karnataka, Madhya Pradesh, Andhra Pradesh and Maharashtra have already begun implementing rain water conservation measures on a large scale. Roof water harvesting is very common in north-eastern states, Andaman Nicobar, Lakshadweep Islands, Rajasthan and southern parts of Kerala and Tamil Nadu, It is worthwhile to mention that in desert areas of Rajasthan, rain water harvesting structures are being developed by the Central Arid Zone of Research Institute, Jodhpur. The state of Mizoram which receives around 2500 mm of rainfall every year began to realize the value of rain water harvesting since the late nineteen hundred eighties.

Centre of Science and Environment has started rain water harvesting in Tughlakabad over its building so that sufficient quantity of rain water can be stored in the underneath of the building. Even this practice has been started in schools. For instance, Shri Ram School, Vasant Bihar North Delhi, has started this activity in its school building by involving students. The school students are now very much concerned with environmental issues, and are spreading the message of water harvesting and its conservation. Various other schools have also started water harvesting technique by excavating trenches and making bunds to regulatory the flow of rainfall run-off.

Now-a-days, a number of new water harvesting techniques are being used in a number of cities like Chennai, Delhi, Jodhpur, Meerut and many others. The method which is quite specific consists of a number of steps like collection of rain water, separation of first rain flush, filtration and storage of rain water and finally its distribution (Gupta, 2009e).

4.6.1 Construction of Check Dams

Construction of ground water dams is not a new concept. Various small scale ground water damming techniques have been developed and adopted all over India. This technique has, however, proved very useful for arid and semi-arid regions in hard rock areas. Low rainfall belt in Western India, which receives monsoon rains during 4 months in a year, forms a typical area in which check dams are constructed for water conservation. Rainfall in most of this region lies between 300-700 mm annually.

Construction of check dams of suitable height (5 to 6m) is also found useful in Southernmost Himalayan Region or Siwaliks to conserve rain water including *Kandi* belt of Jammu (Jamwal and Gupta, 2007), Punjab, Haryana, Uttaranchal and Himachal Pradesh. For example, over 200 check dams have been constructed by the Haryana Forest Department in the Siwalik foot hills for providing irrigation facilities to the farmers of the area during the lean months and to recharge the water table in the sub-soil. These check dams were full to the brim due to good monsoon rainfall during 2003, rainy season. The water of these dams percolates into the substrate and recharges underground water table on the downstream side. Actually, runoff in the catchment is impounded in these water reservoirs which are subsequently used by the farmers for irrigation on the downstream side.

Prior to constructing these check dams the team of the Haryana Forest Department identified the barren area - the village Sukhomajari falling in the Sukhna catchment as the primary source of erosion. This area was badly cut and almost devoid of vegetation. The cattle had free access and used to feed on whatever scanty vegetation was available, leading to further acceleration of soil erosion. After convincing the villagers for the need to conserve water and soil, the forest department constructed earthen embankments to conserve rain water and prevent soil erosion. As the cultivable area of this village is rain-fed so the villagers like to utilize this water for irrigating their fields. The Forest Department agreed to this proposal with the commitment from the villagers that they would help the department in protecting the catchment by not sending the cattle for grazing there but resort to stall-feeding. This concept of Joint management has strengthened.

Check dams of varying design are constructed for rain water harvesting even under arid conditions. For example, over two hundred check dams constructed by the Haryana Forest Department in Siwalik foot hills for providing irrigation facilities to the farmers of the Sukhomajri area during the lean months and to recharge the ground water. The water collected in these dams percolates into the substrata and recharges underground water table on the down stream side.

The area which was hitherto badly eroded and almost unproductive has now become productive. It is because the peasants utilize the stored water of the check dams for irrigating their fields.

The construction of check dams in foot hill Siwaliks of Jammu, Jammu and Kashmir, *Kandi* belt, have also proved boon for the farmers of this belt. Check dams are made of locally available materials like loose rocks, poles, woven wire etc.

4.6.2 Other Ways to Tide over the Scarcity of Water

To make the judicious use of the fresh water, the following other methods would help to tide over the water scarcity:

4.6.2.1 Revival of Ground Water Resources

Revival of ground water resources particularly wells is another area which needs special attention at the macro-planning and management. It is worth mentioning that many people of Jammu city utilized well water for drinking purpose. While going to Pir Kho via Pir Mitha, one can come across a few wells the water of which used to utilize even by the Dogra rulers for home consumption. Such wells if are made functional, their water can be used both for drinking and irrigating purposes (Gupta, 2009e).

The scanty rains during monsoons for the past two years had started creating grave situation vis-a-vis availability of drinking water in Jammu (Verma, 2010). It was because the water level of tube wells had depleted. This required installing of more tube wells. Although there was a proposal to install deep tube wells in kandi districts of Jammu i.e., Jammu, Samba and Kathua to overcome the problem of water scarcity (Anonymous, 2010a) vide which tenders had already floated. There are also upcoming schemes for augmentation of drinking water facilities in the border districts of Kupwara and Rajouri of Jammu & Kashmir State.

4.6.2.2 Roof Water Harvesting

As already stated that due to rapid urbanization, infiltration of rain water into the subsoil has decreased drastically and recharging of ground water has almost diminished. Construction activities in and around the cities, is resulting in the drying up of water bodies and reclamation of these tanks for construction into plots for houses. Thus, the system of rain water harvesting from roof tops has been found useful mainly for drinking water purposes. In this system rain water falls on the roofs which can be collected through a system of pipes and semicircular channel of galvanized iron and is stored in tanks in suitably sized storage tank for providing sufficient water reasonably long duration. On the roof top as well as at the entry to the tanks, simple filters are placed to remove dust and other particles.

4.6.2.3 Revival of Ponds

The huge ponds or tanks which stored rain water in old days are nowhere seen today. As a consequence in comparison to ancient times, water scarcity is a more blatant misery for the Jammuites of the present generation.

During Dogra rulers there were number of ponds for storing rain water. Rani Ka Talab, A Jabghar Talab and Talab near Raghunath temple are glaring examples in this respect which now stand covered and parks/secretariat have come up. Before 1947, the water of these ponds is used not only for washing clothes but for drinking purpose also. So wherever possible the government should think about the construction of new ponds on scientific lines and existing ones must be renovated. Ponds of Kandi belt Jammu, Punjab, Haryana and Himachal Pradesh as well as of other states should also be revived.

4.6.2.4 Digging of Trenches

Digging of trenches activity has turned to be very popular among the beneficiaries as it fulfills the twin objectives of: i) water harvesting and ii) plantation. Standard size of the trench under Kandi soil conditions is 1.5 m x 0.45 m x 0.45 m. As water is a forest produce and we must realized that without vegetative cover on the soil, there can not be any assured water supply so plantation of trees is a must. Native tree species such as Acacia nilotica. Acacia modesta, Acacia catechu, Dalbergia sissoo etc. should be planted. Among the various fruit trees mango (*Mangifera indica*), jamun (*Syzygium cumini*) and phalsa (*Grewia asiatica*) should be preferred to plant as reported by Gupta et al. (2016).

4.6.2.5 Digging of Ponds Around Field Crops

Interestingly, people in rural areas in Kashmir have appeared to realize the importance of rain water harvesting. And most of them follow a procedure in a wonderful manner. Farmers dig ponds around their rice and vegetable fields and capitalize on the reserved water when surface water or canals dry up (Parvaiz, 2009). They also make their cattle to quench their thirst on these ponds which harbor harvested rain water. What is more important is that rural people have often been seen using harvested pond water for construction purposes.

4.7 Change of Crops to Cropping System

Since the sustainable agriculture is the need of the hour, so the focus has to change from crops to cropping systems which are more input use efficient and going for resource conservation technologies. Out of more than 250 cropping systems in India, 30 are the most common ones in use. However, in irrigated agriculture a limited number of cropping systems are prevalent in the country, with the most popular ones being rice- wheat and rice-rice.

4.7.1 Eastern States

In the eastern states like Orissa, West Bengal, Bihar and Assam, where an average annual rainfall varies from 1000 to 1500 mm and wherever sufficient water is available, rice-rice crop sequence is dominant. However, this cropping system is now facing problems owing to deterioration of soil health, increasing pressure of diseases and pests vis-a-vis increasing water scarcity for the winter rice cultivation. For example, long term fertility trials in acid soils of Orissa have revealed drop in sulphur content from 11 to 4 mg ha^{-1} due to continuous rice-rice cropping pattern (Ghosh, 1999). In the rice, potato and maize undertaken in terraced upland of upper Shilong (1850); it was found that percentage of water use from rainfall under the *Rabi* crop of potato was higher than under *Kharif* crop of potato and maize. Run-off constituted only 6.1 and 3.8 per cent of the rainfall in the *Kharif* crop of potato and maize, respectively. No run-off occurred in *Rabi* crop of potato. Deep percolation losses constituted 46.7, 52.4 and 32.3 per cent of the rainfall in *Kharif* potato, *Rabi* potato and maize, respectively (Borthakur, 2001)

4.7.2 North Indian States

The availability of water has increased considerably in both Haryana and Punjab during the last three to four decades. As for example in Punjab, gross irrigated area increased from 3136 x 10^3 ha during 1965-66 to 7315 x 10^3 ha during 1994-95. Similar was the situation in Haryana, where these figures were 1736 x 10^3 ha and 4515 x 10^3 ha during the same periods. Consequently, the main crops like rice, wheat and cotton, are almost totally irrigated in these states, while the crops such as chickpea and pearl millet still remain largely un-irrigated (Anonymous, 1998). The number of tube wells has increased tremendously (Table 4.1) during the last three decades. This has resulted in over exploitation of ground water in a number of areas leading to decline in ground water more than 2 m during last 20 years depending upon the area. This decline of water table has enforced the farmers not only to lower the pumps further deeper in the wells, thereby, enhancing the costs of pumping and energy use, but the water also brings underground salts and certain metals to the soil surface (Singh, 2004). Falling of water table has compelled the farmers in Punjab to go for deeper submersible tube wells with cemented structures at the base (Bariana, 2009). Although it was found in select areas of Bathinda and Mansa districts where water table has gone below 90 m (300 feet) yet experts feel that other areas will also have deep tube wells very soon.

Table 4.1: Number of tube wells which increased from Green Revolution Period (1966-67) to 1994-95 when rice-wheat yields started decline (Adapted from Anonymous [1998])

States	Years	Tube wells
Haryana	1966-67	12,570
	1994-95	5,47,216
Punjab	1966-67	16,932
	1994-95	7,40,000

It is worthwhile to mention that following of continuous rice-wheat cropping sequence has deteriorated and is still deteriorating the soil health. As for example, in the major rice-wheat regions including southern parts of Haryana, the organic carbon content of soils was less than 0.5 per cent during sixties, which now has gone down to 0.2 per cent. Lack of organic carbon in these soils has, thus reduced water holding capacity. Available nutrient status has also changed. The soils with low phosphorous content have now revealed increase from 3.5 per cent in 1975 to 73 per cent in 1995. Due to increasing input costs, the N: P2O5 and K2O ratio being used by the farmers' has changed from 3.2: 1 to 4.2: 1. About 20 per cent of the Haryana soils are deficient in Fe and 33 per cent deficient in sulphur. In Punjab several important crops are now suffering from micro-nutrient deficiencies like Zn, Cu, Fe and Mn.

4.7.3 North-Eastern Hills

The soils of whole of the north-eastern region covering the states of Meghalaya, Nagaland, Manipur except those of Tripura, are acidic in nature (pH 4.5 to 5.5), but are high in organic matter content (2.1 to 4.6%). Agriculture in this region is carried out under varying conditions of slope (0-100%), altitude (50-3000 m), climate (tropical to temperate), and rainfall (1500 to 3000 mm). Accordingly, three cropping systems exist in the region namely Jhum or shifting cultivation, valley and terrace systems of cultivation. Rice is the principle crop (60% of total net sown area) grown both in upland terraces and valley areas. Potato and maize are next in importance to rice production. However, in the upper reaches of Meghalaya, rice and maize are not profitable but cultivation of these crops is crucial to the agricultural economy of this region for ensuring Subsistence. Inclusion of groundnut through intercropping and, thereby, raising cropping intensity without any crop replacement, is a useful proposition in this situation. The maize fallows in Meghalaya and Nagaland, and river beds in Tripura can prove an asset for groundnut cultivation.

From water balance studies based on cropping system replacement of winter rice by groundnut has advantages as: i) the groundnut can be grown in this cropping system only with 450 mm water, whereas, about 1500-1800 mm water is required for winter rice, and ii) groundnut being a leguminous crop involves the soil fertility by fixing atmospheric nitrogen. Further in the upland rice ecosystem in these states its yield is not remunerative as it is very low i.e., about 0.65 t ha^{-1} as compared to the national average of 1.75 t ha^{-1}. But rice cultivation is crucial to the agricultural economy of this region for ensuring subsistence of large population. Hence, a quite feasible proposition would be the inclusion of groundnut, a protein energy rich crop through intercropping with rice (to be sown during mid January to mid July). This grain legume in cereal based cropping system will thus, play a significant role in sustaining crop production and fulfilling the dietary protein requirement. Riverbed ecosystems of Assam and Orissa, where lot of silt deposits due to occurrence of flood almost every year, have become ideal for cultivation of groundnut, especially for HPS cultivar on residual moisture only or with one or two productive irrigations.

4.8 Food Security and Nutritional Security

In the light of the afore said facts as stated above, it becomes an imperative to follow these methods or technologies of water management to ensure the food security and nutritional security which stand defined as under:

4.8.1 Food Security

Conventionally, food security may be defined as the balanced supply of food, especially supply of cereals and effective demand for food. Otherwise, it would be termed food self sufficiency within a territorial entity, usually the country.

4.8.2 Nutritional Security

Nutritional security in terms of adequate protein, energy, micro-nutrients and minerals for all households' members as balanced diet and safe drinking water promising long ago still remains unchanged. Thus, it must be one's sincere effort and desire to take agriculture to a new level of knowledge based development. A development should be equitable, a development that is environmentally sustainable and development that is regionally balanced. It is our goal to ensure both livelihood security and food security for all people, paying particular attention to the needs of the small and marginal farmers.

More than 850 million people worldwide suffer from hunger; about 820 million of them are in developing countries (Mukherjee, 2009). Among 118 countries, India is 94th in global hunger index prepared by International Food Policy Institute (IFPI). The poor donot have the capacity to get protein rich food like milk, fish, meat, egg etc. and prices of pulses reaching ridiculous heights, these have become beyond their reach.

Iron rich variety of rice has been developed by Hari Krishan of the University of Missouri (US) and P. Sukumar of Tamil Nadu Agricultural University, Coimbatore. The variety is a cross of normal rice variety of *Oryza salva* with a wild variety of *Oryza nivare.* This rice variety not only cooks well but has an average protein content of 12 per cent. About 3.5 billion people in the developing countries are lack of enough Fe in their diets.

Amidst impending food crises in the world and reports of large quantities of food grain like maize/corn being diverted for bio-fuels production. United Nation FAO has urged the world to view challenges facing global food security and cumulative actions to deal with the emergency situation (Anonymous, 2008). Diversion of maize towards bio-fuel was having an effect on food prices. All dimensions have to be analyzed before taking decisions. A decision needs to be taken out at the structural level across the world. In United States alone 100 million tonnes of cereals were diverted for bio-fuel production and the quantity was estimated to increase 12 fold by 2017 (Anonymous, 2008).

4.9 Conclusion

Water is the essential commodity for survival of mankind and other living organisms including crops. It is one among the fabulous gifts of nature for supporting its creatures. Although about 75 per cent of earth's surface is covered with water yet out of this only about 2.7 per cent is fresh water. Availability of the fresh and clean water on Earth planet is, thus, very limited in quantity. Of this limited supply, about 23 per cent is utilized for industry and only 8 per cent is used for domestic purpose. Approximately 1.3 billion people lack access of potable water worldwide and 800 million people don't have enough to eat. Since there is limited supply of water for irrigating the crops to sustain their productivity so for increasing growing demand of water both for drinking and irrigating purposes, rain water harvesting is very essential. The same has been described in details in this chapter of the book. Apart from this the solution to meet future food and water requirement for ameliorating the agricultural productivity and water use efficiency has also been given.

4.10 Future Research Strategies/Priorities

- Participatory irrigation management is the need of the hour as the farmers can use water prudently and economically if the management of irrigation system is handed over to them.
- The farmers must be encouraged to do voluntary service for maintenance of the irrigation system at micro level. They will be willing to accept the responsibility for management if they realize that they had a voice in the decision making.

- Trainings have to be imparted to the officials of irrigation agency, farmers, functionaries of WUA's and Non-Government Organizations.
- Apart from the support from the Central and State Governments, the other components necessary for successful participatory irrigation management programme, is creating awareness among the farmers about the benefits of this irrigation management system.
- To impart training; is a key input for implementation of the ideas of participatory irrigation management.
- Mouths of storm water drains while harvesting the rain water needs to be raised above ground level to prevent water from draining away.
- The structure made within the building during water harvesting required to be inspected frequently. To prevent leaves and debris from entering the system mesh filters must be supplied at the mouth of the drain pipe. Sand filters should be provided over the tanks meant for storing the water.
- Development of fertigation techniques for fertilizer use efficiency is the need of the hour.
- Integrated water management studies to work out the water requirement at different stages of crops growth for optimum utility of water should be undertaken on priority.
- Rain water from ground catchments should be harvested and used to supply water, for crops, wild life and livestock as well as for domestic use.
- With growing population and ever increasing water needs of household agriculture and industry, we may face a major water trouble in near future in case of failure of monsoon. The policy planners should, therefore, undertake a comprehensive decision to earmark a fixed portion of budget to be used for rain water harvesting.

Study Questions

1. Define the following:
   i. Food security
   ii. Nutritional security
   iii. Rain water harvesting
   iv. Irrigation management

2. What is meant by participatory irrigation management? Describe its various classes in details.
3. How we can improve irrigation efficiency to improve agricultural productivity?
4. What are the other ways to tide over the scarcity of water? Write their brief accounts?
5. What is meant by cropping system approach? How it has proved useful from change of crops to this system?
6. Why water table is declining year after years in Punjab and Haryana?
7. Write in details about the benefits of mulching?
8. Differentiate between diggings of trenches and check dams?

References

Abrol, V., Sharma, P., Sharma, V. and Kumar, R. (2013). Upgrading dryland farming for future challenges. Indian Farmers' Digest 46 (4): 41-42.

Anonymous. (1998). Decline in crop productivity in Haryana and Punjab: Myth or Reality. Report of the Fact Finding Committee (May, 1), ICAR, New Delhi, India.

Anonymous. (2008). Food crises due to wrong policies says FAO. The Tribune 128 (100): 15 [April 11, Friday].

Anonymous. (2010a). Deep tube wells in kandi districts soon. The Tribune 130 (61): 3 [March 4, Thursday].

Anonymous. (2014d). World water day - 2014: The water energy nexus (2014). Agriculture Today 17 (5): 42-43.

Bariana, S.S. (2009). Fall of water table compels farmers to go for deeper tubewells. The Tribune 129 (196): 10 [July 7, Friday].

Borthakur, D.N. (2001a). Agricultural Research for North East Hill Region. ICAR, Research Complex for North East Hill Region. Ex-Vice Chancellor, Assam Agricultural University, Jorhat, Assam, India (Personnel communication).

Ghosh, S.P. (1999). Citrus germplasm in India and its conservation strategy. Intensive Agriculture 36 (11-12): 8-10.

Gupta, R.D.. 2016. Wildlife and Wetland Ecosystems of Jammu and Kashmir & Hamalayas. NR. Book International, Pacca Danga, Jammu Tawi - 180001, J&K State, India

Gupta, R.D. (2005c). Environmental degradation of Jammu and Kashmir Himalayas and their control. Associated Publishing Company, New Delhi (India).

Gupta, R.D. (2006a). Environment Pollution: Hazards and Control. Concept Publishing Company, New Delhi (India).

Gupta, R.D. (2009e). Scarcity of water in Jammu. The Daily Excelsior 45 (203): 6 [July 23, Thursday].

Gupta, R.D. (2013f). Rain water harvesting recalls-Indus valley civilization. The StateTimes 16 (312): 6 [November 12, Tuesday].

Gupta, R.D., Gupta, S.K. Bhardwaj, S.D. 2016. Agrotechniques and uses of medicinal plants. Associated Publishing company-A division of Astral International Pvt. New Delhi, India

Jaggi, R.C., Mahajan, A., Choudhary, A.K. and Dogra, R. K. (2001b). Plant nutrient deficiency symptoms are partly dependent. Indian Farmers' Digest 3 (49): 9-11.

Jamwal, G.S. and Gupta, R.D. (2007). Watershed management project - A case study of Bari-Badhori, Jammu and Kashmir State (pp. 226- 231). In Natural Resource Management for Sustainable Hill Agriculture (Eds. S. Arora, S.S. Kukal and V. Sharma). Soil Conservation Society of India, Jammu Chapter SKUAST Jammu, J&K State, India.

Jhunjhunwala, B. (2011). Judicious use of water. The Daily Excelsior 47 (128): 6 [March 9, Monday].

Kumar, P. And Kher, S.K. (2012). Rainfed agriculture in India: Constraints and priorities. Agriculture Today 15 (2): 57-58.

Mahajan, A. (2004). Watershed - Concept and management. Agrobios Newsletter 3 (1): 28-31.

Mahajan, A., Bhagat, R.M. and Sharma, S. (2004). Conservation tillage - An ecological approach for better soil management to improve crop productivity. Indian Farmers' Digest 37 (3): 16-20.

Makwana, J.J., Tiwari, M.K. and Pampaniya N.K. 2019. Impact assessment of land use and coverchanges on resources using R.S. and GIS techniques in watershed area. Soil and water conservation 18(1): 11-23.

Mukherjee, D. (2009). Matured strategy imperative. The Kashmir Times 66 (182): 9 [July 2, Thursday].

Munshi, M.M. (2005). Water scarcity may lead to wars. The Daily Excelsior 41 (294): 6 [October 25, Tuesday].

Musaida, P.K. (2011). Micro-irrigation. Agriculture Today 14 (9): 32-34.

Parvaiz, A. (2009). Rural way of rain water harvesting: A lesson town dwellers should learn. The Kashmir Times 66 (186): 7 [July 6 Monday].

Rao Sarvotham, C. (2000). A need for conservation of the water resources of the country. Environment and People 7 (1): 6-8 [June issue].

Shirgure, R.S., Srivastava, A.K. and Singh, S. (1999). Water management in citrus. A timely demand. Intensive Agriculture 36 (11-12): 12- 16.

Singh, A.K. (2002). Water - The key to sustainability. Indian Soc. Soil Sci. Newsletter 13, pp. 1-2.

Singh, H.R., Gangwar, K.S. and Tomar, O.K. (2010). Water management for enhancing crop productivity. Indian Farmers' Digest 43 (9): 5-8 & 14.

Singh, T. (2004). Jalandhar water table hits new low-down 8 m in a decade. The Sunday Hindustan Times [July 18, Sunday], Chandigarh, India.

Suryakumar, P.V.S. (2014b). What our farmers want. Agriculture Today 17 (8): 58.

Vaidyanathan, A. (1999). CSE Conference on potential of water harvesting. The Catch Water 1 (1): 2 [February Issue].

Verma, M. (2010). Water level depleting in tube wells due to dry spells. The Daily Excelsior 46 (20): 1 and 4 [January 20, Wednesday].

5

Watershed Concept and Management

Abstract

The term "watershed" refers to the divide, separating one drainage basin from the other. It is, infact, a hydrological unit which comprises combinations of arable/non arable lands and drainage lines and is utilized by permanent and transient populations with varying skills and commitments to long term husbandry or thrifty management. Watershed management has come not a day too soon. It was long over due, especially in a country of poor millions, who are mostly dependent upon the vagaries of weather there is no alternative but to manage our water properly- And there is no better method for hill country side to harvest its water wherever.it falls. Management of water will automatically prevent the soil erosion and run-off losses. The water so conserved can be used not only for agricultural production but also for olericulture/ horticultural production and livestock rearing as well as for forestry raising. Watershed management thus, constitutes an holistic approach which requires to be followed in all the deforested catchments and one must fully know about the details of "Watershed Management". This chapter of the book is, therefore, devoted in this respect. Concept of watershed, problems of watershed, necessary points/ factors for selection and operation of watersheds, causes of watershed deterioration, watershed planning, meaning of watershed development or watershed management and its objectives are described in this chapter. Apart from this, watershed development programmes, success obtained in few crop production programmes through watershed management are the other topics which have been dealt with. Future research strategies/priorities are also given.

Keywords: Watershed; watershed management; agricultural production; holistic approach; water; olericulture/horticultural production; soil erosion; run-off; livestock; forestry; deforested catchments

Now-a-days, a watershed management or watershed development has been fully recognized as a unit for over all development of rainfed areas (Gupta, 2006a and b). Hydrological problems such as excessive runoff, large scale soil erosion, inadequate or faulty drainage and occurrence of droughts are the most significant problems in almost all the rainfed areas. These problems, however, can be alleviated through running watershed based development programmes that recognize the natural topography and water movement patterns of the land.

A watershed is the land area from which surface water drains to a single outlet. Thus a watershed is bounded by a ridge line and constitutes an independent land unit for surface hydrology. It is infact, a natural unit which is meant for designing and implementing practices for resource development and integrated land use management. All soil and water conservation measures involving treatment of agricultural vis-a-vis non- agricultural lands are to be planned to conserve soil and moisture for enhancing not only crop productivity but is also essential for increasing the productivity of horticulture, olericulture, forestry, sericulture and animal husbandry. It, therefore, becomes an utmost important to know more about watershed planning and management. It is because India requires about 325 million tonnes of food grains to feed an expected population of about 1.5 billion by 2025 (Choudhary et al., 2005)

In this chapter of the book, the concept of watershed, its present scenario, the soil survey and land use capability classifications, success obtained in a few crop production programmes, factors distinguishing small and large watersheds, points for selection of watershed, have been detailed. Apart from this, watershed management, its objects and steps in watershed management have also been given.

5.1 Concept of Watershed

The concept of watershed is quite old yet its application has been largely since two and half decades only i.e. from early nineteen hundred ninety. It is worthwhile to mention that all round development of the farmer's activities is the order of the day. As such various line departments like raising of livestock and fisheries, development of horticulture, olericulture, sericulture, forestry and agrostology etc. involved in their activities have to be made rapid in for undertaking the programme in selected areas. The watershed frame work therefore provides the necessary inter-ecological linkages and is a logical unit for the integration of sustainable use of land and water vis-a-vis other resources for the production of agriculture and allied sectors as stated above. It helps in augmenting soil and water resources. It also enjoins the biophysical, social and economical inputs which if optimally managed, lead to diversified high agricultural production. Watershed management helps to control environmental degradation and provides a mechanism for the recharge of ground water aquifers. In this way watershed development ensures water adequacy to crops and reduces the risk of yield loss occurred due to water related stress.

5.2 Definition of Watershed

"Watershed is defined as a hydrological unit of land draining to a common outlet". The terms catchment, watershed and drainage basin are synonymous, but are used depending on the size of the drainage area. If large, it is called catchment, if small - watershed. A watershed reflects the interactions of vegetation, soil and ground water.

5.3 Problems of Watershed

The main problem in watershed is deforestation which has resulted into the high rate of soil erosion by rainfall and run-off losses as well as soil erosion caused by landslides/landslips. The run-off water transports the sediments, which blocks the channel head, dam, reservoir and storage structures. The degradation processes in the catchment areas may result in an alarming rate of sedimentation in the reservoirs, structures and at the river mouth. Nizamasagar reservoir in Andhra Pradesh is an example of an alarming rate of sedimentation. It has lost 50 per cent of its capacity to provide irrigation to the command areas. Similar is the fate of Bhakra Dam in Himachal Pradesh and/or Salal Project in Jammu and Kashmir.

5.4 Factors Distinguishing Between Small and Large Watershed

The following factors that distinguish the small and large watersheds are as follows:

5.4.1 Size

The size of small watershed varies from few hectares to 100 hectares, while in large watershed the size exceeds 100 hectares.

5.4.2 Overland Flow

The overland flow is more predominant in small watershed as there is less network of drainage system, while in large watershed channel flow is more predominant.

5.4.3 Channel Storage

The channel storage is significant in large watershed, while it is insignificant in small watershed.

5.4.4 Land Use

It affects the overland flow, hence significant to small watersheds, while in large watersheds there is channel flow predominant. It is therefore the land use in large watersheds is not effective.

5.4.5 Intensive Rainfall

In small watershed, intense rainfall causes rapid peak flow, while in large watershed, it is not so prominent.

5.4.6 Roughness of Soil Surface and Channels

The surface roughness of soil or land creates effect on overland flow i.e. there would be less overland flow when surface roughness is more., but roughness of channel causes precedental effect on channel flow.

Table 5.1: Sub-division of watersheds (operational sizes for agricultural purposes)

Sr.No.	Sub-division	Hectares	Base map size
1.	Sub-watershed	10000-50000	1: 50000
2.	Milli watershed	1000-10000	1: 50000
3.	Micro-watershed	100-1000	1: 8000
4.	Mini-watershed	1-1000	1: 4000 or larger

5.5 Points for Selection of Watershed

The following points should be considered for the selection of watershed:

- Watershed which has acute shortage of drinking water.
- Watershed that has a large population of scheduled castes or scheduled tribes depend on it.
- Watershed that has a preponderance of wastelands.
- Watershed that has a preponderance of common lands.
- Watersheds where actual wages are significantly lower than the minimum wages.
- Watershed that is contiguous to another watershed that has already been developed or is selected for development.

- In order to check the siltation, it is intrinsic to conserve the soil of catchment area by following agronomical, mechanical and biological methods (Gupta, 2016). This will help in checking water scarcity of the catchments (Anonymous, 2018).

5.6 Factors Affecting the Operation of the Individual Watershed

The factors that affect the operation of the individual watershed are discussed here under:

5.6.1 Size of Watershed

Both run-off volumes and rates increase as the size of the watershed increases. However, both rate and volume per unit of watershed decreases as the area increases. The size, namely area of watershed, is an important parameter in determining the peak rate of run-off. It is essential for designing erosion control structures and channels to carry maximum run-off.

5.6.2 Shape of Watershed

The shape of watershed is divided into two types - fan and fern. Fan shaped watershed produces higher peak rate of run-off in shorter duration than fern shaped watershed because in fan shaped watershed, all the parts of the watershed contribute the run-off to the outlet simultaneously in little time period

5.6.3 Time of Concentration

The time taken by water to travel from the most distant point of the watershed to the outlet or some other down-stream point of reference is called as *"time of concentration"*. The longer the watershed, the greater is the time of concentration. The longer the time of concentration, the greater is the time available for the water to infiltrate, evaporate and get utilized by the growing vegetation.

5.6.4 Drainage Density

The drainage density (length of drainage channels per unit area) is an important parameter of drainage that influences the watershed behaviour. Higher the drainage density more is the run-off yield, as the rain water enters the drains immediately and reaches the outlet.

5.6.5 Slope of the Land

The run-off water from steep slopes reached the main channel quicker than the gentle slope because in sloppy lands, the run-off velocity is higher and infiltration loss is less resulting into higher peak rate of run-off and vice-versa.

5.6.6 Land Use

Land can be used in such a way as to provide high degree of watershed protection. Land use affects the rates of water infiltration and run-off, types and nature of vegetation e.g. an area that is under forest is covered with a thick layer of mulch of leaves and grasses. The surface run-off therefore, becomes too less in the forest area, as huge rainfall amount is absorbed by the soil due to increase in infiltration rate and formation of resistance in the flow path of water over the ground surface while in barren lands just reverse trend is found.

5.6.7 Soil Types

The soil texture determines the infiltration rate of rainfall e.g. in light texture soils like sandy or loamy sand soils, the soil particles are coarser and have large pore size, resulting into rapid absorption of rainwater and thus it has less run-off potential. But in heavy texture soils like clay or clayey soils, the soil particles are finer and have small pore size, resulting into less absorption of rainwater and thus it has huge run-off potential.

5.6.8 Rainfall

The amount of pattern of rainfall is the most important factors in determining what happens in a specific watershed.

5.6.9 Vegetation

The type of vegetative cover plays an important role in watershed management. Vegetative cover controls infiltration, erosion, sediment production and evapotranspiration.

5.7 Causes of Watershed Deterioration

The watershed deterioration takes place due to unplanned, uncontrolled and unscientific land use activities:

5.7.1 Agricultural Lands

Cultivation on sloping land without precautions; cultivation along nallahs or stream banks; cultivation of erosion permitted crops; over cropping areas with out replenishing soil fertility are the main causes which deteriorate the watersheds.

5.7.2 Forestlands

Clear-felling on steep slopes; drastic thinning of plantation along slopes; faulty logging roads and disturbance of forest floors during extraction are the conspicuous factors responsible to create disturbances in the watershed areas.

5.7.3 Grasslands

Excessive grazing results in disappearance of protective cover; development of cattle tracts into channels, gullies etc.; causes compaction of soil resulting thereby in lower infiltration rates etc.

5.7.4 Fire Consequences

Intentional or accidental fire produces loss of vegetation, organic matter and micro-organism. Fire also disturbs the hydrological behaviour of the watershed very seriously for a long period.

5.7.5 Unscientific Mining and Quarrying

This practice results in exposure and digging up of slopes that causes considerable damage to the landscape by destroying vegetation. Haphazard disposal of the mine spoils and blocks drainage channels, covers up pastures and agricultural fields.

5.7.6 Alignment and Construction of Roads in a Faulty Way

Alignment and construction of roads in a faulty way can and is causing a lot of dislocation of life in watersheds located in hilly areas. This is a precisely major problem of Himalayan watersheds where many new roads (length about 10,000 km) have been made. These contribute largely coarse sediments to the drainage channels causing, thereby, lot of blockade to flow.

5.7.7 Shifting Cultivation

Shifting cultivation which is mostly common in North Eastern States destroys protective and productive vegetation in preference to a very brief period of immediate crop production and results in soil loss and consequential damages. This practice can cause wide spread destruction in the watershed.

5.7.8 Non-Cooperation of the Peoples

This is another most important factor which may deteriorate the watershed considerably. As a matter of fact without people's cooperation, no watershed can remain undisturbed and it is the people's non-cooperation that results in people's ills.

5.8 Consequences of Watershed Deterioration

The deterioration of watershed causes the following effects:

- There is low productivity of land with respect to production of food, fibre, fuel, forage and fruits.

- There is reduction in biomass production.
- There is quick siltation of various reservoirs, lakes, dams, canals etc as observed by Gupta (2016).
- Reduction of water quality and quantity, both due to heavy sediments which are brought by the flash floods from the deforested/denuded areas of watersheds.
- Erosion and denudation within and outside the watershed is a common phenomenon.
- There is occurrence of frequent floods and droughts.
- Poor returns from agriculture, grassland, forests, fruits etc. are obtained.
- Creation of poverty in the area is likely to occur.
- Quality of life becomes poor.

5.9 Watershed Planning

Currently Watershed Development programmes are implemented by taking a revenue district a unit. This is more so in Jammu and Kashmir State under the single line administration programme. The main objective behind this concept is involvement of the people in the planning process through the Institution of District Development Boards. The planning unit has lately been brought down to the block level. However, this system pre-supposes uniform conditions over an area, be it district or a block which is not so in the actual field. These areas are heterogeneous in nature and extent of their problems and constraints in the process of their development. The present system would, therefore, compartmentalize various sectors and activities.

One of the most essential aspects of an Integrated Area Planning is the spatial distribution of socioeconomic facilities. In formulating such policy the planner is often confronted with two conflicting demands viz. a) villagers want immediate approach to all other facilities, b) the government owing to its limited financial resources wants to limit the investment on these facilities to the minimum possible extent. Many approaches viz., central place theory, the theory of growth poles and various other theories have been developed from time to time in different contexts. Under most of these approaches a village is generally considered as the unit of investigation, planning and development.

5.10 Micro-Watershed and Planning

A micro-watershed usually has discernible drainage lines, wherein land use system of whatever nature - good or bad effects, its hydrological behavior is measurable. It is also convenient unit for design development with the people that live in it. As such a village unit is an integral part of a Micro-watershed. Apart from it, any scientific and technological issues of development are simpler and much easy to explain and demonstrate and also act upon in "Watershed" nicely.

5.11 Planning and Participation of Local People

For specific interaction "Watershed Planning" must be according to the needs of the people with due consideration to the level of ecological degradation which already stands set in the area. Moreover, it should be in accordance with the prevailing socio-economic and cultural demands for continued adoption of the developed/tested technologies. Such investigations besides identifying agro-climatic and physiographic parameters that are to be dealt with must quantify the level of interactions required for balancing the disturbed ecosystem in a watershed.

5.12 Comprehensive Treatment Plans

Based on a realistic baseline data, comprehensive treatment plans should be drawn in consultation with multidisciplinary team of experts and local people for implementation in a integrated manner. A team of experts should consist of soil scientist, agronomist, horticulturist, a man of animal husbandry, forestry and engineering etc. Need based menu of treatments should be predetermined duly quantified to meet the resources gaps required to be covered under the programme. Treatments must be technically effective with principles of low cost, but must be more sustainable in the long run and wide scale replicability. Referring specially to our conditions, site specific models have to be developed for treatment of different land use units for increasing productivity. These land use units comprise a number of enterprises like fuel wood, fodder, cropland, fruit and vegetable production with an inbuilt provision for soil and moisture conservation. Improving the quality of the livestock population and forestry within watershed are also the components of land use units.

5.13 What is Meant by Watershed Development or Watershed Management?

Watershed management refers to a scientific use of all the resources i.e. land, vegetation and water of watershed to achieve maximum production with minimum hazard to the natural resources to meet the basic needs of the people. Watershed management provides protection against land deterioration, loss of productivity and flood hazards. The task of watershed management includes the treatment of land by using most suitable biological and engineering measures in such a manner that, the management work must be economical and socially acceptable.

In nutshell, in the watershed management concept we must adopt the right land use pattern with emphasis on production of horticulture, olericulture, animal husbandry and silviculture instead of agriculture alone as well as adoption of suitable soil conservation measures. Development of grasslands for raising livestock is also a major component (Gupta, 2009a). Whatever mixed farming we adopt, it must be coordinated and all the agencies must work together.

Thus, ***"Watershed development or watershed management is the rational utilization of land and water resources for optimum production of food crops, fodder, fuel wood, timber, fibre and fertilizers with minimum hazards to the natural resources".*** It essentially results in conservation of soil and water in watershed leading to proper land use i.e. protection of land against deterioration, building up soil fertility, conserving water for farm use and ground water recharge, loss of run-off control, improving drainage, flood control and ultimately increasing productivity.

5.14 Objectives of Watershed Management

Watershed management is aimed at the following objectives:

- To control the damage caused by run-off and, thereby, conservation of soil and water.
- To manage and utilize the run-off for useful purposes or soil moisture conservation purposes.
- To protect, conserve and improve the land of watershed for more efficient and sustained production basis.
- To rehabilitate the deteriorating lands.
- To enhance the ground water recharge, wherever it is possible and acceptable.
- To expand electricity programme by way of storing water in check dam.
- To intensify various income generating agricultural activities like bee keeping, sericulture, mushroom cultivation etc.
- To protect and enhance the water resources originating in the watershed.
- To control of flash floods.
- To render regulation of water supply.
- To sustain food, fibre, fuel, forage and fruit production.
- To have general development of the area.
- To ameliorate quality of life.

5.15 Steps in Watershed Management

The major steps or requirement for formulation of watershed management plan is to have basic information on human and livestock population as well as land and soil characteristics of the watershed, which are as follows:

5.15.1 Basic Information or Watershed Characteristics

In order to start the project in a systematic way, the following basic information is needed:

- Statistics of human population and livestock population.
- Pattern of land ownership.
- Topography, land use, crop rotations and cropping systems and potential capability of land classification.
- Data on natural vegetation, rainfall and temperature/type of climate, erosion problems and ground water.
- Information on existing water sources like tanks, wells etc.
- Service facilities such as banking, schools, market, extension agencies, health and veterinary facilities

5.15.2 Soil and Land Characteristics

Apart from morphometry of watershed, geophysical aspect, hydrological, vegetation and climatic factors, the data on soil and land characteristics are also one of the major components of watershed management planning.

The scientific way of acquiring data on soil and land characteristics lies with soil survey. Before taking up soil survey work, one must be fully aware of the objective of soil survey. According to soil survey objective, the scale of mapping base map and level of mapping are taken into consideration. Of the various Intensity/Level of Soil Survey and Mapping Details, for watershed management planning soil's information requires to be collected on large scale at least on 1:15,000 scales for effective planning.

After performing soil survey, the data on soil and terrain characteristics are so generated need to be interpreted for various agricultural and engineering applications, forest and pasture development, and crop cultivation. This could be a very useful tool for preparing "watershed Management Plan". Therefore, different thematic deduced from soil maps viz., suggested measures on soil conservation, land irrigability, water conservation and suitability for different land use practices would through valuable information about the land use potential of the watershed.

5.15.3 People's Participation

The success of any Watershed Management is wholly solely dependent upon people's participation, it is because the people living in the watershed especially peasants are well versed with land use, water and forest resources and investment in long range programme. Hence, people's participation should be part and parcel in watershed management planning. It will also encourage the people to share their innovative ideas and to assist the planner and manager to find new ways to cut cost and increase the efficiency.

5.15.4 People's Awareness Programme

This programme will not only help to educate the farming community but also bring awareness among the general public living in the area towards adoption of scientific technologies generated for cultivation of crops, soil and water conservation measures, afforestation techniques etc. For this purpose, it is required to involve the students of school/college and university and media personnel's to improve communication between researchers and farmers, foresters, soil scientists and engineers.

5.15.5 Development Component

The following components must receive attention in any watershed development project:

- Soil conservation measures such as moisture retention, safe disposal of run-off, various mechanical measures etc.
- Storage and recycling of run-off through ponds, storage reservoirs and conveyance.
- Improvement of fuel fodder production including horticulture and change in land use pattern whenever necessary.
- Optimal land use and cropping systems including mid-season correction and appropriate.
- Ground water recharge and development.
- Water management such as drainage, lining of watercourses, proper field layout, land leveling, crop saving irrigation methods etc.
- Development of livestock, poultry and other associates activities.

5.16 Watershed Development Programmes

Watershed Development Projects have been taken up under different programmes launched by the Government of India. The Drought Prone Areas

Programmes (DPAP) and the Desert Development Programme (DDP) adopted the watershed approach during 1987. The Integrated Watershed Development Project (IWDP) was started in 1989 by the National Wasteland Development Board with an aim to develop wastelano on watershed basis. The concept of National Watershed Development Programme in Rainfed Areas under the Ministry of Agriculture (NWDPRA) was started during 1990-91. These programmes laid down their own guidelines, norms, technical components and funding pattern based on their respective and specific aims While DDP focused on reforestation to arrest the growth of hot and cold deserts, the DPAP concentrated on non-arable lands and drainage lines for in situ soil and moisture conservation, agroforestry, pasture development, horticulture and alternate land uses. The IWDP, on the other hand, had made silvipasture, soil and moisture conservation on wastelands as the predominant activities. The NWDPRA combines the features of all these three programmes with an additional dimension of improving arable lands through better crop management technologies.

5.16.1 The Technical Aspects

Scientists, after working in various watershed projects of the country especially in Hill States, have developed appropriate technologies to find solutions to most of the problems related to watershed treatment. Such technologies range from simple "Check dam to large percolation/irrigation tanks, from vegetative barriers to contour bunds".

5.17 Success Obtained in a few Crop Production Programmes through Watershed Development

Success obtained in Rain-fed Agricultural Production through watershed management in Jammu Region of Jammu and Kashmir State and parts of Uttaranchal has been highlighted. In Jammu Region, watershed management was started on the lines of National Watershed Development Project for Rain-fed Area (NWDPRA) in 1990-91. Prior to intercepting NWDPRA, Mir Watershed Development Project was started during 1984-85 as an operational Research Project, the progress of which stands discussed hereunder along with that of NWDPRA (Gupta et al., 2009):

5.17.1 Multiple Cropping in Watersheds

After introducing the watershed development programme in some of the areas of J&K, Himachal Pradesh and Uttarakhand, multiple cropping systems have been started.

5.17.1.1 Keri Watershed

Farmers of Kandi belt of Jammu were earlier growing only one crop either maize or wheat. But the Department of Agriculture under watershed programme in Keri (Bhalwal block of Jammu district, Jammu & Kashmir State) introduced multiple cropping as demonstration plots. Under this cropping system third crop of wheat (IWP-72) was sown in the field of Mr. Bansi Lai. This helped in raising Agricultural Productivity for unit of land, better utilization of land/cropping intensity and bettering the lot of farmers. About 15 q yield of maize was obtained ha^{-1} and 2 q of toria ha^{-1} a cash crop and oilseed Zaid crop for the first time during 1992-93. Now many farmers are growing toria apart from maize and wheat as recently reported by Gupta (2006a and b).

5.17.1.1.1 Strategies for Rainfed Agro-ecosystem

A Sub-mountainous area of Siwaliks is locally known as ***Kandi.*** The ***Kandi*** belt of Jammu and ***Karewas*** of Kashmir are characterized by low rainfall i.e. less than 950 mm and 750 mm, respectively, undulating topography, low to medium soil organic matter content with light coarse textured soils and frequent occurrence of drought (Gupta, 2013h). Not only this, low and uncertain rainfall varying in time and space accompanied with droughts owing to change in climate cause low agricultural productivity. The growing of crops by the farmers in this area is not only risky but also pays very less and sometimes even the investment of inputs cannot be met with. This, therefore, necessitates following up new strategies for the rain-fed agro-ecosystem of ***Kandi*** and ***Karewas*** dwellers. Such strategies consist of adoption of agroforestry, promotion of horticulture, olericulture and floriculture. Shifting from chemicalised farming to organic farming, watershed management, performance to grow pulses and adoption of green manuring are the other strategies to follow.

5.17.1.2 Mir Watershed

Before the initiation of the watershed development in Mir (Udhampur District, Jammu and Kashmir), the area was mono- cropped. The maize was the major crop of the area. During Rabi season usually most of the area used to leave fallow and very few farmers grew wheat during Rabi. Efforts were therefore made to increase the area under Rabi crops. This increased the intensity of cropping from 103 per cent during 1984-85 to 183 during 1987-88. The yield obtained by improved methods in different years is summarized here under.

Table 5.2: Yield of crops obtained through years (Adapted from Gupta [1994) - Scientist Incharge, National Watershed Mir Udhampur)

Crops	Yield (q ha'¹)						
	Traditional	1984-85	1985-86	1986-87	1987-88	1988-89	1989-90
Wheat	5	14	31	30	25	26	24
Barley	8	15	15	18	13	15	14
Mustard	2	4	4	4	F*	8	F*

'Crop failed due to hail storms.

5.17.1.3 Devak Watershed

The Devak watershed spread over an area of 9,733 ha has been divided almost into 15 micro-watersheds. The work of this Integrated Watershed Development Project was completed by 1999. Over 481 ha of land not suitable for agricultural production were brought under silvi-pasture and horticulture rejuvenation. Since, inception of project during 1992 and up to 1998, 6,000 q of grass supplied to the people. An area of 1354 ha of non-arable land was utilized for afforestation. Species like khair (*Acacia catechu*), sissam (*Dalbergia sissoo*) and kikar (*Acacia nilotica*) were planted. These species were planted in the contour trenches. A number of nurseries were also developed.

5.17.1.4 Integrated Watershed Management and Eco-development Project

With the commencement of "Integrated Watershed Management and Eco-Development Project" in 1998 village Sah near Samba, Project Authorities laid greatest emphasis on various components. These were planting of herbs, trees and shrubs, promotion of agriculture, horticulture, natural regeneration, afforestation of degradated area, soil and water conservation works, and promotion of non-conventional source of energy. By the end of March 1996, the Eco-Task Force accomplished the gigantic task of covering 3,929 ha where up on 22.2 lakh plants of local species were planted which were socially acceptable to the people (Jasrotia, 1996). Besides 8.58 lakh saplings were distributed among different agencies so as to further increase the cause of forestry. A large number of plants are going on very well.

Owing to adequate protection and seeding the fodder production has been improved significantly and, thus, fodder deficient area has now become self reliant in fodder production rather, locals have started exporting surplus grass to adjoining areas. Emphasis has also been laid on promotion of agroforestry system (Gupta, 2005a and b) viz., agriculture - horticulture, agriculture - sericulture - silviculture and non-conventional source of energy vis-a-vis apiculture and mushroom production.

The physical achievements of the Eco-Task Force and Integrated Watershed Management ending 1996 are given here under:

Table 5.3: Physical achievements relating to agriculture/allied disciplines (Adapted from Jasrotia [1996])

Sr.No.		
1.	Planting of trees	22.24 lakh plants
2.	Planting of grass slips	2.07 lakhs
3.	Fruit plants distributed	6,000
4.	Fodder grass distributed	4.47 lakh quintals
5.	Fuel wood distributed	69,720 quintals

In addition to the above, some income generation through forestry development work, rope, basket and doona making selling of milk and milk products, collection of honey, has also been occurred.

5.17.1.5 Fakot Watershed (Uttarakhand)

Like Mir watershed after preparation of soil and water conservation work plans and implementation of the measures, the yield of local crops of Fakot Watershed in the Hill Region of Garhwal erstwhile Uttar Pradesh, now Uttaranchal Uttrakhand was very much improved which hitherto was very low. Crop improvement programmes were taken up to improve the yield of various varieties of crops. For this purpose, crop demonstration was conducted on farmer's fields. As a result of the work done, the area under assured irrigation and rainfed agriculture sustainability increased. Besides crop intensity as well as average yield of major crops also went up.

The technologies developed at the Watershed Management Project Fakot are now being adopted by the Directoratc of Watershed Development of Uttar Pradesh for application in the hilly region of Garhwal (now in Uttarakhand. These are terrace improvement, bench terrace and shoulder buildings. The Directorate has also undertaken crop management by use of high yielding varieties of crops, line sowing, and balance use of fertilizers.

In Jammu region, a total of 13,000 ha of cultivable land have been undertaken in watershed management with 26 micro watershed, each covering an area of 500 ha since 2005 (Kaul, 2005). This watershed management programme has started in six districts viz., Jammu, Udhampur, Doda, Kathua, Rajouri and Poonch. There is every possibility that with the fully starting of the work of watershed management programme, it will bring a "silent revolution".

5.17.2 Vegetable and Fruit Production through Watershed Management Areas of Jammu and Samba Districts of Jammu Region

The construction and upgrading of check dams, earthen dams, ponds, baulies through Village Development Committees have provided better irrigation facilities to the farmers. As such they have developed more interest in the production of vegetables. Not only in the semi plain areas but also in the hilly and partially hilly areas in the Siwalik range of Akhnoor, the farmers have started producing vegetables. The production of tomatoes and peas has brought an additional income for the farmers of these areas. They are sending these vegetables to Akhnoor and Jammu markets for the last 2-3 years (Sharma, 2004) The areas such as Dori Dager, Ghar Majoor, Sungal, Panjgran Maira, Pallwan etc., are now yielding good quality and quantity of vegetables. Advanced technology and hybrid seeds are being utilized to improve the quality of vegetables. In horticulture sector orchards of citrus fruits, mangoes, litchi, have been raised. Apart from citrus fruits, mangoes and litchi, pomegranate and local fruits like amla *(Phyllanthus emblica)* and lasoora *(Cordia dichotoma)* were also introduced in the villages of Keri watershed which are now in the bearing stage (Gupta et al., 2009) and as such the peasants of the area are obtaining lot of income. Similarly many of the farmers of Bari-Badhori Watershed Project sanctioned in Samba District by the CAPART for Siwaliks of Jammu are getting an income of Rs. 1,000-1,500 or even more tree^{-1} year^{-1} from local variety of amla known as amli (Jamwal and Gupta, 2007). Jamun *(Syzygium cumini)* and phalsa *(Grewia asiatica)* have also been planted. Phalsa is yielding an average yield of 4 kg per plant in the month of June which is sold at Rs. 100 kg^{-1} i.e., Rs. 400 per plant that works out to be Rs. 4 lakhs ha^{-1}. Even if the farmers get Rs. 50 kg^{-1} he would earn 2 lakhs every year besides providing Rs. 60,000 to the local landless labour particularly women who pluck the fruit at Rs. 15 kg^{-1} at village Suchani (Samba) as reported by Jamwal (2009).

In light of the above, the farmers must grow phalsa wherever it is possible. Growing of phalsa not only provides handsome income to the farmers but helps in recharging ground water by harvesting rainwater. Moreover, phalsa is a medicinal plant with following characteristics:

- Leaves are known to have antibiotic actions which are applied on skin eruptions.
- Fruits are cooling tonic, and allay thrust and burning sensations.
- The phalsa seeds produce approximately 5 per cent yield of bright yellow oil.
- The fruit is also astringent and stomachic.

5.17.3 Work on Soil and Water Conservation

The various kinds of works or activities undertaken to conserve soil and moisture in the above said watersheds consisted of building check dams for storing rain water, gully plugging, land leveling, bunding etc. The details of building of check dams has been described in chapter 3 of this book along with rain water harvesting technique and other ways to tide over the scarcity of water. Various measures viz., land capability classification, use of agronomic and engineering practices involved in controlling soil and water conservation are given elsewhere (Gupta and Sharma, 2007).

5.18 People's Participation in Watershed Project

Study conducted in two watersheds viz. Akhnoor and Bari Badhori in Jammu (now Samba) to find out the role of people's participation in watershed management (Sharma et al., 2008). It was observed that participating farmers had higher level of adoption in latest technology like soil water conservation, fruit trees plantation, vegetables and crop production than those of non-participating.

The innovative approach of the Forest Department has not only ensured active involvement of the local people but also helped in achieving the target of the beneficiary contribution for the Rs 337 crore by World Bank Funded under Mid Himalayan Watershed Development Project in Himachal Pradesh. The seven year project is slated for completion in March 2013 and accordingly local community was required to contribute Rs 27.5 crore for various activities. However, the contribution of the beneficiaries has already crossed 28 crore mark and with two years to go it is likely to overshoot the target by a significant margin.

A host of works were undertaken under the project. They comprised mainly of construction of water harvesting structure, foot bridges, roads, water storage tanks and micro-irrigation schemes etc.

5.19 Women's Participation in Watershed Management

As women constitute the backbone of a number of agricultural, horticultural and animal husbandry operations, and majority of agricultural labourers are women, so their participation is necessary for implementation of watershed management/development programme. It is attributed to their contribution in every stage of food production such as growing and gathering, preparing, selling, processing and storing as well as food processing, water management and managing dairy animals/poultry birds. There are certain agricultural operations which totally belong to the women's domain. Such operations mainly consist of weeding, picking, winnowing, grading and packing as well as to manage resources stocks particularly soils, firewood for fuel, water for drinking, irrigating the crops and other purpose for their households (Gupta, 2015b).

5.20 Conclusion

Watershed management aims at managing water, land and vegetation for increased production of agricultural crops, grasses and trees on sustained basis, conserving soil and water, reducing run-off (thus floods), and decreasing sediment in run-off.

5.21 Future Research Strategies/Priorities

- To develop the horticulture in the watersheds more but dedicated horticulture extension staff must be deployed with proper accountability. This is very essential in the Siwaliks of Jammu and Kashmir, Punjab, Himachal Pradesh, Haryana, Uttarakhand and Uttar Pradesh especially in the *Kandi* belt which is not suitable for agriculture.
- Indigenous fruits depending up on the area must be grown. These may consist of shatoot (*Morus alba*), phalsa (*Grewia asiatica*), jamun (*Syzygium cumini*), amla (*Phyllanthus emblica*), girna (*Carissa spinarum*), karonda (*Carrisa carandus*), amb (*Mangifera indica*), ber (*Zizyphus jujuba*) etc.
- The aforesaid states must develop their own fruit nurseries. Only then our watershed management concept will succeed.
- While planting bushes/trees in silviculture, local species like *Acacia, Albizia, Terminalia, Cassia fistula, Pinus roxburghii, Cedrus deodara, Quercus* spp. etc. should be preferred to grow.
- One way to develop watersheds is to start with biogas units. By standing such units will not only provide biogas slurry to serve as manure for various crops but also constitute non-conventional source of energy.
- It is also suggested that series of food security systems can be designed and implemented in various watersheds of the country to boost agricultural production. A number of Agroforestry systems like Agriculture-Horticulture, Agriculture-Silviculture, Agriculture-Horticulture-Silviculture, Agriculture-Silviculture-Fishry and Agriculture-Silviculture-Animal Husbandry can be established.
- The details of these systems have been given in separate chapter i.e., Agroforestry of this book.

- It is stressed up on involvement of people through Panchayats and other Rural based village development committees.
- It is called upon people to make the best use of Integrated Watershed Development Projects.
- Wherever, it is possible the small ditches be constructed like Dumka district of Jharkhand as these have been found fruitful for the farmers (Raghav et al. 2016), especially for lowland areas of the country to conserve soil and water.

Study Questions

1. Define the following:
   i. Watershed
   ii. Watershed development or watershed management
2. How the watershed management proves boon not only for the farming community but also for the general public?
3. Describe in details the main objectives of watershed management?
4. Write the various factors responsible for causing deterioration in watershed.
5. Write a short note on the participatory approach on integrated watershed management.
6. Comment about the saying "have watersheds and save lands".
7. What are the basic requirements of watershed management planning?
8. Explain the role of women's participation in watershed management.

References

Anonymous. 2018. Water Scarcity in Anantanag. The Daily Excelsior 54(88)6

Choudhary, A.K., Thakur, R.C. and Shobhna (2005). Integrated plant nutrient supply system and sustainable agriculture - A review. Farmers' Forum 5 (5): 23-24.

Gupta, R.D. 2005. Environmental Degradation of Jammu and Kashmir Himalayas and their Control. Associated Publishing Company-A division of Astral International Pvt. Ltd. New Delhi, India

Gupta, R.D. (2005a). Cover crops enhance food production-l. The Kashmir Times 62 (293): 13 [October 22, Saturday].

Gupta. R.D. (2005b). Cover crops enhance food production-ll. The Kashmir Times 62 (300): 13 [October 29, Saturday].

Gupta, R.D. (2006a). Upliftment of rural people through technologies. Technology for dryland farming - I. The Kashmir Times 63 (53): 13 [February 23, Thursday].

Gupta, R.D. (2006b). Upliftment of rural people through technobgies. Technology for dryland farming - II. The Kashmir Times 63 (80): 13 [March 23, Thursday].

Gupta, R.D. (2009a). Agroforestry for farmers. The Daily Excelsior 45 (180): 6 [June 30, Tuesday].

Gupta, R.D. (2013h). Strategies for rain-fed agroecosystem. The Daily Excelsior 49 (342): 8 [December 14, Saturday].

Gupta, R.D. (2015b). Women's participation in watershed management. The Daily Excelsior 51 (88): 8 [March 31, Tuesday].

Gupta, R.D. and Sharma, P.D. (2007). Soil erosion and conservation in North West Himalayas (pp. 205-214). In Natural Resource Management for Sustainable Hill Agriculture (Eds. S. Arora, S.S. Kukal and V. Sharma). Soil Conservation Society of India, Jammu Chapter, SKAUST-Jammu, J&K State, India.

Gupta, R.D., Sanjay, A. and Sharma. V. (2008). RairvTed agricultural production watershed management approach in Siwalik region of Jammu. J. Soil and Water Conservation 8 (1): 19-24

Gupta, R.D. (2016). Wildlife and Wetland ecosystems of Jammu and Kashmir Hamalayas: N.R. Books International Pacca Danga jammu Tawi- 180001, J&K, India

Gupta, S.C. (1994). Experience of Watershed Development Programme with special reference to Mir Watershed. Paper presented by the author in the seminar held in Department of Agriculture, Jammu and Kashmir, Talab Tilo Jammu on March 13,1994.

Jamwal, G.S. (2009). Harvest phalsa potential. The Daily Excelsior 48 (241): 4 [August 30, Sunday].

Jamwal, G.S. and Gupta, R.D. (2007). Watershed management project - A case study of Bari-Badhori, Jammu and Kashmir State (pp. 226- 231). In Natural Resource Management for Sustainable Hill Agriculture (Eds. S. Arora, S.S. Kukal and V. Sharma). Soil Conservation Society of India, Jammu Chapter SKUAST Jammu, J&K State, India.

Jasrotia, R.S. (1996). Forest Newsletter. 16 Inagural Issue, May-June, 96. Jammu and Kashmir Forest Department, Exchange Road, Jammu.

Kaul, Y. (2005). Watershed management long way to go. The Kashmir Times 62 (154): 9 [June 5, Sunday].

Lohumi, R. (2011). People's participation in watershed project pays off. The Tribune 131 (261): 7 [September 21, Wednesday].

Mahajan, A. (2004). Watershed - Concept and management. Agrobios Newsletter 3 (1): 28-31.

Raghav, D.K., Thakur, A. and Singh, R.K. (2016). Impact of small ditches (Dhobas) in rainfed areas in Dumka district of Santhal Pargana, Jharkhand-A case study. Soil and Water Conservation 15 (4): 356-60.

Sharma, G. (2004). Watershed project a big boon for socioeconomic development of Kandi people. The Daily Excelsior 40 (200):4 [July 11, Thursday].

Sharma, R., Chauhan, J., Meena, B.S. and Chauhan, R.S. (2008). Impact of participatory approach in management of watershed practices in Jammu. Indian Res. J. Ext. Edu. 8 (1): 19-21.

6

Organic Farming for Sustainable Agriculture

Abstract

Organic farming relies upon ecosystem management and attempts to reduce external agricultural inputs, especially fertilizers and pesticides. It is a holistic production management system that enhances health of agro-ecosystem including bio-diversity, biological cycle as well as soil biological activity. As a matter of fact, the food farming system based upon chemical fertilizers and inorganic pesticides has started to peak out. It was quite evident from the average food grain production which was estimated at about 201 million tonnes for the last twelve years except for the figures of 201 and 227 million tonnes obtained during 2005-06 and 2008-09. Thus, in view of all ill effects created by the chemicalised farming like degradation of environment, soil, air and water pollution, food unsafely and energy shortage, there is an increasing awareness of "organic farming" practices the world over. In this chapter of the book, concept of organic farming along with its definition, principles, benefits, scope of organic farming in respect of Indian and world scenario, traditional Views of Indian farming and International competence for organic agriculture, have been narrated. Options of organic farming, its constraints, various methods of organic farming like crop rotations, soil tillage, use of organic manures including green manuring, Panchgavya and Jaivic and Vedic inputs, have also been described.

Keywords: Organic farming; agro-ecosystem; ecosystem management; production management; food grain production; environment degradation; chemical fertilizers; inorganic pesticides; soil pollution; air pollution; water pollution; crop rotations; soil tillage; organic manures; green manuring; Panchgavya; Jaivic and Vedic inputs

There is no doubt that after independence, particularly during 1966-68, the use of hybrid and composite seeds of various crops, chemical fertilizers and pesticides resulted in an increase in the Indian agricultural output leading to "Green Revolution". But quite soon, the deleterious effects of farming based on chemical methods came to fore. These deleterious effects are: (i) Accumulation of excess nitrate in the underground water which caused an infant disease, known as Methaemoglobinaemia or blue baby disease; (ii) Nitrosamines formed by the reaction of excessive amount of nitrate with secondary amines serve as carcinogenic agents; (iii) Most of the pesticide and herbicide residues which affect the central nervous system, respiratory system and gastrointestinal system of human beings were observed. There was also depletion of ground water, ecological imbalance in soil flora and fauna, resistance in pests to chemicals and elimination of natural enemies of pests (Ramanathan, 2006).

The excess nitrate leached to water bodies along with phosphate increases algal growth in their surface, lead to accumulation of considerable masses of dead organic material which produces deficiency of oxygen supply and finally influences the other aquatic life i.e., "eutrophication" (Gupta, 2017). Accumulation of heavy metals like Cd, Pb and Cr in soils and water, are the principal causes of environment concern due to excessive use of fertilizers. Excessive use of nitrogen causes more production of gases in the form of NO, NO2, and N2O which adversely affect the ozone layer. Details of ill-effects caused due to "Green Revolution" are described elsewhere (Gupta and Singh, 2006). There is increased resistance among pests and weeds which is becoming difficult to control with the use of indiscriminate pesticides. There is also evidence that many of these chemicals have brought alteration in the soil biological ecosystem like bacteria, actinomycetes, fungi, micro-fauna and macro-flora like earthworms, centipede, millipede etc.

The harmful effects of agricultural chemicals based farming as stated above, forced us to return to "Organic Farming" so that the land/soil remains more fertile and productive for longer period, and our health and environment are protected from the harmful effects (Mahajan et al., 2007a and 2007b).

The primary concern of Indian agriculture in the coming years is the challenge for increasing food production to keep pace with the nutritional needs of the growing population of the country. India's population is expected to increase from 1125 million during 2011 to 1500 million by 2020, which will require 300 million tonnes of food grains yearly. Keeping in view, the current food production level, India has to raise food production by 5 million tonnes as against 3.1 million tonnes per year achieved over the past 50 years. Thus, the farming system's approach alone, may not be considered as a food production system, but may be considered to resource management, ensuring soil fertility, ecosystem management, water management and pest management. With this background, the concept of organic farming as the basis of "Sustainable Agriculture" came into existence for overall healthy agricultural development in the country (Singh, 2012).

The present chapter dedicates on the concept of organic farming and the methods by which the sustainability of agricultural production can be achieved.

6.1 Concept of Organic Farming

Organic farming means farming in the spirit of organic relationship. It is holistic production management system which avoids the use of synthetic fertilizers and pesticides, and promotes and enhances health of agro-ecosystem including biodiversity, biological cycles and soil biological activities. It emphasizes the use of management practices in preference to the use of off-farm testing carried out under local specific and regional conditions (Mahajan et al., 2007c) and use of bio-pesticides (Rishipal et al., 2013).

The concept of organic farming was started with the establishment of International Federation of Organic Agriculture Movement (IFOAM) on 6th of November, 1972 in France. Today nobody can deny that organic trade is a growing reality all over the world (Gupta et al., 2005). There are about 17,000 organic farmers in Germany and in Switzerland the organic share has reached about 7 per cent with the largest share being 30 per cent in Kant on Graubunden. Italy has about 45,000 organic farms, 6 per cent agriculture land in Czech Republic is farmed organically. Austria boasts of a 13 per cent share of organic farms (Geier, 2005). In Austria, there were more than 20,000 organic farmers. Sweden and Finland have arrived at the level of Switzerland. In Uganda, 7,000 farmers have chosen to cultivate organic crops, whereas in Mexico innumerable small farmers have started to produce organic coffee and other staple foods. Though worldwide markets for organic foods are expanding, the concentration of growth has been mainly in the Europe, the USA and Japan. All these three markets have recorded annual growth rates of 15 to 30 per cents in the past 5 years. The overall European market is in the range of 13 billion US Dollars and worldwide organic market reached 25 billion US Dollars in 2003. According to the International Trade Centre projections, the organic market size would be around 46 billion U.S Dollars in the European Union, 45 billion Dollars in the U.S.A and 11 billion U.S Dollars in Japan during 2010. According to IFOAM, by late 2006, organic farming accounted for a total surface area of 30.5 m ha, representing almost a three-fold increase in the worldwide certified surface area between 1999 and 2006, covering 7,19,000 organic producers (Anonymous, 2009a).

In India National Standard Committee also drafted the concept and principles of basic standards of organic agriculture in 1996. More than 20 Inspectors were trained to carry out inspection of organic food by the Spices Board, Government of India during 1998. During May, 2000, the Ministry of Agriculture, Government of India declared 21st century as an era of Organic Farming.

Accordingly, the information with regard to the research conducted in the field of organic agriculture, was collected to initiate further activities from this angle.

Further in order to promote organic farming in the country in view of its eco-friendly nature, the National Steering Committee (NSC) under the chairmanship of Secretary Commerce approved National Programme for Organic Production (NPOP) at its meeting held on 2nd May, 2001. Under NPOP, documents like National Standards Accreditation Criteria for accrediting, inspection and certification agencies were prepared and approved by the NSC. The NSC has designated Agricultural and Processed Food Products Export Development Authority (APEDA), Coffee Board, Spices Board, Tea Board, Coconut Development and Directorate of Cashew nut and Coca Development as Accreditation Agencies. APEDA has been assisting the Commerce Ministry in preparing various documents under the NPOP and has also developed the National Organic Logo. There were 107 certified organic producers in the country for the products covered under APEDA. APEDA has also setup 4 model organic farms in North Eastern Region (NER) for Joha rice, sugarcane, passion fruit and pineapple. Organic farming has the potential to generate over 8 million jobs in Rajasthan (Anonymous, 2013b)

The concept of organic agriculture for food production and biodiversity conservation is considered universally at International level. However, to the farmers of Developing countries, the concept of organic farming is dubious. In India, it implies the use of organic manures and no use of synthetic fertilizers and pesticides. International Union for Conservation of Nature (IUCN) recognizes the potential of organic agriculture in several protected area categories. Now organic agriculture is being encouraged in several national and regional areas. Protocol to produce organic Kalanamak aromatic rice variety of north eastern part of Uttar Pradesh is one of the example in this regard (Chaudhary and Mishra, 2012).

From the aforesaid literature concerning with the concept of organic agriculture, it is inferred that organic farming is one of the most widely practised, diversified conventional farming system to render the agriculture sustainable. In agriculture, sustainability refers to "development not only in terms of output but also the socio-economic and ecological parameters." According to United Nation Development Programme (UNDP), safeguarding national food security means improving the quality of life of farming community by ensuring no further depletion of natural resource base. Though we have inherited organic farming as a tradition, still the question is whether we should go for it or not. But, in the long run to meet the food grain requirements, the most desirable way is to maintain soil health, stability in production through use of organic and biological resources (Gupta, 2012a).

6.2 How Organic Farming is Defined?

Organic farming is the "Rishi Agriculture" of our ancient tradition. It is also called "Natural Farming", "Biological Farming". The other terms which are used to signify organic farming, are "Green Food", "Organic Food", "Biodynamic Farming" and "Ecological Agriculture" or "Eco farming". Definitions of some of these terms as reported by Gupta (2019a and 2019b) are as follows:

6.2.1 Organic Farming

It is defined as a system of agricultural production that encourages healthy soils and crops through such practices as nutrient recycling of organic matter, crop rotations, proper tillage and avoidance the use of synthetic fertilizers and pesticides. Organic farming as enunciated by IFOAM, covers all aspects of operating farms that promote environmentally, culturally and economically sound production of food, fibre, herbs and fuel wood. Organic farming is, in fact, an important form of eco-friendly practices of food production (Mohan and Maheshwari, 2013)

Organic farming is infact, the science and art of producing crops and rearing of livestock. It possesses three distinct features of appropriate cropping patterns: (i) crop rotations, intercropping, relay cropping, alley cropping, mosaic king with multipurpose tree species and livestock integration; (ii) use of organic materials; and (iii) use of non-chemical methods for control of pests (Angelina-Briones, 1999). Organic materials here refer to use of FYM, compost, green manure, crop residues, bio-fertilizers etc. (Gupta, 2014). Thus, organic farming may be defined *"comprehensively as environment friendly ecological production system that promotes and enhances biodiversity, biological cycles and biological activities".*

6.2.2 Natural Farming

It is a system of agriculture devised by Masanobu Fukuoka (Japanese Plant Pathologist) that seeks to follow nature by minimizing human interference: no mechanical cultivation, no synthetic fertilizers, no weeding by tillage or herbicides, no dependence on chemicals.

6.2.3 Ecological Agriculture or Eco-Farming

Farming practices that enhance or, at least, do not harm the environment and are aimed at minimizing the use of chemical' inputs, rather than completely avoiding them as in organic farming, is known as "eco- farming".

6.2.4 Biodynamic Farming

Holistic system of agriculture devised by Rudolph Steiner, seeks to connect nature with cosmic creative force. An attempt is made to create a whole form organism in harmony with its habitat. Compost and special preparations (e.g., plant derived sprays) are used. It consists of using "cow-dung horn technology", which means composting of dung from lactating cow in hollow of cow horn after burying in soil at an auspicious moment. Thereafter, applying it in pellet form according to Homeopathic doses (Menon and Karamarkar, 1994). Other details of biodynamic farming have been given elsewhere (Gupta et al., 2005). Synthetic fertilizers and pesticides are avoided.

6.2.5 Sustainable Agriculture

Management of resources for agriculture to satisfy changing human needs while maintaining or enhancing the quality of environment and conserving the natural resources is called sustainable agriculture.

6.2.6 Organic Agriculture

Codex Alimentarius (FAO/WHO) Commission defines organic agriculture as a holistic food production management system which promotes and enhances agro-ecosystem's health including biodiversity, biological cycles and soil biological activity.

6.3 Principles of Organic Farming

The basic principle of organic farming is to enhance organic matter content of the soil which has a profound impact on soil quality, by way of enhancing soil structure, soil fertility, water infiltration and its storage capacity. Moreover, soil erosion is reduced due to bacterial activity in producing humus in soil. Organic farming maintains soil biological activity at high level. Above all, organic farming maintains environment quality (Mahajan et al., 2007a and b; Mahajan and Gupta, 2009).

6.4 Benefits of Organic Farming

Experiments from around the world proved merits of organic farming. A scientific study by the American National Academy infers that application of fertilizers and pesticides does not necessarily lead to better farming. Credit for putting natural farming on global scale goes to Japanese, Plant Pathologist - Masannobu Fukuoka, who secured Food and Agriculture Organization (FAO) award during, 2002 for signal service, rendered to the cause of natural farming. He advocated that soils nourishing qualities stem from organic matter that returns to soil periodically. Fukuoka took 25 years to develop rice variety known as "Fukuoka rice" is the best genetically modified by natural breeding using organics. This variety has given promising yield in the coldest places all over the world and can yield 3 times more yield than present varieties.

In Southern India organic farming is of immense importance these days. Fertility of the soil is sustained as it is free from the bad effects caused by the Green Revolution like deterioration of soil health and environment pollution. Harmful chemicals do not enter the food grains and other farm produce. Occurrence of blue baby disease caused due to excessive amount of nitrate can be minimized.

Organic agriculture emphasizes the use of management practices in preference to the use of farm inputs taking into account that regional conditions require locally adapted systems. This is accomplished by using wherever possible agronomic, biological and mechanical methods as compared to using synthetic materials to fulfill any specific function within the system (Mahajan and Gupta, 2009). Also, organic farming increases soil biological activity, enhances biological diversity within the whole system, maintains long-term soil fertility and re-cycles wastes of plant and animal origin in order to return nutrients to the land/soil.

"Agnihotra" (ashes left over after performing *yagna)* is considered as the complete or perfect plant food (Pathak and Ram, 2003), though it mainly contains potash. This defies scientific logic. Agricultural scientists must not tolerate such absurdities and treat them with contempt they deserve. Moreover, the organic produce is more nutritious and tasty, and as such fetches more income. The European Union has passed strict regulations on organic products and farmers have to observe the same to be labeled this produce as "organic product". The regulations apply to milk and meat products also. According to these regulations the farmers are not allowed to use any chemical on the crop including fodder.

Today, organic foods or green foods are in more demand globally. Such foods are supposed to be healthier for humans as well as for animals and hence command a high premium in the international market. These foods are also free from all harmful chemicals. Though our farmers have been in practice of producing green foods since ages, our faster growth in population has created new challenges. Health conscious has increased and organically produced foods are sought after everywhere.

Since most of the fruits and vegetables are taken fresh and raw, the residues of pesticides present in such products, therefore, may lead to various kinds of health hazards (Gupta and Singh, 2008). Hence green food production offers a better possibility in horticultural and olericultural crops rather than field crops. Besides, green food production is a holistic production management system, which promotes and enhances agro-ecosystem health including biodiversity, biological cycle and soil biological activities.

The increased demand of organically grown foods is due to perpetual consumer's demand and its positive environmental effects. Many aspects of organic farming are important elements of a system approach to sustainable food production both for domestic and export markets. Nevertheless, there is a possibility that products are falsely claimed to be organic, may be presented to the consumers. Organic farming not only enables to capture foreign market but also helps in preserving the ecosystem and sustainability of farming.

The quality of horticultural produces like flowers, vegetables and fruits is improved when the nutrients are supplied through organic manures than in the form of fertilizers. It is due to the supply of all the growth principles like enzymes, hormones and vitamins in addition to all the essential plant nutrients. As a result, the metabolic functions get regulated more effectively resulting in better synthesis of proximate constituents and consequent improvement in the quality of the produce.

The cultivation of almost all horticultural and vegetables round the year, is an advantage for Indian conditions to compete in the world market. Major export markets for organic foods exist in cocoa, spices, cashew nut, fruits, vegetables, herbs. Another area having good market potential is off season cultivation of fruits and vegetables. In the next few years, organic food production could jump from roughly 1 per cent (Dutt, 2002) to 5-10 per cent of food sales in the major markets. Besides fresh fruits and vegetables, spices, cashew nuts and coffee, tea, rice, flowers, saffron and many others organically grown products found in grocery stores, are part of the broad and growing market for developing countries to export organic food.

6.4.1 Environmental and Economic Benefits of Organic Farming

The FAO Report points out to further benefits such as better animal welfare, wildlife protection, avoidance of pesticides and less energy use (Burcher, 2007). Results from the studies carried out by the US Department of Agriculture support the FAO findings showing that organic crops are worth more than conventional crops in the market. Organic agriculture has the potential to reduce carbon dioxide, nitrous oxide, methane and other greenhouse gases which contribute to "Global Warming". Organic agriculture could double soil carbon sequestration in livestock based systems and decrease greenhouse gases by 48-60 per cent (Burcher, 2007).

6.4.2 Health Benefits of Organic Farming

Organic foods tend to have higher micro-nutrient content that contributes to better health, lower incidence of non-communicable diseases and boosts plant and animal immunity against diseases. The UK Soil Association carried out a systematic review of the evidence comparing trace minerals in organic and non- organic food and found that on an average, organic food contains higher levels of Vitamin C and essential minerals like Ca, Mg and Fe.

6.4.3 Organic Farming Provides Long-Term Solutions

A broad scale shift to organic farming can produce enough food on a global per capita basis to feed the world's population over the next 50 years. Therefore, as the myth of "low yield organic agriculture recedes", it is up to the agricultural researchers, officials and governments to invest in long-term alternative agricultural systems. One such system consists of green manure that can provide enough biologically fixed nitrogen to replace the artificial nitrogen currently used on the planet (Gupta, 2014). The results would strongly benefit farmers, consumers and the environment.

"The strongest benefits of organic agriculture are its reliance on fossil fuel independent as it is dependent on locally available resources that incur minimal agro-ecological stresses and cost-effective", said by Nadia Scialabba (FAO official). She described organic agriculture as a "neo-traditional food system which combines modern science and indigenous knowledge." The FAO report strongly suggests that a worldwide shift to organic agriculture can fight world's hunger and at the same time tackle climate change (Burcher, 2007).

6.5 Scope of Organic Farming

Organic farming has lot of scope in the rain-fed or un-irrigated areas where the farmers mostly grow their crops with organic manures. It has also a very much scope in the irrigated areas where chemicalised agriculture has degraded the lands.

6.5.1 Indian Scenario

Indian agriculture needs new directions in order to improve socio-economic conditions of the peasants. This can be done to great extent by diversifying agriculture (Anonymous, 2013a). New opportunities have been created and made available due to globalization of the trade. In fact, liberalization and globalization have brought forth opportunities and threats.

Many states of India like Jammu and Kashmir, Punjab, Himachal Pradesh, Karnataka, Andhra Pradesh, Maharashtra, Kerala, Tamil Nadu, Gujarat, can create an international market for vegetable, fruits, flowers and milk products. But we can compete with the world market only if we produce quality products. In order to produce for the world market, a dynamic agricultural production system is required. This will also require a suitable technology, effective communication and delivery system and supportive Government policies.

As per the study conducted by FAO during mid-2003, India had 1426 certified organic farms producing 14,000 tonnes of organic produce annually. During 2002, the organic produce amounting to 11,925 tonnes was exported which mainly consisted of tea, coffee, fruits, herbal plant products, cotton (Mahajan et al., 2007b). The area under organic farming was more than 75,000 ha during 2001 as reported by Chhonkar (2001) and Chhonkar (2003). Tea and medicinal plants covered under 10,000 ha each and 12,000 ha was under cotton. Pine apple, pepper, turmeric, walnut were the other products which are being produced organically. Upto November, 2005, India had 11 accredited agencies that used to certify organic farm produce for export (Anonymous, 2005).

There are large number of farmers in India who have either never cultivated their lands chemically or have come back to organic farming due to the beliefs or purely for reasons of economics. These farmers are cultivating hundreds of thousands hectare of land but have not classified as organic growers though they are. Their produce is generally sold in open market along with the other grown food. These farmers are not keen to opt for certification because of more involvement of cost vis-a-vis extensive documentation required for such certification.

It is the best time for those farmers who can produce for processing international market and the worst for those who do not like to participate in the world market. It is worthwhile to mention that our products are of poor quality and have been found to possess traces of chemicals used to control pests and weeds. For example, 130 containers of fresh grapes of India were rejected by the Netherlands during July, 2003 because of high pesticide residue levels in them (Naidu, 2003). In the wake of this rejection, the Bio- Agriculture Inputs Producers Association of Andhra Pradesh is now seeking to promote farmers awareness of alternate methods for tackling pests i.e., use of bio-agriculture inputs. The farmers require to be made aware of about international standards and maximum tolerance for pesticides so that those exporters donot suffer in future. For the development of organic farming, the Union Budget 2014-15 had proposed Rs.100 crores for the north eastern region of India (Singh, 2014). This is, in fact, a very good proposal upon implementation, will set out an environment for an organic revolution in India.

6.5.2 World Scenario

Asia has about 13 per cent of world's organic land, dominated by China, ranking second highest in the world after Australia and then followed by India and Russia (Mahajan et al., 2007a, b and c). According to world market survey conducted during 2001, organic agriculture was being practised in about 17 m ha of which 7.7 m ha was in Australia alone (Magar, 2004), 2.8 m ha in Argentina and about 4.2 m ha in European countries. Countries like UK, Switzerland, Denmark and Sweden have achieved the organic growth rate to the range of about 30-50 per cent. In Germany, it has already reached a share of 80 to 90 per cent (Geier, 2005). More than 30 per cent of the daily bread in and around Munich and 30 per cent of all milk in Denmark are certified organically. Even in a country like Egypt, organic products are becoming main stream.

These days nearly 130 countries in the world produce organic products covering an area of 24 million hectare, which works out to be 1.6 per cent of agriculture area (Dahatonde and Khambalkar, 2007).

However, according to FAO report, organic agriculture is no longer to be considered a niche marked within developed countries. But, a vibrant commercial agriculture system which is practised in 120 countries covering 31 m ha of cultivated land and 62 m ha of certified wild harvested areas (Burcher, 2007). Growing consumer demands are also reported in countries like Brazil, Poland, Uganda and Mexico, where an impressive mostly small farmer produce organic coffee, coca, fruit, vegetables and spices.

The estimated global market for organic produce (fresh and process) is 13 billion U.S. Dollars and is still growing, which reached 21 billion during 2002 (Magar, 2004). The U.S. market for organic products alone is estimated to be about 9.5 billion U.S. Dollars during 2002. Up to November, 2005, the global market for organic products was over 30 billion US Dollars with 13 billions in European market alone (Anonymous, 2005). The organic market was worth 40 billion US Dollars in 2006 and expected to reach 70 billion US Dollars by 2012 (Burcher, 2007). Retail sales of organic foods and beverages worldwide were projected to grow at 5-15 per cent over the next five years. Worldwide markets for organic foods are expanding. The past five years i.e., from 1998-2003 have witnessed an annual growth rate of 15-30 per cent in the U.S., Europe and Japan.

Despite the demand, less than 0.1 per cent of the farmers in the country use bio-pesticide because of lack of awareness. But use of bio-pesticides can result internationally acceptable quality organic production within 3 years. So some re-orientation of farming practices and honest commitment from the farmer in this regard is required. In brief, it is concluded that organic farming is of much significance as it provides eco-technological stability, sustainable pest management and an alternative to in-organic fertilizers.

6.6 Traditional Views of Organic Farming in India

Organic farming is very much native to India, but it is almost forgotten. Any farmer having interest in organic farming would therefore, first look at the ancient Indian agricultural practices which are still being practised in many parts of the country. Ladakh cold arid region of Jammu and Kashmir is following traditional method of organic farming. In vegetable and crop production, use of fertilizers and pesticides is mostly avoided. For enriching the soil, processed night soil, animal dung and, kitchen and farm wastes are used. The recycling of human and animal excreta in soils of Ladakh is an old and time immemorial practice (Gupta and Upadhaya, 2006) but modern practices are slowly corrupting the ancient sound practices.

Modern agriculture in India is hardly 50 years old and "Green Revolution" was introduced during 1966-68. In comparison to this, agricultural practices of India date back to more than 4,000 years, which maintained the soil fertility over a long period of time. Most of the traditional practices are transmitted into a number of forms like maxims, proverbs, rituals, folk songs and riddles for the guidance of the peasants. *"A field without manure is as useless as a cow without a calf and "no fodder, no cattle; no cattle, no manure; no manure, no crop", are some of the Tamil proverbs and "un poshi talli, yalli van poshi" of Kashmiri, meaning agriculture is subservient to forests.* The farmers of ancient India, while developing farming systems and practices, were always supported with natural products for enhancing soil fertility. As for example, traditional farmers are still using cow's urine to combat crop diseases (spray with 10% concentration, the urine turned the diseased plants to healthy ones). Till the middle of the twentieth century, the farmers farmed with traditional practices to maintain the soil fertility and protect the crops. However, with an inception of modern agricultural practices from late 1960, a number of such practices were given up. Another important factor that has affected old practices is that agriculture is used as an occupation and not a business. But now it is a business. Non-cultivating proprietors are increasing in number, i.e. Zimidari system.

6.7 Agriculture in Vedic Periods

Agriculturists during the Vedic periods possessed a lot of knowledge about soil fertility. According to its fertility, soil was mainly divided into two classes: *urvara* (fertile) and *unurvara* or *usara* (sterile). Classification of soil was based on two grounds - medicinal and economic. The medical authorities e.g., Charaka and Susruta had in view the efficacy of vegetables drugs which was thought to depend on the nature of the soil, where the medicinal plants grow. And politicians like Chanakya cared for the productively of different types of soil, attention to which is necessary on the part of a good government to prevent famine (Raychaudhuri, 1982).

The ancient Indian cultivators were a wealthy and respectable section of the people, and had also good knowledge of climate, plant physiology, rotation and protection of crops. Indeed, one will be filled with surprise and admiration if one cares only a look into the detailed accounts as have been stated in Arthashastra, and Brhat Samhita and Agnipurana regarding the selection and treatment of seeds and use of animal excreta, fish and bones, beef and fish washings, and various kinds of mixtures and decoctions as fertilizers. Smearing of seedlings with mixture of sesamum, ghee, honey, cow dung and milk prior to transplanting, treating of seeds with a paste of vidanga, ghee and silt to control diseases caused by cold and hot sun, are the examples of traditional agricultural practices. Indian Plant Science, i.e. Vriksh-Ayurveda, is of immense relevance in agriculture, horticulture that provides knowledge about agronomical practices, soil classification and testing, pest identification and control measures, methods of cultivation and grain storage practices (Ghosh et al., 1999).

6.8 Guidelines

Recently, FAO in collaboration with World Health Organization (WHO) has published some guidelines (FAO Document NO. CACGL 32-1999) to provide agreed approach to the requirements which underpin production of, and the labeling and claims of, organically produce foods.

These guidelines set out the principles of organic production at farm, preparation, storage, transport, labeling and marketing stages, and provide an indication of accepted permitted inputs for soil fertilizers and conditioning, plant pest and disease control, food additives and processing aids. For labeling purpose, use of terms inferring that organic production methods have been used, are restricted to products derived from operators under the supervision of a certified body or authority.

These guidelines do not prejudice the implementation of more restrictive arrangements by member countries in order to maintain consumer credibility and prevent, fraudulent practices, and to apply such rules to organic products from other countries on the basis of equivalency to such more restrictive provisions.

6.9 GEO Farm Approach

The word Geo means Earth. It also stands for Global Ecological Organization. GEO Farm Approach addresses 12 Integrated Appropriate Technologies which are ecologically sound, simple, low cost, culturally appropriate, sustainable, adaptable and integratable. The goal of GEO farm was propounded by Guevara (1999) with an aim to live in harmony with the Nature. The objectives and methods of GEO Farm Approach are:

Table 6.1: Objectives and methods of GEO Farm Approach

Sr.No.	Objectives	Methods
1.	Natural food production	Integrated farming
2.	Alternate/renewal energy	Biogas, solar, windmill
3.	No waste	Total waste management
4.	No pollution	Biological water treatment
5.	Promote health	Herbal medicine
6.	Combat malnutrition	Algalculture-Spirulina
7.	Technology transfer	Education and training

6.10 International Competence Centre for Organic Agriculture (ICCOA)

Organic farming is an innovative of contemporary agriculture in India. ICCOA is one of the primary organizations in the field of contemporary agriculture supporting the cause of organic farming and developing newer concept in creating a sound market base for organic products in India. ICCOA is, infact, an interface organization working as knowledge centre. It caters to the needs of all organic stakeholders by providing information service, advice and marketing linkages. The mission of this organization, which was created in 2003, is to help in building the competence of organic stakeholders so as to build ecologically, economically and socially sustainable agricultural systems (Sarkar, 2007).

ICCOA also assists in arranging Indian Organic Trade Fair. Arranging of this fair is a land marker event for the organic farming sector. From the previous two such fairs held in 2005 and 2006, the statistical highlights are: More than 45,000 visitors, over 100 overseas delegates, 330 exhibitors representing the entire spectrum of the organic business. Six exhibitors from overseas participate in 102 meetings of trades with organic farming producer groups.

6.11 Options in Organic Farming

Following three kinds of options in organic farming are available (Palaniappam and Annadurai, 1999):

6.11.1 Pure Organic Farming

This excludes the use of in-organics both chemical fertilizers and pesticides, but advocates the use of organics i.e. organic manures and bio-pesticides including biological control techniques/ methods.

6.11.2 Integrated Green Revolution Farming

In this type of option, the basic trends of the Green Revolution such as intensive use of external inputs, increased irrigation, development of high yielding crop varieties and hybrids are retained. But much greater efficiency on the use of these inputs is obtained to limit the damage to the environment and human health. For this purpose, some organic techniques have been developed which are combined with high input technology. These are known as "Integrated Nutrient Management (INM)" and "Integrated Pest Management (IPM)".

6.11.3 Integrated Farming System

This is a low input organic farming in which farmers depend on local resources and ecological processes, recycling agricultural wastes and crop residues. To ensure sustainability in small holding, age-old mixed farming systems are prudently integrated with cropping system.

6.12 Identification of Crops Used for Organic Farming

As catalogued by Mahajan and Gupta (2009), the following crops are being currently cultivated under organic farming in India:

Table 6.2: Crops used for organic farming (Adapted from Rajendran [2005]; Mahajan and Gupta [2009])

S.No.	Type of products	Products
1.	Cereals	Rice, wheat, maize, bajra, sorghum, pearl millet
2.	Pulses	Pigeon pea, chick pea, green gram, black gram, Bengal gram, red gram, white gram
3.	Oil seeds	Groundnut, castor, mustard, sesame, sunflower
4.	Spices	Ginger, turmeric, chilies, cumin, cardamom, black pepper, vanilla, cinnamon, mace, nutmeg, tamarind, cloves
5.	Plantation crops	Tea, coffee, cardamom
6.	Fruits	Banana, sapota, custard apple, pineapple, papaya, mango, passion fruit, orange, walnut, cashnewnut
7.	Vegetables	Tomato, brinjal, okra, onion, tomato, potato, cucurbits, cole crops, leafy vegetables
8.	Commodities	Sugarcane, cotton
9.	Others	Herbal extracts

6.12.1 Organic Vegetable Production

One of the biggest advantages of a kitchen garden is that one can grow vegetables and fruit trees organically i.e. without using inorganic or chemical fertilizers and pesticides. The vegetables and fruits so obtained are called greens (Gupta, 2011a). These greens are freshening, tastier and contain more nutrients, minerals and vitamins.

If there is any choice the kitchen garden must be close to the house so that odd hours could be utilized conveniently. The selected side should be quite sunny but could be located in northern side. Vegetables and fruit trees should be planted in way that they do not obstruct sun rays and ensure that kitchen garden gets plenty of light as otherwise in the shade vegetables will grow pale and silky (Gupta, 2011a).

Apart from growing of food grains production, vegetables and fruit trees cultivation organically medicinal plants can also be grown using organics. Such medicinal plants mainly consist of Aonla, Harbal and Bahera which are mostly grown in Siwaliks of Jammu, Punjab, Haryana and Himachal Pradesh (Gupta, 2017).

6.13 Constraints in Organic Farming

The organic farming is expensive and labour intensive. It is restrained by a number of constraints-availability of organic manures, transport from source to site and lack of price premium. Such organic matter (organic manures) is bulky and huge quantities, are required to be generated. In situations where on farm generation of organic manure is not possible commercially the cost of, transporting the same from its source to the users site is prohibitive.

Biomass generation as a source of organic input is feasible in high rainfall coastal and hill ecosystem. However, there is serious limitation in areas, where dry land agriculture is in vogue. Although propagation of suitable quick growing tree species in common property lands, and farm borders can augment organic source yet collection, processing and utilization of biomass is cumbersome and requires coordinated efforts.

No doubt, total organic farming is the most desirable proposition but it is not possible in modern commercial agriculture which can not be sustained at high levels of production. It is because (as stated above) we do not and can not have enough organic materials for all our arable lands. Total organic farming would be possible only under subsistence farming situations as had been practised in India a few decades a back. It is also possible under special conditions like horticultural/olericultural/floricultural farming where sufficient organic material would be available for recycling.

Although in India, total availability of plant nutrients from different organic sources has been projected about 7.75 million tonnes by 2025 (Tandon, 1997) yet it would not fulfill their total uptake which has been found to the order of 35.5 million tonnes, requiring 301 million tonnes food grains production for increasing population of 1504 million (Sekhon, 1997). It is obvious from the above said data that if all the organic sources are harnessed (Table 6.3), even there would be its deficiency of nearly 30 million tonnes. Hence, the alternative is to supplement the organic nutrients with those of inorganic or chemical fertilizers. This is known as *Integrated Nutrient Management.* However, in our country the contribution of organic sources which was 80-100 per cent in 1949-50, reduced to about 32 per cent in 1993-94 and 20 to 25 per cent in 2004-

05 (Gupta and Sharma, 2006). So, need of the hour is to enhance the contribution of organic sources in total nutrient consumption.

Table 6.3: Projections on the availability of organics for Indian agriculture during 2000-2025 Adapted from Tandon (1997)

Resources	2000	2010	2025
I. Generators			
Human population (millions)	1000	1120	1300
Livestock population (millions)	498	537	596
Food grain production (million tonnes)	230	264	315
II. Nutrients (theoretical potential in million tonnes of N+P2O5+K2O)[1]			
Human excreta	2.00	2.24	2.60
Livestock dung	6.64	7.00	7.54
Crop residues	6.21	7.10	22.27
Total	14.85	16.34	32.41
III. Nutrients (considerable tapable potential in million tonnes of N+P2O5+K2O)[2]			
Human excreta	1.60	1.80	2.10
Livestock dung	2.00	2.10	2.26
Crop residues	2.05	2.34	3.39
Total	5.05	6.24	7.75

[1]All data pertaining to nutrients in dung and residues are counted twice the extent these are fed to the animals. [2]Tapable = 30% of dung, 80% of excreta, 30% of crop residues.

Food security of our ever increasing human population has become possible by growing high yielding varieties of various crops responding to higher doses of fertilizers and pesticides. So at this juncture, it would be unwise and un-remunerative proposition to grow crops especially food crops without fertilizers. Hence a conjunctive use of both manures including bio-fertilizer and chemical fertilizers as an integrated approach, and not exclusive use of either alone, has been suggested by some of the scientists (Kumaraswamy, 2002; Dutt, 2002; Reen, 2003). However, the experiences of Danish and other European countries farmers as well as those of the U.S.A confirm that it is possible to raise crops, vegetables and fruits without the application of chemical fertilizers.

The challenge, before us, therefore, is to develop suitable technology packages following all ways and means such as cultural methods, application of manures including green manuring and bio-fertilizers, use of small quantity of fertilizers and bio-pesticides as is being done by the south Indian farmers. They will facilitate restoration, improvement and maintenance of soil fertility and not harm the environment either. At the same time, they will guarantee production at high levels with high quality produce as well. Agriculture at high levels of productivity could be sustainable only through such an integrated approach. Top rich forest soil known to Europe for centuries, is an important resource of inputs. "Let us learn from Kashmiri farmer who applies this input literally with high profits".

6.14 Methods of Organic Farming

Since the main aim of organic farming is to develop viable and sustainable agriculture so number of methods are available from this angle. They mostly consist of cultural methods like crop rotation, mixed cropping or inter cropping, soil tillage, use of organic manures etc. which stand discussed hereunder:

6.14.1 Crop Diversification Methods

Crop diversification method mostly consists of crop rotation, mixed cropping and multiple cropping (Kang, 2013). These methods not only help in supplementing the income of the farmers but also assist in checking soil erosion, improving soil fertility and, thereby, sustaining the crop productivity. Use of trap crops and tillage operations are the other cultural or diversified methods which have proved beneficial in checking crop insects and pests. Cultivation of marrigold along with other crops has been found beneficial for the farmers (Gupta, 2019a and 2019b).

6.14.1.1 Crop Rotations

The value of cultivating crops in a judicious crop rotation and inclusion of legumes as means of improving soil fertility appears to have been known in Vedic age (Raychaudhuri, 1982). In India, the first trial was probably conducted by J.W. Leather during 1903-06 at Kanpur (Uttar Pradesh) and Pusa (Bihar) who observed the highest loss of drainage water and nitrate from fallow land. This loss could be minimized to about one half by cropping once in a year, two thirds by a two cropping system and still further by including legumes and grasses (Sarkar et al., 1998). Agroforestry is another effective system (Gupta et al., 2012).

Both Haryana and Punjab have significantly contributed to food grain procurement, especially of wheat and rice with 53.6 and 84.4 per cent, respectively of the total rice and wheat procured during 1994-95 (ICAR, 1998). Thereafter, in areas where rice and wheat had been growing continuously, their yield either became stable or began declining. It has been ascribed to narrowing down of crop diversity in both the states. Therefore, Haryana Agriculture Department prepared an Action Plan in 2003 to divert areas from rice and wheat to encourage diversification of crops like Kharif pulses (black gram, green gram, pigeon pea), sugarcane, oil seed crops (groundnut, mustard/rape seed) and cotton. Cultivation of fruit and vegetable would be given thrust in the entire state (Kang, 2013).

Like Punjab and Haryana, the yield of rice and wheat has also sufficiently reduced in areas of Jammu region (Jammu and Kashmir) where rice-wheat crop rotation had been following since late 1960. Similar was the situation at

Krishi Vigyan Kendra farm of Jammu district at Ranbir Singh Pura so "On Farm Testings" were laid out with regard to change of cropping pattern from rice-wheat to rice-berseem during 1997-98. It was found that rice-berseem cropping pattern not only brought a revenue of Rs. 1,60,000 ha^{-1} during 1998-99 through sale of berseem as fodder and increased the yield both of rice and wheat but berseem also fixed atmospheric nitrogen in the range of 110-135 Kg ha^{-1} $year^{-1}$ (Sharma and Gupta, 2003 and Gupta and Sharma, 2002). Apart from fixation of atmospheric nitrogen, cultivation of berseem also increased the exchangeable status of Ca, Mg and K by 13, 11 and 19 per cent, respectively as compared to control. It was attributed to enhance amount of organic carbon in soil due to berseem cultivation which in turn increased the exchangeable amount of said cations. Further, rice-berseem crop rotation helped in controlling obnoxious Phalaris minor weed. In another change of crop rotation from rice-wheat to green manure-rice-wheat, reduction of nitrogen content by one-third of the recommended dose in rice and wheat was observed besides enhancing soil fertility and yield of rice and wheat (Gupta and Sharma, 2002).

Indian farmers have learnt them selves with their long experience to reduce the pest problems by carefully rotating the crop belonging to one family with one of a different family. This practice is still in vogue. Rotating groundnut with maize or smaller millet reduces an attack of white grubs. Similarly rotating of pigeon pea or chickpea with non- leguminous crops is one of the practice for combating fusarium wilt and nematode problems.

6.14.1.2 Mixed or Intercropping

Multiple cropping is the intensification of cropping in time and space dimension. It refers to growing of more than two crops (routine practice) in well planned sequence (rotation) in a fixed period.

A way to multiply the yield further is "Mixed" or intercropping i.e. growing more than single pure crops together in the same piece of land as maize + black gram or wheat + gram and wheat + mustard. This covers the risk of crop failures, reduces the cost of cultivation per unit of yield, renders economy in inputs and meets various needs of farm families and animals. Mixed cropping is the common practice of the farmers of dry land agriculture, particularly of Kandi belt of Punjab, Haryana and Jammu, Jammu and Kashmir, and Changer area of Himachal Pradesh. They either grow maize and black gram in the same field during Kharif or black gram + sesamum. Sometimes black gram is grown on the bunds of maize fields. Whichever practice is followed its main aim is to enhance the soil fertility through atmospheric nitrogen fixation by black gram.

The yield potential and economies of multiple cropping sequences worked out under irrigated conditions of Jammu indicated that on the whole the rice-wheat fallow is the best followed by rice-wheat-green gram in terms of total food production, food quality and energy output value as well as net profit (Gupta et al., 1992). However, in terms of economic return, rice-wheat-sorghum-cowpea fodder is the best. The study conducted under KVK for 2 years in 8 different locations (4 each year) showed more yield in the black gram and sesamum intercrops (4:2 rows) than mixed cropping (Gupta and Sharma, 2002).

Intercropping is also being practised in irrigated areas of Bhalwal and Marh blocks of Jammu district. Sugarcane + bitter gourd, sugarcane + tomato, cauliflower + tomato, mustard + radish and mustard + peas, are the main intercrops being followed by high tech farmers. The details of income generated by one high tech farmer, Mr. Gandharv Singh of village Batera (Bhalwal block), is presented in Table 6.4. The data presented in Table 5.4 indicate that he usually earns an amount of Rs. 3,000 and 18,000 from bitter gourd and sugarcane intercrops, respectively, in an area of 0.1 ha (Sharma and Gupta, 2003; Anonymous, 2013a).

Table 6.4: Details of income generated by high-tech farmer from Batera, Bhalwal block of Jammu district (Adapted from Sharma and Gupta [2003])

Sr.No.	Crop	Income qenerated (Rs.)	Cost: benefit ratio
1.	Sugarcane (0.1 ha)	18,000	1:4
2.	Bitter gourd as intercrop in sugarcane	3,000	1:6
3.	Okra (0.1 ha)	6,000	1:4
4.	Cauliflower (2 crops) 0.05 ha	12,000	1:3
5.	Potato (0.1 ha)	7,500	1:3
6.	Wheat (0.7 ha)	1,800	1:3

The high tech farmer also grows arum and colocacia using organic manures only which are more tasty and sweet. He further adds that organic farming provides eco-technological stability, sustainability in agriculture, pest management and an alternative to inorganic fertilizers. Probably, he was the first farmer in the area, who started using biogas manures and found that these manures could substitute the entire NPK requirement of tomato, chilies, arum and colocacia.

6.14.1.3 Trap Crops

In certain cases, use of trap crops has been found suitable method for diverting the attention of the pest from the main crop. Mustard which is preferred by aphids, diamond back moth and cabbage leaf Weber is successfully used as a trap crop in checking pests of cabbage (Saxena, 1996). These pests are allured by mustard crop and cabbage grows almost free from the attack of any pests.

Similarly castor is used as trap crop for attracting tobacco caterpillar and white flies in tobacco fields.

6.14.2 Tillage in Relation to Sustainable Crop Production

In relation to sustainable crop production, there are two schools of thoughts on the appropriateness of tilling soils in tropics (Laryea et al., 1994). Some scientists are of the firm view that tillage has beneficial effects on soil and sustainable crop production. It is because tilling the soils is essential for controlling weeds, for loosening compacted and crusted soils, and for increasing the rooting depth of shallow soils. Others believe in that by cutting, pulverizing and inverting, tillage operations in the long run do more harm than good to soils and, therefore, should be discontinued. A closer glance of the research experimental data (Nicou and Charreau, 1985) indicate that tillage is not very essential for the soils having clay content more than 20 per cent with dominance of smectite group of minerals. It is because in these soils, phenomena of swelling, shrinking and their resultant inversion of soil as it wets and dries, regenerate a good desirable structure and tillage may not be required. However for soils showing less than 20 per-cent clay with non shrinking minerals (Kaolinite group) require some form of tillage to loosen the hard and compact soils. This in turn, will provide adequate and desirable soil-water-ratio for growth and easy movement of both of the crop roots, soil macro and micro-organisms. On crusting and hard setting Alfisol in India, however, Laryea et al. (1994) reported that biological practices such as application of straw, farm yard manure and pasture rotations could ameliorate the structurally unstable soils rather than following the tillage operations. Absence of tillage leaves the soil surface covered with previous crops residues if not removed grazed by animals or burnt. This promotes the environmental conditions conducive to the prolific growth of fauna and flora in the surface soil underneath the litter (Laryea et al., 1994).

The types of implements used in tillage also play an important role. The mould board ploughs are generally found useful in killing hibernating insects, exposing eggs and larvae to the surface and destroying weeds which provide food and breeding sites for many pests. Ploughing during summer kills a number of cut worms, white grubs, grass hoppers and inhabiting beetles and caterpillars.

6.14.3 Use of Organics

Use of organics involves incorporation of crop residues as a source of plant nutrients, use of organic manures like FYM, compost including vermicompost, green manures as well as use of organic pesticide (Mehta, 2014). The details of these organics are as follows:

6.14.3.1 Incorporation of Crop Residues

Residues of various crops are a good source of plant nutrients. In India, nearly 400 million tonnes of such residues are available having about 7.5 million tonnes of NPK nutrients. So their incorporation will increase total and available NPK content of the soils. The crop residues incorporation into the soils inhabits 5-10 times more aerobic bacteria, 5-9 times more actinomycetes and 1.5 to 11 times more fungi. Various enzymatic activities of surface soil like urease, phosphates, nitrogenase and dehydrogenase also get increased with crop residues incorporation. Hence, such crop residues, if managed properly, have great potential to be utilized as source of plant nutrients in achieving sustainable crop productivity and for substituting a part of fertilizer requirement of the field crops. However, results of research trials conducted at various places and reviewed by Bhardwaj et al. (1998) indicate their beneficial effects to the subsequent crops rather than standing ones. It is ascribed to higher C: N ratio of the residues ranging from 40 to 84 or even more which results in immobilization of nitrogen immediately after their incorporation. Therefore, narrowing down of C: N ratio of the organic residues, is necessary for their adequate decomposition to obtain real benefits not only in crop nutrition but also in amelioration of soil fertility and tilth. Further, decomposition of plant residues under anaerobic conditions produces phytotoxic level of organic acids, nitrite, ammonia and possibly Fe and Mn.

It has also been reported that about 30 days could be an appropriate pre-cropping period for the decomposition of crop residues from crop production point of view. Earlier Indian scientists i.e. before 1980, used to incorporate organic materials in the soil about 6 to 8 weeks in advance of crop planting with application of 50 to 60 Kg P_20_5 ha^{-1} as rock phosphate or basic slag. The position in respect of the management of the legumes residues however, seems to be some what comfortable. These materials have generally been found to have beneficial effects on crop yields even when incorporated in the soil at the time of sowing the crops (Gupta and Sharma, 2006).

A field experiment was conducted on Inceptisol under subtropical irrigated conditions of Jammu (J&K) to see the effect of rice residue management in wheat crop and soil properties (Sharma and Bali, 1998a). The results showed significant effect of urea application with rice stubbles at the time of field preparation on growth and yield attributes, probably due to proper decomposition of rice stubbles. Grain and straw yields were significantly influenced by each other treatment (Table 6.5). The treatment (T_4) produced significantly higher grain yield which was statistically at par with T_3 treatment. The values of pH and EC were slightly increased with incorporation of rice stubbles in combination with urea, while organic carbon per cent and available NPK contents increased.

In another study (Sharma and Gupta, 1998b) carried out under intermediate or sub temperate soil conditions of Rajouri district (Jammu and Kashmir State), it was found that application of leaves of Leucaena leucocephala and farm yard manure had residual effect on succeeding wheat crop by producing equivalent yield as 100 per cent NPK. Among different organic sources, farm yard manure proved superior to black gram straw followed by leucaena leaves, while wheat straw proved inferior source of N substitution.

Table 6.5: Yield of wheat as influenced by rice residue management (pooled data of two years) (Adapted from Sharma and Bali (1998))

T reatment*	Grain yield (t ha^{-1})	Straw yield (t ha^{-1})
T_1	2.80	3.91
T_2	2.50	3.60
T_3	3.12	4.21
T_4	3.17	4.50
LSD (0.05)	0.25	0.28

*T,. removal of rice stubbles + 120 Kg N ha t T_2, no urea application to decompose rice stubbles +120 N Kg ha^{-1}, T_3, urea @ 30 Kg N ha^{-1} to decompose rice stubbles + 90 Kg N ha^{-1} and T_4 urea @ 30 Kg N ha^{-1} to decompose rice residues + 120 Kg N ha^{-1}.

Effect of compost prepared from paddy straw on the yield of wheat crop was studied under KVK farm Jammu district, during 1999-2000 and 2000-01. The data (Table 6.6) indicate that use of compost prepared from paddy straw and treated with DAP at the time of composting, along with 1/3 recommended dose of NPK has given the highest increase in wheat yield as compared to control followed by the compost which was treated with urea. The yield of both of these treatments was found to be superior to the yield obtained with recommended NPK dose.

Table 6.6: Effect of paddy straw compost on the yield of wheat (Adapted from Gupta and Sharma [2002])

Treatment	Yield (q ha^{-1})		Average	% increase over control
	1999-2000	2000-2001		
Ti=1/3 NPK + Urea treated compost	26.8	25.9	26.4	98.5
T_2=1/3 NPK + DAP treated compost	27.2	26.5	26.9	99.2
T_3= Recommended NPK	26.0	25.4	25.7	95.4
T_4= Control	14.3	12.3	13.3	-

Data generated under KVK Farm.

6.14.3.2 Use of Organic Manures

Organic manures are valuable by products for sustainable farming. These are generally consid l ered inferior in performance to chemical fertilizers. This may be true in shorter term objectives but in longer term and total usefulness gained, organic manures far exceed the short term gains of chemical fertilizers. Regular addition of organic manures gives consistent output unlike application of NPK fertilizers, where the results are spectacular in the first two decades (Balasubramanian, 2000; Mehta, 2014) and decline later.

The results of several experiments compiled by Gaur (1982) regarding response of added organic manures indicated:- (i) rice and wheat responded to application of 12.6 tonnes of FYM or compost ha^{-1}, in the range of 1.0 to 2.2 and 2.1 to 2.5 q ha^{-1}, (ii) application of rape cake and FYM were effective in maize crop (iii) addition of 25 tonnes of FYM or compost gave an average response to the order of 8 tonnes of sugarcane, (iv) millets showed significant response to FYM which was comparable to NPK application, (v) additional yield of potato to the range of 7.7 to 19.1 t ha^{-1} was obtained with application of 50 to 100 tonnes of FYM.

While reviewing the interaction of organic matter with nutrients, Sarkar et al. (1998) reported that;

- From a long term field study in a sub-mountainous soil (Himachal Pradesh), application of FYM had a better cumulative effect on potato tuber yield than the combined application of P and K as in organic fertilizers at the same level of N application in potato-wheat-peas rotation. Continuous application of FYM also gave higher yields of maize and other crops.
- With the residual effect of FYM in Kharif economy of NPK application in growing of Rabi crops could be achieved,
- Application of FYM and NPK over a period of 20 years has helped in maintaining and improving physical properties, organic carbon content and enzymatic activities of soils whereas application of nitrogenous fertilizer alone deteriorated soil physical properties. It also improved P use efficiency.
- Addition of FYM along with pyrite increased the chlorophyll content in leaf, S concentration and its uptake by lentil.

Studies carried out on farmer's fields showed that combined use of 12 tonnes of FYM and 4 tonnes of poultry manure gave rice yield equivalent to that obtained with 120 Kg N ha^{-1}. Integrated use of manures and fertilizers is responsible in increasing protein and oil contents of groundnut. A recent study was undertaken to see the influence of commercial as well as locally available organic manures like oilcake pellets, FYM, and vermicompost on the improvement of soil properties and yield of rice var. Pusa Basmati (Prakash et al., 2002). The results indicated that grain yield with FYM was significantly higher than chemical fertilizers and other different manures. In general, availability of almost all major nutrients was higher in treatment with organic nutrient source as compared with chemical fertilizers and untreated control.

A long-term field experiment conducted at Regional Research Station, Bajaura (Kullu, Himachal Pradesh) with maize-wheat rotation under rain-fed condition during 1994-98 (Parmar and Sharma, 2002), revealed that 50 per cent N applied to both the crops and 10 tonnes of FYM to only maize can sustained the yield at par with 100 per cent NPK. This saved 50 per-cent N and 100 per cent P and K fertilizers. Katyal (1992 and 1994) have already given much emphasis on the use of organic manures and bio-fertilizers to meet nutrient needs of the crops grown under rain-fed areas. Besides supplying a host of nutrients, improving soil physical and biological environment, also help in preventing hidden deficiencies and maintaining sustainability of an agricultural system. However, due to lack of organic manures and inability of bio-fertilizers to meet total nutrient requirements to sustain large scale productivity goals, their integrated use with fertilizers is advocated. Consistent with this, incorporation of FYM, green manure and Azotobacter inoculants in combination with nitrogenous fertilizers increased the grain and straw yield of rice (Singh et al., 2002). Green manure (20 q ha^{-1} fresh weight) with 30 Kg N ha^{-1} resulted in an increase of grain yield equivalent to 60 Kg N ha^{-1}. Combined use of N fertilizer, green manure, Azotobacter and FYM gave an additional increase in rice grain and straw yields of 11.2 and 19.9 q ha^{-1}, respectively over control. The supply of nutrients through incorporation of green manure, FYM and Azotobacter created conducive physical environment leading to better aeration, rooting activity and nutrient absorption and the consequent complementary effect in economic grain and straw yields of rice. Low cost production technologies i.e., green manuring and efficient use of N-fertilizer in rice-wheat cropping system has given good results (Bansal et al., 2007)

Gupta et al. (2001) found that application of FYM @ 10 t ha^{-1} alone gave lower yield but when it was supplemented with 30 or 40 Kg N ha^{-1} the yield of maize and wheat increased over control vis-a-vis farmer's practice under dry land conditions. The results of "On Farm Testings" in Kandi belt of Jammu on maize showed that when FYM was applied with N, P_2O_5 and K_2O fertilizers there was improvement in the yield of maize. To maintain sustainability of maize production, half of the recommend N, P_2O_5 and K_2O through fertilizers and half through FYM, is desirable. In irrigated belt of Jammu, 50 to 75 per cents of the recommended doses of N, P_2O_5 and K_2O can be saved through FYM and green manure in case of rice-wheat cropping system.

Farmers of Himachal Pradesh invariably apply 10-15 tonnes of FYM ha^{-1} to maize crop in maize-wheat cropping sequence and to wheat crop in rice-wheat cropping sequence (Acharya et al., 2001). In ginger cultivation, about 50-60 tonnes of FYM are applied with good amount of mulch materials. Incorporation of Eupatorium @ 20 t ha^{-1} on fresh weight basis was as effective as FYM; however, their combined use was better. Spreading of Eupatorium twigs on the soil surface favoured conditions for the growth of earthworms just before sowing of wheat. Acharya and Kapaur (2001) suggested that due to paucity of FYM for potato cultivation for shallow depth soils of hilly areas can be mitigated through addition of compost prepared from abundantly growing wastes like Eupatorium spp. A number of biological materials like kudzu, wild sedge and Eupatorium can be used to prepare compost in the Kumaon and Garhwal divisions of Central Himalayan Region of Uttar Pradesh, now Uttarakhand (Singh et al., 2001). In a study after 3 years of experimentation to assess the performance of 2 organic and 1 inorganic sources towards sustainability of upland agriculture in hills of Uttarakhand, it was found that all sources were equally effective in affecting yield of wheat (Singh and Chauhan, 2002). The residual effect of organics i.e., FYM and kudzu compost was, however, prominent in affecting the ragi yield. The soil properties also improved under organic nutrition.

The research work on organic farming in the Department of Agronomy and Agro-meteorology, Punjab Agricultural University, Ludhiana (Punjab) was started in 1995. The results showed that during initial three years, yields of organically grown crops were generally low as compared to inorganically grown crops (Kler and Walia, 2004) as has been mentioned below. Later on, the organic farming showed slight edge than chemicalised farming alone.

Table 6.7: Effect of organic and chemical farming on grain yield of maize-wheat systems (q ha^{-1}) (Adapted from Kler and Walia [2004])

Crop	1st to 3rd year (Mean)		4th and 5th year (Mean)	
	Organic	Chemical	Organic	Chemical
Maize	41.3	43.3	51.3	48.5
Wheat	40.8	49.0	52.8	52.3

Apply 20 tonnes, well rotten FYM on dry weight basis to each crop during the first 5 years at the start of organic farming and later on, reduce by 50 per cent of the dose in maize and 25 per cent in wheat. Incorporate residues of maize and wheat in the field. The data (average of 6th and 7th year) presented below indicated increasing trend in yields of organic maize and wheat over chemical farming system:

Table 6.8: Effect of organic and chemical farming on the grain yield of maize-wheat system (6^{111} and *V* year) (Adapted from Kler and Walia [2004])

	FYM (t ha^{-1})		Grain yield (Mean q ha^{-1})	
	Maize	Wheat	Maize	Wheat
20	20	Incorporation	52.0	58.0
10	15	Incorporation	50.0	55.8

6.14.3.3 Use of Green Manures

Among the various crops used as green manures like cowpea, cluster bean, green gram, black gram, dhaincha and sunnhemp, the performance of dhaincha has proved very effective in parts of Jammu and Kashmir, Punjab, Himachal Pradesh, Uttar Pradesh, Gujarat, Orissa, Tamil Nadu and Andhra Pradesh (Acharya et al., 2001). In Gujarat alone, green manuring with dhaincha (Sesbania aculeata) in transplanted rice is practised in more than 88000 ha area in its south and middle parts. Madre tree (Glyricidia sepium), pongan (*Pongamia pinnata*), neem (*Azadirachta indica*), babul or kikar (*Acacia nilotica*) and subabul (*Leucaena leucocephala*) are the main tree species, whose loppings are also used as green manure. In Poonch, green leaves of drek (*Melia azadarch*), Cedrela serrata and khrik (*Celtis australis*) make excellent manure applied fresh in the rice fields before transplanting. Use of dhaincha in rice-wheat cropping system as green manure has shown higher yield of these crops under Uttarakhand soil conditions (Bansal et al., 2007).

In plains and sub-mountainous areas of Himachal Pradesh like Una, Amb, Gagret and Palampur, green manuring with dhaincha can compete well with optimum doses of nitrogenous fertilizers in wetland rice and meet the requirement of various rice varieties which demand up to 120 Kg N ha^{-1}. In Bharmour area of Chamba district and in some rice growing areas of Kangra, Kullu and Mandi, green manuring is done with leaves/twigs of wild bushes like basoti or branker (*Adhatoda vasica*), ban mar or Eupatorium (Eupatorium odoratum). The green material is added before transplanting. In direct seeded rice it is a practice to do wet puddling with animal drawn plough, locally called "halod" in which weeds are uprooted and buried underneath while rice seedlings are uniformly spaced. In Uttar Pradesh increase in rice, wheat and sugarcane yield has been noticed 51, 42 and 30 per cents, respectively. In Kandi belt of Jammu, green manuring sunnhemp (Crotalaria juncea) has also been proved useful. In this practice, the green manure crop is grown with 70 kg DAP. Sowing is done with the onset of monsoon and green manure is incorporated 45 days after sowing by about middle of August. It helps in increasing about 30 per cent wheat yield without any application of additional fertilizer to wheat crop. "On Farm Testings" conducted for three years under KVK Jammu, revealed that 60 per cent of N, P_2O_5 and K_2O requirements of rice IET-1410 variety could be met through green manuring done with dhaincha. Sarkar (1998) reported that green manuring with 40 Kg N ha^{-1} was comparable to application of 120 Kg N ha^{-1} alone. Also the residual fertility due to green manuring increased the yield of succeeding wheat crop significantly. In Punjab, about 60 days old green manure crop saved 75 Kg N ha^{-1} in maize and 40 Kg N ha^{-1} in succeeding wheat crop. The maximum saving was accomplished with cowpea followed by Sesbania and Cluster bean.

In Andhra Pradesh, leguminous crops like cowpea can be used as green manure very profitably. When 10 Kg N was combined with 5 tonnes cowpea green manure ha^{-1}, it was proved as good as 40 Kg N ha^{-1}. Green leaf manuring with *Sesbania rostrata* (5 t ha^{-1}) increased the yield of rice which was at par with *Sesbania aculeata*. In coastal Andhra Pradesh, the green manure crops like dhaincha, sunnhemp and cowpea, are grown after the harvest of paddy. These are partly used for green or dry fodder and partly buried into the fields at the time of puddlings. In Tamil Nadu, besides growing dhaincha, sunnhemp and cowpea, cluster bean and green gram are also grown as green manuring. They are grown in the fields and incorporated in situ or they are brought from out side and applied to soil.

Sunnhemp and leucaena, when used as green manure resulted in substantial increase in the yield of sorghum and safflower in Madhya Pradesh. Sesbania rostrata was found to produce the highest biomass (20-25 t ha^{-1}). Green gram and black gram are the main green manuring crops used in areas of Chhatisgarh. Pods of babul or kikar tree are also buried into the soil as manure. Green manuring is common practice in all the irrigated areas of Orissa during Kharif. This practice helps in better growth of the rice crop and results in higher yields. In red lateritic region, some parts of alluvial Terai region of West Bengal, farmers grow green gram during summer season as catch crop between the harvest of previous Rabi crop and planting of Kharif crop. After 3-4 hand pickings of pods, the left over crop residues are incorporated in situ with a summer ploughing. In rice-wheat system of Bihar, Sesbania or Crotolaria for green manuring during summer season has shown poor results. A short duration crop like summer green gram after picking of pods and turning the biomass into soil when green, gives dual profit by providing grain for consumption and straw for green manuring. This practice is very common in the hills of Chotanagpur plateau.

As already stated, lopping of various tree species can also be used as green manure. Leucaena loppings (75 Kg N equivalent) treatment was supplemented with 25 Kg N ha^{-1} (Katyal, 1994). Increase in organic matter content and simultaneously rise in water retention capacity of soil profile was the additional benefits arising from Leucaena treatment. In Goa, the farmers use the lopping of *Glyricidia sepium*, *Cassia tora* and *Cassia auriculata*. *Sesbania aculeata*", *Calotropis gigantea*, *Leucaena leucocephala* and *Glyricidia sepium*, are the common green leaf manuring species, whose loppings are used by the farmers in Andaman and Nicobar Islands.

6.14.3.3.1 Shifting Cultivation and Green Manuring

North-eastern hilly region of the country consisting of Assam, Arunachal Pradesh, Manipur, Mizoram, Tripura, Nagaland and Meghalaya, is still lagging behind, especially in agriculture production despite being endowed with rich natural resources. Low productivity of various crops is attributed to a number of factors viz. prevalence of shifting cultivation, hilly terrain, un-predictable climate changes and low levels of inputs.

Shifting cultivation locally known as Jhum is the main form of agriculture in north eastern hilly regions. Jhum cultivation is practicing in these areas since long with traditional knowledge. Shifting cultivation drastically declines their soil fertility and became unsustainable primarily due to increase in population which led to increase in food demands. Practising green manuring has proved restoration factory to maintain soil fertility for sustainable agriculture (Kumar and Kumar, 2014).

6.14.3.4 Use of Bio-Fertilizers

This topic has been described in a separate chapter.

6.14.3.5 Use of Bio-Pesticides

Topics relating to use of bio-pesticides have been dealt with separately.

6.14.3.6 Use of Panchgavya

It is a natural indigenous organic pesticide which can be used by many organic farmers in various parts of the country. The 'Panchgavya" is normally prepared by using ghee, milk, curd, cow dung and cow urine. In modified formulations, besides five basic ingredients, sugarcane juice, tender coconut water and ripe banana are also added. The entire mixture is, then fermented for 22 days. The fermented solution after filtration was found to be very effective in control of various pests of sunflower and green gram.

6.14.3.7 Use of Jaivic and Vedic Inputs for Crop Production

At Indian Grassland and Fodder Research Institute, Jhansi, use of various Jaivic and Vedic Krishi inputs such as Angara, Amritpani, Panchgavya and -Gomutra indicated improvement in crop productivity, soil microbial population and soil biological activity. Foliar application of "Gomutra" and "Panchgavya" increased the growth of mustard. Similarly, foliar spray of Angara and Amritpani increased sorghum green and dry fodder production.

6.15 Conclusion

It is concluded from the foregoing literature that organic farming system which means farming in the spirit of organic relationship usually shuns the use of chemical fertilizers and pesticides. However, prior to using various organics, these must be well decomposed to prevent immobilization of nitrogen. This can be judged by determining the C: N ratio of the used organics. Use of organics and adoption of other natural methods of farming will not only help in checking toxicity of the chemical fertilizers/pesticides to the human, animal and soil's health but also maintain the sustainability of the crops. Sustainability in agriculture here refers to development in terms of output as well as socio-economic and ecological parameters. According to UNDP, safeguarding National food security means improving the quality of life. The farmers should, therefore, follow this farming system to increase organic matter and microbial biomass in their fields to support agroecosystem.

6.16 Future Research Strategies/Priorities

- Preservation of indigenous techniques and their further improvement in the context of quality and efficiency of output is needed. Infact, traditional technologies must be upgraded for their market competitiveness without losing social, ecological and economic advantage by integrating them with modern or frontier technologies.
- Creative blending of new and traditional techniques can assist in achieving required improvement in agricultural production and economic growth. There is generation of employment avenues also.
- To bring out main development in the major areas of agriculture, it is argued that carve out the knowledge of traditional agriculture that is still practised by the Indian farmers in some parts of the country.
- In many states of India, majority of the farms are small and marginal and as such organic farming can not be taken up to individual farmers. There has to be a cluster farm approach to maintain sustainability.
- The farmers need to be made aware of about the use of bio-pesticide and bio-fertilizers for organic food production.
- India's strength in organic food production could lead the country to becoming a major player internationally besides making this segment of exports a main foreign exchange earner, provided the government and

non-government organizations give an adequate and full support to the farmers, fruit and vegetable growers. Lack of an attractive market premium for organically grown products, an important factor that limits wider adoption of organic farming by farmers, must be strengthened.

- The process of preparing compost, FYM and vermicompost at the farm level needs to lie standardized.
- Improved methods including vermicomposting and Japanese techniques should be tested and adopted to up grade the quality of compost.
- Many non-chemical organic sources are being made available as organic pesticides for controlling plant diseases, but awareness about their use is limited. Hence awareness training programmes are required to be organized. Neem based derivatives are such emerging examples.
- Usefulness of allelo-chemical requires to be studied in detail.
- Indigenous technical knowledge (ITK) is also a farce to be used. Although many traditional methods of cropping, especially in containing damage due to pests and diseases are used to some extent but few attempts have been made to prove them scientifically and exploit their eco-friendly commercial potential. Therefore, this untapped area calls for attention.
- The techniques of organic farming should be incorporated in the package of practices published jointly by State Agricultural Universities and State Department of Agriculture.
- Niche markets for organically produced food should be identified.
- To meet the fast paced demand of developing countries exporters need to adopt this kind of farming, processing and marketing techniques and build good connections with importers and distributors in Europe, the U.S.A and Japan.
- In the Indian context, compulsions of food and nutrition security, and the need to restore soil health and quality would require a strategy that integrates many useful components of organic farming into Green Revolution technology. For example, continuation of Green Revolution is inevitable in resource endowed regions. Negative effects of Green Revolution are required to be rectified.
- The challenge before us is to develop suitable technology packages to make full use of organic residues of the farms, to improve food production without impairing soil health which would ensure sustainability in agriculture production.

- The estimated nutrient potential from organic sources is approximately 20 million tonnes year"[1] which is sufficient for the present output target, but these organic sources are not fully utilized for agriculture/horticulture/olericulture production. Hence awareness must be created among the farming community to exploit these sources, after conducting appropriate/suitable research. It needs a strong political will
- All said and done, rely on agroforestry practice i.e., Agriculture and Silviculture should be done jointly on farms. Remember that Agriculture is meant for the plains while Silviculture is for mountains.
- Large scale plantation of quick growing leguminous trees on roadsides, wastelands and field bunds should be encouraged.
- There should be a ban to use cow dung as fuel as has been done in China.
- Bio-fertilizer technology needs to be refined for easy adaptability by the farmers.
- Developing simple techniques to handle bulky organic manures so that their processing, use and application become as simple as chemical fertilizers.
- A belief that organic food is tastier. Research in this line should be carried out in details as has been done by the Soil Scientists of the UK Soil Association.

Study Questions

1. Define the following:
   i. Organic
   ii. Natural farming
   iii. Biodynamic farming
   iv. Sustainable agriculture
2. Differentiate between mixed and intercropping?
3. Write in details about the concept of organic farming.
4. What are the various advantages or benefits of organic farming?
5. Write briefly the various diversification methods which are used in organic farming.
6. Name the crops of India which are currently cultivated following organic farming.

7. Describe the traditional views of organic farming in India.
8. Write brief notes on the following:
    i. Pure organic farming
    ii. Integrated farming system
    iii. Integrated nutrient management

References

Acharya, C.L. and Kapaur, O.C. (2001). Using organic wastes as compost and mulch for potato in low water retaining hill soils of North West India. Glimpses of the published work in Indian Journal of Agricultural Sciences (Personnel Communication).

Acharya, C.L., Sood, M.C and Sharma, R.C. (2001). Indigenous nutrient management practices in Himachal Pradesh (pp. 121-146). In Indigenous Nutrient Management - Wisdom Alive in India (Eds. C.L. Acharya, P.K. Ghosh and A. Subba Rao), Indian Institute of Soil Science, Nabibagh, Bhopal, Madhya Pradesh (India).

Angelina-Briones. (1999). Organic farming: Root of sustainability in Asia (pp. 73-76). In Food Security in Harmony with Nature. Proceedings of the 3rd IFOAM-Asia Scientific Conference and General Assembly, UAS, Bangalore.

Anonymous. (2005). Indian organic products certifying bodies to get EC equivalence status. Agrinews - Farmers' Forum 5 (11): 28.

Anonymous. (2009a). Raising a toast to the organic sector. Agriculture Today 12 (1): 46-47.

Anonymous. (2013a). Crop diversification: An ideal way to deal with food insecurity. Agriculture Today 19 (4): 50-51.

Anonymous. (2013b). Organic farming can create more jobs in Rajasthan. Indian Farmer's Digest 16 (1): 13.

Balasubramanian, C. (2000). Friendly alternatives to chemical fertilizers and food grain production. Himsa Research Foundation Ahims Sthal Mehrauli, New Delhi (India).

Bansal, S.K., Shekhar, K.S. and Ram, S. (2007). Affordable Technologies for sustainable agricultural production. Indian Farmer's Digest A0 (3): 32-34.

Bhadauria, B.P. and Singh, Y. (2013). Exploration of bio-fertilizers and bio-pesticides in IPM for sustainable agriculture. Indian Farmer's Digest 46 (7): 14-17.

Bhardwaj, K.K.R, Kalyana Sundaram, N.K. and Khan, H.H. (1998). Management of organic materials for field and plantation crops. Bull. J. Indian Soc. Soil Sci. No. 19, pp. 122-134.

Burcher, Sam (2007). FAO promotes of organic agriculture. The Kashmir Times 64 (320): 10 [November 24, Saturday].

Chaudhary, R.C. and Mishra, S.B. (2012). Kalanamak the Heritage rice of eastern U.P. and its likely role in farmers' prosperity. Agriculture Today 15(12): 36-37.

Chhonkar, P.K. (2001). The state of Indian soil science and challenges to be faced during twenty first century. J. Indian Soc. Soil Sci. 49 (4): 532-536.

Chhonkar, P.K. (2003). Organic farming: Science or belief. J. Indian Soc. Soil Sci. 51 (4): 365-377.

Dahatonde, S. and Khambalkar, V.P. (2007). Use of organics in Indian agriculture. Agrobios Newsletter 5 (12): 38-39.

Dutt, S. (2002). Organic farming for sustainable development. The Daily Excelsior 38 (251):7.

Gaur, A.C. (1982). Organic manures and bio-fertilizers. Review of Soil Research in India. Part 1, 12th International Congress of Soil Science New Delhi (India), 8-16 Feb. 1982.

Geier, B. (2005). An overview on worldwide organic agriculture and markets. Farmers' Forum 5 (11): 5.

Ghosh, S.K., Dakshina Murthy, K.M., Ramesh, G. and Palaniappan, S.P. (1999). Need for organic farming. Employment News 24 (22): 32-33.

Guevara, Edgard. (1999). GEO farms: Integrated appropriate technologies (pp. 63-72). In Food Security in Harmony with Nature. Proceedings of the 3rd IFOAM-Asia Scientific Conference and General Assembly, UAS, Bangalore.

Gupta, D.K., Gupta, R.D. and Singh, H. (1992). Cropping patterns in the Kashmir Himalayas (pp. 78-91). In Conserving Indian Environment (Ed. S.K. Chadha). Pointer Publishers, Jaipur, Rajasthan, (India).

Gupta, R.D. (2011a). Grow organic vegetables in kitchen garden. The State Times 16 (161):6 [June 12, Sunday].

Gupta, R.D. (2012a). Bio-fertilizers for sustainable agriculture in north western Himalayan soils (pp. 142-154). In Model Training course on agroforestry in sustainable farming and natural resources management in hills of north western Himalayas (Eds. S. K. Gupta, M.D. Saleem, K.K. Sood, N.S. Raina, L.M. Gupta, S. Sehgal and M. Gupta). Sponsored by Department of Agriculture and Cooperation, Ministry of Agriculture, Government of India, Division of Agroforestry FOA, SKUAST-J, Chatha, Jammu, J&K (India).

Gupta, R.D. (2014). Food security through green manure. The Daily Excelsior 50 (285): 8 [October 16, Thursday].

Gupta, R.D. and' Sharma, A.K. (2002). Organic farming in Jammu and Kashmir. The Daily Excelsior 38 (330): 6 [November 27, Wednesday].

Gupta, R.D. and Sharma, S.P. (2006). Soil pollution and its solution through integrated nutrient management (pp. 32-38). In Balanced Fertilization for Sustainable crop Production (Eds. S.P. Sharma, S.K. Sharma, R.P. Sharma and T.S. Verma). Proceedings National Seminar on INM for Balanced Fertilization held on November 17-18, 2006.

Gupta, R.D. and Singh, H. (2006). Indiscriminate use of fertilizer poses health hazards. Farmers' Forum 6 (9): 20-24.

Gupta, R.D. and Singh, H. (2008). Indiscriminate use of pesticides in agriculture: Public health issues and their control. Indian Farmers' 'Digest 41 (1): 8-13.'

Gupta, R.D. and Upadhaya, V. (2006). Ladakh farmers nourish the soils with night soil compost for agriculture sustainability. Agrobios Newsletter 4(11): 32-33.

Gupta, R.D., Khajuria, M.R., Gupta, J.P. and Bali, S.V. (2001). Indigenous technology for nutrient management in Jammu and Kashmir (pp. 147-164). In Indigenous Nutrient Management practices - Wisdom Alive in India (Eds. C.L. Acharya, P.K. Ghosh and P. Subba Rao). Indian Institute'of Soil Science, Bhopal (Madhya Pradesh), India.

Gupta, R.D., Kher, D. and Jalali, V.K. (2005). Organic farming: Concept and its prospective in Jammu and Kashmir. J. Res., SKUAST-J 4 (1): 25-37.

Gupta, R.D. 2019a. Organic farming need of the day. The State Times.24(5):6 [January 5, Saturday]

Gupta, R.D. 2019b. Organic farming need of the day. The State Times. 24(6):6 [January 6 Sunday]

Gupta, R.D. 2017. Medicinal Plants for Greening the Siwaliks. Natural Resources Management for Climate Smart Sustainable Agriculture (Edited by Sanjay Arora, Sanjay Swami and Suraj Bhan), Soil Conservation Society of India NASC Complex, Pusa, New Delhi – 110012, India

Gupta, S.K., Saleem, M.D., Sood, K.K., Raina, N.S., Gupta, L.M., Sehgal, S. and Gupta, M. (2012). Model Training course on agroforestry jn sustainable farming and natural resources management in hills of north western Himalayas (October 12-19, 2012). Sponsored by Department of Agriculture and Co-operation, Ministry of Agriculture, Government of India, Division of Agroforestry FOA, SKUAST-J, Chatha, Jammu, J&K State, India.

Indian Council of Agricultural Research. (1998). Decline in crop productivity in Haryana and Punjab: Myth or Reality. New Delhi, (India).

Jillani, M.H., Singh, C.B. and Tripathi, R.K. (2005). Role of women in livestock farming in Uttaranchal. Indian Farmer's Digest 38 (7): 40-44.

Kang, M.S. (2013). Achieving crop diversification in Punjab. The Tribune 133 (62): 9 [March 4, Monday].

Katyal, J.C. (1992). In Balanced Fertilizer use for Increasing Food grain Production in Northern States (Ed. G. Dev). Proceedings of the Zonal Symposium. Potash and Phosphate Institute of Canada India Programme Gurgaon (Haryana), India.

Katyal, J.C. (1994). Nutrient management in soils of arid and semiarid regions. In Soil Management for Sustainable Agriculture in Dry land Agriculture (Eds. T.D. Biswas, G. Naryanasamy, J.S.P Yadav, G. Dev, J.C. Katyal and P.S. Sicfhu). Bull. Indian Soc. Soil Sci. No. 16, 78-91.

Kler, D.S. and Walia, S.S. (2004). Organic farming for ecologically sound maize-wheat production. Indian farming for ecologically sound maize-wheat production. Indian Farmer's Digest 37 (10): 43-44.

Kumar, R. and Kumar, S. (2014). Green manuring: Effective approach to sustain soil fertility in Jhum area of north eastern hilly state, Indian Farmer's Digest 47 (2): 42-44.

Kumaraswamy, K. (2002). Organic farming - Relevance Prospects. Indian Soc. Soil Sci., Newsletter 12 (March Issue): 1.

Kumari, M.S.S. and Kumari, K. (2002). Effect of vermicompost enriched with rock phosphate on growth and yield of cowpea. J. Indian Soc. Soil Sci. 50 (2): 223-224.

Laryea, K.B., Gogle, A.K. and Pathak, P. (1994). Tillage Problem in Rain-fed Soil: relevance to sustained productivity. In Soil management for Sustainable Agriculture in Dry Land Area (Eds. T.D. Biswas, G. Naryanasamy, J.S.P Yadav, G. Dev, J.C. Katyal and P.S. Sidhu). Bull. Indian Soc. Soil Sci. No. 16, pp. 119-133.

Magar, S.S. (2004). Organic farming: Technical feasibility, economic viability and social acceptance. J. Indian Soc. Soil Sci. 52 (4): 374-379.

Mahajan, A. and Gupta, R. D. (2009). Integrated nutrient management (INM) in a sustainable rice-wheat cropping system (pp.1-267). Springer Science + Business Media (www. Springer.com).

Mahajan, A. Gupta, R.D. and Sharma, R. (2007a). Organic farming - A need for all times. Gram Vikas Jyoti (July-September Issue): 7-8.

Mahajan, A., Sharma, R., Gupta, R.D. and Trikha, A. (2007b). Organic farming and sustainable agriculture. Agriculture Today 10 (3): 48-50.

Mahajan, A., Sharma, S.K., Gupta, R.D. and Sharma, R. (2007c). Morphological, physical and chemical properties of soils from North West Himalayas. Bulg. J. Agric. Sci. 13 (5): 607-618.

Mehta, M.H. (2014). Clean village - Eco-village. Agriculture Today 17 (1): 36-38.

Menon. T.G.K. and Karamarkar, V.B. (1994). Biodynamic agriculture (pp. 16-18). In Organic Farming (Eds. Shroff et al.). Jawahar Lai Nehru Agricultural University, Jabalpur (India).

Mohan, L. and Maheshwari, R.K. (2013). Sustainability of organic farming for future perspectives. Agrobios Newsletter 12 (6): 41.

Naidu, S. (2003). Uses of bio-pesticides can control residues levels. The Economic Times, New Delhi (11th July issue, 2003).

Nicou, R. and Charreau, C. (1985). Appropriate technologies for farmers in the semi-arid-West Africa, International Programme in Agriculture, Agricultural Administration Building, Purdu University West Lafayette, Indians, U.S.A.

Palaniappam, S.P. and Annadurai, K. (1999). Organic Farming: Theory and Practice. Scientific Publishers, Jodhpur (India), pp. 18.

Parmar, D.K. and Sharma, V. (2002). Studies on long term application of fertilizer and manure on yield of maize under rain-fed conditions in Western Himalayas. J. Indian Soc. Soil Sci. 50 (3): 311-313.

Pathak, R.K. and Ram, K.A. (2003). Biodynamic Agriculture Bull. No. 14. Central Institute for Subtropical Horticulture, Lucknow (India), pp 42.

Prakash, Y.S., Bhadoria, P.B.S. and Rakashit, A. (2002). Comparative efficiency of organic manures on the changes in soil properties and nutrient availability in an alfisol. J. Indians Soc. Soil Sci. 50 (2): 219-221.

Ramanathan, K.M. (2006). Organic farming for sustainability. J. Indian Soc. Soil Sci. 54 (4): 418-425.

Raychaudhuri, S.P. (1982). Host country, its people history, land and soils. Souvenir, 12[h] International Congress Soil Science, New Delhi (India) 8-16 February 1982.

Reen, A.S. (2003). Organic farming for sustainable agriculture. The Daily Excelsior 39 (56): 8 [February 26, Wednesday].

Rishipal, Malik, Y.P. and Prasad, C.S. (2013). Bio-pesticides: An Eco-friendly approach for insect pest management. Indian Farmer's Digest 46 (7): 14-16.

Sarkar. A. (2007). India organic 2007: Curtain raiser event suggests a whole new approach to organic farming in India. Agriculture Today 10 (11): 16.

Sarkar, M.C., Sachdev, M.S. and Datta, S.P. (1998). Interaction of soil organic matter with nutrients. In Bulletin Indian Soc. Soil Sci. No. 19, pp. 90-102.

Saxena, R.D. (1996). Report on awareness programme on pesticides and sustainable agriculture. 228, Sidharth Enclave, New Delhi (India).

Sekhon, G.S. (1997). Nutrient needs of irrigated food grain crops in India and related issues (pp. 78-90). In Plant Nutrient Needs, Supply Efficiency and Policy Issues: 2000-2025 (Eds. J.S. Kanwar and J.C. Katyal). National Academy of Agricultural Sciences, New Delhi (India).

Sharma, A.K. and Gupta, R.D. (2003). Only Effective legume association fixes atmospheric nitrogen. Co-operative Bulletin 46 (4): 12-14.

Sharma, M.P and Bali, S.V. (1998a). Effect of rice residue management in wheat yield and soil properties in rice-wheat cropping system. Indian J. Agril. Sci. 68 (10): 695-696.

Sharma, M.P and Gupta, J.P. (1998b). Effect of organic material on grain yield and soil properties in maize-wheat cropping system Indian J. Agril. Sci. 68(11): 715-717.

Singh, R.D. and Chauhan, V.S. (2002). Impact of inorganic fertilizers and organic manures on soil productivity under wheat-ragi system. J. Indian Soc. Soil Sci. 50 (1): 62-63.

Singh, R.D., Rao, K.S. and Bisht, K.K. (2001). Indigenous nutrient management technology in Central Himalaya (pp. 165-176). In Indigenous Nutrient Management - Wisdom Alive in India (Eds. C.L. Acharya, P.K. Ghosh and A. Subba Rao), Indian Institute of Soil Science, Nabibagh, Bhopal, Madhya Pradesh (India).

Singh, S., Singh, R.N., Prasad, J and Kumar, B. (2002). Effect of green manuring, FYM and bio-fertilizers N on yield and major nutrient uptake by upland rice. J. Indian Soc. Soil Sci. 50 (3): 313-314.

Singh, V. (2014). Organic farming: A Cornucopia of benefits. Indian Farmer's Digest 47 (8): 2.

Singh, Y. (2012). Biological aspects of organic farming for sustainable agriculture. Indian Farmer's Digest 45 (5): 35-39.

Tandon, H.L.S. (1997). Organic resources an assessment of potential supplies, their contribution in agricultural productivity, and policy issues for Indian agricultural from 2000-2025 (pp. 15-28). In Plant Nutrient Needs, Supply Efficiency and Policy Issues: 2000- 2025 (Eds. J.S. Kanwar and J.C. Katyal). National Academy of Agricultural Sciences, New Delhi (India).

7

Integrated Nutrient Management (INM) System and Sustainable Agriculture

Abstract

Continuous use of high levels of fertilizers over a prolonged period, ignoring totally the use of organic manures after "Green Revolution" brought about in rice-wheat cropping system under irrigated conditions affected the sustainability of these crops adversely. Evidences showed that grain yields under rice-wheat cropping system have either plateaued or are registering a declining trend (Mahajan and Gupta, 2009). The results from long-term fertilizer experiments at Pantnagar (Uttarakhand), Palampur (Himachal Pradesh), Ranbir Singh Pura (Jammu region of Jammu and Kashmir), Barrackpur (West Bengal) and Ranchi (Jharkhand) have revealed a decline in organic matter status of soils with application of fertilizers alone. This, therefore, warrants mobilizing and conserving all the locally available organics i.e., plant or crop residues, organic manures including green manuring, bio-fertilizers etc. In this chapter of the book, concept and aim of integrated Nutrient management (INM) system i.e., use of both organics and in-organics has been described. Besides this, components of various INM system viz., organic manures like bulky and concentrated manures and green manuring, and their advantages vis-a-vis biofertilizers have also been dealt with in detail.

Keywords: Green Revolution; long-term fertilizer experiments; sustainability; rice-wheat cropping system; organic manures, bulky manures; concentrated manures; fertilizers; crop residues; green manuring; bio-fertilizers; INM system

Today a number of countries have moved away from the inorganic or chemicalised farming to organic farming yet it is not possible for India to depend entirely upon this system of farming due to increase in human population. As total availability of plant nutrients from different organic sources is projected to 7.75 MT by 2025 against the total requirement of 37.50 MT for 1504 million populations, so it means if we depend on organic sources there would be deficiency of about 30 million tonnes of nutrients. Hence, it becomes an imperative to use both organic and inorganic sources of plant nutrients i.e. Integrated Nutrient Management (INM) System to enhance the food production for the increasing population (Gupta and Sharma, 2006). However, in our country the contribution of organic sources which was 80-100 per cent in 1949-50, reduced to about 32 per cent in 1993-94 and 20-25 per cent in 2004-05. So, the need of the hour is to enhance the contribution of organic sources in total nutrient consumption.

The INM system does not restrict its activities to low crop productivity. Necessary plant nutrient supply is identified according to the targeted yields and different nutrient sources are managed in an integrated manner to achieve the specified target. Long term fertilizer experiments involving intensive cereal based cropping systems reveal a declining trend in productivity even with the application of recommended levels of N, P and K fertilizers (Mahajan et al., 2008a). Thus INM system holds a promise in sustaining higher crop yields besides improving soil health which has been dealt herewith in this chapter of the book.

7.1 Definition of INM System

Integrated nutrient management (INM) or integrated nutrient supply (INS) system that aims at achieving a harmony in the judicious and efficient use of chemical fertilizers in conjunction with organic manures, use of well decomposed crop residues, green manures, inserting of legumes in cropping systems, use of biofertilizers and other locally available nutrient sources for sustaining soil health and ameliorating of environment as well as crop productivity on long term basis (Mahajan and Sharma, 2005). Long term fertilizer experiments involving intensive cereal based cropping systems reveal a declining trend in productivity even with the application of recommended levels of nitrogen, phosphatic and potassic fertilizers. The increase in crop productivity results in from the combined effect of chemical fertilizers and organic manures which helps in the improvement of physical, chemical and biological properties and consequently the soil organic matter and nutrient status and eventually the crop production.

7.2 Concept and Aim of INM System

The basic concept underlying INM system is the maintenance or adjustment of plant nutrients supply to achieve a given level of crop production by optimizing the benefits from all possible sources of plant nutrients in an integrated manner, appropriate to each cropping system and farming situation in its ecological, social and economical possibilities (Jaggi et al., 2001). The main aims of INM system are:

- To manage the soil fertility.
- To sustain the agricultural productivity and improve the farmer's profitability through the judicious and efficient use of chemical fertilizers, organic manures, crop residues and bio-fertilizers.

- To have practical and efficient blend of diverse nutrient sources which can produce desired yields and maintain soil health on long term basis.

7.3 Need of INM System in Sustainable Agriculture

INM system is an age-old concept but its importance was not realized earlier as nutrient removal by the crops was very low due to subsistence farming. At present, INM system has a great significance because of intensive farming which is now being practised on commercial basis. (Mahajan and Sharma, 2005). The need of INM system in modern agriculture has arisen due to:

- Escalating prices of chemical fertilizers.
- Imbalances in the ratio of NPK consumption.
- Imbalances between consumption and domestic production.
- Deterioration of soil health.
- Inorganic fertilizers consume non-renewable energy sources.
- Pollution hazards of chemical fertilizers.
- Loss of chemical productivity.
- Deterioration in soil physical properties.
- Deterioration in biological activity.
- Organic and mineral fertilizers show additive effects.
- Organic materials as a source of secondary and micronutrients.
- Reduction in crop productivity.

7.4 Components of INM System

Green revolution in India during 1966-68 has no doubt brought about self-sufficiency in food production. However, excessive use of inorganic fertilizers and plant protection chemicals for maximizing crop yield has resulted in deterioration of physical, chemical and biological health of cultivable soils due to imbalance use of N, P2O5 and K_2O fertilizers for which their an ideal ratio is 4: 2: 1. However, if we have a glance of N, P2O5 and K_2O fertilizers ratio it has never been found an ideal as was evident from the consumption ratios of N/ P2O5/K2O in the range of 5.9: 2.4: 1 during 1991-92 to 5.7: 2.2: 1 during 2004-05 (Mahajan and Gupta, 2009). It is therefore imperative to supply the plant nutrients in balanced ratio as well their use in an integrated manner to maintain the soil fertility for sustaining the desired level of crop production and productivity. This can be done through a judicious combination of chemical fertilizers, organic manures, crop residues, green manures, recyclable waste, compost, vermicompost, legumes and bio-fertilizers. It will also ensure increased crop productivity in an efficient and environmentally benign manner, without diminishing the capacity of the soil to produce crops for present and posterity. In other words, integrated nutrient management system is an ecologically, socially and economically viable approach, which is un-hazardous and needs to be followed. Thus the inorganic fertilizers, organic manures (green manures, compost, and vermicompost), crop residues/wastes, legumes and bio-fertilizers are the main components of INM system.

7.4.1 Chemical Fertilizers

Fertilizer contributed to about 50 per cent increase in food grain production of our country. Use of chemical fertilizers in balanced proportions and recommended amounts is the quickest way of boosting crop production (Plate 7.1). Hardly, 30-40 per cent of the applied fertilizer nutrient is utilized by the crop and the rest is lost through various pathways like leaching, surface run-off, volatilization, denitrification, soil erosion and fixation in soil. Therefore, efforts are to be made effectively to increase the fertilizer use efficiency (Raina et al., 2005). Further, chemical fertilizers alone are unable to maintain and sustain the long term soil health and crop productivity because they are unable to supply organic matter and micro-nutrients like organic manures (Mahajan et al., 2007a, b and c). Hence, the integrated use of inorganic fertilizers with organics including bio-fertilizers can greatly increase the efficiency of applied nutrients without any causing ill effect on the soil health.

UREA SSP

MOP

Plate 7.1: Chemical fertillizers supplying NPK Nutrients

7.4.2 Organic Manures

Application of manure alone @ 4, 8 and 16 t ha^{-1} year^{-1} increased the soybean yield by 42, 57 and 75 per cent, respectively and that of wheat by 67, 116 and 143 per cent over control (Reddy et al., 2000). When fertilizer P was also applied with manure, corresponding yield increases were 79, 90 and 93 per cent; and 159, 181 and 197 per cent in case of soybean and wheat, respectively. The treatment with 16 tonnes manure with fertilizer P produced the highest yield of soybean and wheat 2.18 and 4.5 t ha^{-1} (Subba Rao, 2010). Effect of N substitution through different manures on yield and available nutrients in a rice-wheat cropping system in acid soil after 17 and 18 cropping cycles was studied (Subehia and Sepehya, 2012). It was found that though the chemical source of nutrients increased crop yields over control yet these failed to improve the soil properties. However, improvement of soil fertility with respect to available macro and micronutrients was prominent with the application of 50 per cent NPK through fertilizers + 50 per cent N through FYM in Kharif followed by 100 per cent NPK in Rabi.

The organic manures which are mainly prepared from animal's dung/urine and plant residues are referred to as organic manures. Organic manures are generally of two types viz. i) bulky organic manures and ii) concentrated organic manures.

7.4.2.1 Bulky Organic Manures

Bulky manures are bulky in nature and supply plant nutrients in small quantities. They have direct effect on the plant growth, like any other commercial fertilizer. Bulky organic manure contains nutrients in small quantities; therefore large quantities of them need to apply per hectare. Besides the major nutrients, bulky organic manures also contain traces of micronutrients. Common bulky organic manures are as follows:

- Farmyard manure (FYM)
- Compost from farm and town refuse
- Compost from farm litter
- Night soil
- Sewage and sludge
- Green manure
- Poultry manure
- Vermicompost
- Biogas slurry
- Crop residue

7.4.2.2 Concentrated Organic Manures

Concentrated organic manures are those which are organic in nature and contain higher percentage of major plant nutrients compared to bulky organic manure. These concentrated organic manures are made from raw materials of animal or plant origin. The common concentrated organic manures are as follow:

- Oil cakes
- Blood- meal
- Meat meal
- Fish manure

- Horn and hoof-meal
- Chicken manure

FYM is the most important and commonly used organic manure in India. Being a general manure supplying all the major plant nutrients vis-a-vis micro-nutrients, FYM has been commonly used as a source of plant nutrition from time immemorial (Plate 7.2). It is to be kept in mind that organic manures should be well decomposed. Nutrient content of different organic manures are given in Table 7.1. By using this table we can calculate the amount of organic matter to be applied.

Table 7.1: Nutrient content of different organic manures (Adapted from Mahajan et al. 2007)

Sr.No.	Organic manure	Nutrient content (%)		
		N	P_2O_5	K_2O
A. Bulky organic manures				
1.	FYM	0.5	0.20	0.5
2.	Farm compost	0.5	0.15	0.5
3.	Town compost	1.4	1.0	1.4
4.	Night soil	5.5	4.0	2.0
5.	Sewage and sludge	3-6	2.0	1.0
6.	Poultry manure	3.03	2.63	1.4
B. Concentrated organic manures				
1.	Fish manure	4-10	3-9	0.3-1.5
2.	Neem cake	5.22	1.08	1.48
3.	Linseed cake	5.56	1.44	1.28
4.	Groundnut cake	7.29	1.53	1.33
5.	Mustard oil cake	5.21	1.84	1.19
6.	Safflower oil cake	7.88	2.20	1.92
7.	Cotton oil cake	6.50	2.89	2.17
8.	Raw bone meal	3-4	20-25	-
9.	Steamed bone meal	1-2	25-30	-
10.	Blood meal	10-12	1-2	1.0

The advantages of organic manures are as under:

- Keep chemicals off from various food stuffs.
- Protect future generation.
- Protect farm workers' health.
- Prevent soil erosion.
- To maintain the equilibrium and high productivity of the ecosystem.
- Make the plants healthier.

- Organic fertilizers provide many trace elements, major nutrients, as well as organic matter.
- Organic matter in the soil helps to hold the nutrients and water that plant needs.
- Organic matter in the soil provides food for soil organisms that recycle and hold nutrients.
- Save energy.
- Help small farmers.
- Support a true economy.
- Taste better flavour.
- Avoid harms to soil organisms both macro and micro-organisms (e.g. the salts in synthetic fertilizers)

7.4.3 Green Manures

Green manuring is the easiest and quickest way of supplying readily decomposable organic matter to soils and thereby, the release of plant nutrients (Plate 7.3). The magnitude of green manure has been known in periods as far back as 1000 BC because references to the use of stalks and stems of seasmum as manure are found in "Atharva Veda". This practice, however, decreased after late 1960's due to introduction of intensive farming systems and easy availability of chemical fertilizers. But with the energy crisis and increased prices of the fertilizers, green manuring again became popular. Its importance was further realized after Green Revolution and limited supply of FYM (Gupta and Sharma, 2004). Green manuring may be defined as "raising of a quick growing leguminous crop and burying the same in the soil before its flowering stage so as to add organic matter in the soil". Organic matter then decomposes into readily available forms of plant nutrients.

Collecting of green twigs and leaves from all available sources and incorporating them in the puddled lowland rice soils, is an age-old practice in South India. This is known as *Green Leaf Manuring*.

Plate 7.2: Farmyard manure

Plate 7.3: Green manures incorporated in the field

7.4.3.1 Best Known Green Manuring Crops for Indian Soils

For green manuring generally leguminous crops are preferred because they have an additional advantage of fixing atmospheric nitrogen. The crops commonly raised for green manure in many parts of India are dhaincha (*Sesbania aculeata),* sunnhemp (*Crotalaria juncea),* cowpea *(Vigna unguiculata)* and mungbean *(Vigna radiata).* These are all *Kharif season* crops. During *Rabi* season, senji *(Meliotus parviflora syn, alba),* berseem (*Trifolium alexandrinum)* and metha *(Trigonella foenumgraceum)* are found suitable especially for Punjab, Haryana and parts of Himachal Pradesh and Jammu and Kashmir. In Haryana, some of the progressive farmers grow only mungbean and cowpea, and after selling beans as vegetable, the rest of the material is ploughed back to the soil. The suitability of these crops depends upon the amount of nitrogen fixed, type of soil, climate, irrigation facilities and their rate of growth and decomposition. The quantity of organic matter and nitrogen added in soil by different green manure crops has been shown in Table 7.2.

Table 7.2: Quantity of green matter and nitrogen added in soil by green manure crops (Adapted from Pandey [1984-85])

Sr. No.	Crop	Growing season	Green matter ($q\ ha^{-1}$)	N %age on green weight	N - added ($kq\ ha^{-1}$)
1.	Sunnhemp	Kharif	212	0.43	75
2.	Dhaincha	Kharif	200	0.42	68.9
3.	Mung	Kharif	80	0.53	34.5
4.	Lobia	Kharif	150	0.49	50.3
5.	Berseem	Rabi	155	0.43	54.2
6.	Senji	Rabi	280	0.51	120

Green leaf manure trees grown on field bunds and wastelands, their young branches, twigs and leaves are cut and incorporated into the fields during puddling operation prior to rice planting. Biomass of ipomoea (*Ipomoea crassicaulis, Ipomoea carnea),* aak (*Calotropis gigantea),* basuti or branker (*Adhatoda vasica),* water hyacinth (*Eichhornia crassipes*), is also incorporated into the field after chopping/cutting. In hilly areas of North West Himalayas, the leaves of the eupatorium (*Eupatorium odoratum*) or eupatorium (*Eupatorium adenophorum*), are also used as green leaf manuring.

7.4.3.2 Cultivation Practices of Some Green Manuring Crops

Cultivation practices of some of the green manuring crops are as follows:

7.4.3.2.1 Sunnhemp (*Crotalaria juncea*)

It is the most commonly used green manuring crop in India. It grows best in loamy sand and sandy soils where water-logging doesnot exist. It grows rapidly and decomposes quickly. Where irrigation facilities are not available, it must be sown with first shower of monsoon using 60- 75 kg seed ha^{-1} and plough it as green manure prior to flowering i.e. 40-45 days after sowing.

7.4.3.2.2 Dhaincha (*Sesbania aculeata*)

Dhaincha is suitable for sandy loam and clay loam soils. It can even be grown in clay soils also. It can withstand severe drought and also does well in poorly drained land and saline/sodic soils (Najjar and Gupta, 1996). It becomes ready for being ploughed in the soil 1.5-2.0 months after sowing prior to flowering stage. It has been found the most suitable green manure crop for the paddy soils (Gupta and Sharma, 2004). In irrigated areas of Jammu (Jammu and Kashmir), it can be grown in the ending of April or first week of May using 60-75 kg seed ha^{-1}. The crop is then buried into the soil in the mid of June or 3rd week of June before its flowering stage. Four to five days after its burying, the paddy is transplanted.

7.4.3.2.3 Mung bean (*Vigna radiata*)

Being the most popular as *Zaid* pulse, it can also be used for green manure during *Kharif* season (June-July) after taking two-three pickings. The crop can be sown in the second or third week of April i.e. immediately after harvesting of wheat using 20-22.5 kg of seed ha^{-1}. Paddy crop can be transplanted in such fields, after 5-7 days of its incorporation. Where facilities of irrigation are not available, it can be sown in *Kharif* season also.

7.4.3.2.4 Berseem (*Trifolium alexandrinum*)

Berseem is a *Rabi* fodder crop. It can be used as green manure crop also. For this purpose, it can be buried into the soil after taking its 2-3 cuttings as green fodders. For achieving higher yields of green matter, seed must be treated with *Rhizobium* culture.

7.4.3.2.5 Senji (*Meliotus parviflora*)

Senji is a *Rabi* green manure crop and can be grown in irrigated areas using 40-50 kg seed ha^{-1}. It is most suitable green manure for sugarcane and becomes ready for being ploughed in by the end of January. Senji crop can be grown in the middle of December and 1.0-1.5 months after its sowing, it becomes ready for action.

7.4.3.3 Choice of Green Manure Species

In making choice from various crops, the following criteria should be borne in mind:

- A quick growing crop having thick stand, abundant leaf growth and succulent stem should be the main choice.
- Type of the soil is another criterion for making the choice of plant species. Sunnhemp and guara prefer loam and sandy loam soils whereas dhaincha has proved more successful in clay/clay loam soil or loam.
- Cultivation of dhaincha is also recommended for calcareous, saline and sodic soils as well as in the sub-mountainous valleys.
- Berseem and senji can be successfully raised on all types of soils but their growth becomes stunted in sandy saline and sodic soils.
- As legume growth depends upon the presence of suitable Rhizobium strains, so inoculation is essential prior to sowing the seed.
- Plant growth and biological nitrogen fixation can be hindered by water stress, undesirable pH, and lack of nutrients particularly P, Ca, Mo and Zn

and/or Mn toxicity. Application of rock phosphate, lime and ashes can help in improving legumes establishment.

7.4.3.4 Advantages of Green Manuring

As already stated green manuring is a very old farm practice and has been practised from early times a source of supplementing animal manure. It has been found that the yield of rice, wheat, cotton and sugarcane increased from 13 to 100 per cent by using sunnhemp, dhaincha, cowpea, as green manures.

In semi-arid zones, the fertility of the soil can be considerably improved by growing and burying green manuring crops. In tropical regions, the soil organic matter gets oxidized due to high temperature and, as such, it has to be replenished frequently, which can be done by following this practice. In India, on the whole, favourable climatic conditions prevail for extensive cultivation of green manuring, to make up for the shortage of available cattle dung manure, much of which is burnt as fuel. Several trees and shrubs are also available which have been found useful to serve as green leaf manuring. Green manuring is, therefore, an important source of soil fertility, under Indian conditions. The following are the main advantages of green manuring:

- It offers an easy means for meeting our manurial requirements. As most of the Indian soils are lack of organic matter and nitrogen so use of green manure will prove an asset for such soils. The necessity for using organic nitrogenous manures is, therefore, obvious. If the selected crop is legume, nitrogen in the soil is increased considerably. Green manuring crop when mixed with finely divided rock phosphate, can fix more nitrogen from the atmosphere.
- The decomposition of green manures increases the ionic exchange capacity of soils i.e. cation exchange capacity and anion exchange capacity.
- It influences favourably the air cum water permeability, and improves soil tilth and granulation of the soil. It binds the sandy soils and produces crumb structure in clayey soils.
- It takes up soluble substances containing essential plant nutrients, which otherwise might be lost in drainage water. In other words we can say that besides actively fixing nitrogen in the root nodules, green manuring crop also reduces the leaching losses of mineral nitrogen because of its vigorous root system.
- Mineral constituents of soil in combination with humus of green manure are in more suitable form for assimilation by the succeeding crops.

- Solubility of calcium, phosphorous and potassium is increased. It is because the decomposing organic matter liberates carbon dioxide leading to lowering of soil pH especially of calcareous, saline and sodic soils, when help in solubilization of these nutrients. Solubilization of micro-nutrient is also occurring.
- Long/deep roots of green manuring crops carry the plant nutrients upwards from the sub-soil to the root zone of the soil.
- Green manures reduce intercultural operations by keeping down weeds and grasses.
- Dhaincha has proved beneficial in reclamation of alkali and saline soils due to its tolerance to alkalinity and salinity.
- The greatest benefit of green manuring lies in its being a cheaper source of plant nutrients. This kind of manure also helps in keeping the soil well managed.

7.4.3.5 Green Matter Production and Effect on Crop Productivity

The green matter production by various green manuring crops varies according to the type and age of the crop to bury into the soil. The data generated under Jammu soil conditions indicate that dhaincha added to the maximum amount of green matter (300 q ha^{-1}) followed by sunnhemp (284 q ha^{-1}) and minimum by guara (200 q ha^{-1}) when were ploughed into the soil at the age of 60 to 75 days after sowing. However, at 45 days growth stage in alkali soils, dhaincha produced a green matter yield of 200-240 q ha^{-1}, containing about 95-115 kg of nitrogen. In addition to nitrogen, dhaincha also supplied phosphorous, potassium and calcium to the extent of 0.14, 0.61 and 0.44 per cent, respectively.

In situ incorporation of dhaincha before transplanting of rice (PC - 19) variety under Jammu KVK Farm not only showed an increase in yield of the crop but also there was reduction in the application of nitrogen and phosphorous by $1/3^{rd}$ and half, respectively. The yield was to the tune of 34.5 q ha^{-1} with full NPK application, while it 32.5 q ha^{-1} when basal dose of N was missed and one-half of the recommended doses of phosphorous were applied as compared to check (18 q ha^{-1}). The residual effect of green manuring was also seen during *Rabi* crops. In the green manured plot, the yield of wheat, PBW - 343 was observed to be 26 q ha^{-1} compared to control where the yield of the crop was 21 q ha^{-1}.

In Plains and Siwalik hills of Himachal Pradesh, it has been found that growing of dhaincha can meet the nitrogen requirement of various varieties of rice by fixing nitrogen to the extenl of 100 to 125 kg. In Punjab and Uttar Pradesh, the growing of dhaincha as green manure has shown increase in yield of rice by 51 per cent. On the other hand, the sunnhemp showed such an increase in case of sugarcane and wheat by 30 and 42 per cent, respectively.

Since green manuring practice is an easy approach to fulfill the plant nutrients requirement, so the farmers must practise it wherever it is possible. The long run benign influence of green manuring on the "health and wealth of the soil", vis-a-vis its pivotal role in reclaiming saline/sodic soils, call for large scale cultivation of dhaincha. Moreover, this practice will prove an asset in rice-wheat cropping sequence where the yield has either become stagnant or is declining year after year.

7.4.3.6 Role of Green Manuring in Humid Tropics and Sub-tropics

One of the earliest solutions to restore soil productivity in the humid tropics was to use of green manuring (Agboola, 1982). The concept was that green manure increased soil organic matter, maintained soil fertility and, thereby, maintained continuous arable cropping and ultimately sustained crop productivity. It was also highlighted that green manure was capable of recycling of leached plant nutrients. This is accomplished by the absorption of nutrients from the lower depths by the roots of the legumes which translocated them to the leaves and from the leaves to the soil surface (Agboola, 1974), which in turn improve soil structure, water holding capacity, aeration and chemical properties of the soil. It was also explained that after ploughing a green manure crop produces carbonic acid (H_2CO_3) which aids in the liberation of NO_3-N from the soil organic matter. Thus, green manure serves as a catalyst in aiding the decomposition of native soil organic matter, thereby, releasing an amount of NO_3-N.

After the green manure (legume) is turned under, it adds the fixed atmospheric nitrogen also. The amount of nitrogen fixed, however, depends on the availability of other nutrients in soils especially P and micro-nutrients as well as the *Rhizobium* spp. present. This is one of the constraints to the use of green manure as stated by Singh and Balasubramanian (1980), summarizing the use of green manure in Asia:

> "In the flooded rice cultivation of Asia, it has been a common practice to raise green manure crop (using a legume) at the beginning of rainy season, plough it when the crop is 6-7 weeks old and allow it to decompose for another 3-4 weeks prior to the transplanting of rice seedlings. Thus, 10 weeks is needed to practise green manuring in rice culture and farmers may sometimes lose the best part of the rainy season before planting the main crop of rice".

The aforesaid method is, however, traditional one which is growing the green manure crop for more than two months and allows 3 to 4 weeks decomposition before planting the next crop (rice). The new technique used in the studies involved growing the green manure crop for short periods and transplanting of rice immediately after ploughing under the green matter. The modified technique is as follows:

- Sowing date of dhaincha is fixed in such a way in summer so as to give 40-45 days growth period for the green crop to till the transplanting date for rice in the ending of June or fortnight of July.
- 60-75 kg of dhaincha seed soaked in water over night is required for a hectare.
- If possible the seed must be treated with specific *Rhizobium* culture as is done for other leguminous crops.
- 40 to 50 kg of phosphorous as superphosphate or diammonium phosphate is applied to the soil.
- First two irrigations are applied at four or five day's interval while subsequent watering is done at 9 to 10 days interval.
- The green manure is buried with furrow turning plough, power-tiller or tractor disc or any other means an objective to get satisfactory green matter after incorporation.
- Thirty to forty days old seedlings of rice are transplanted either on the same day or after two to three days.
- In situ incorporation of dhaincha before transplanting of rice not only showed increase in the yield of rice but there was also reduction of N by 1/3rd and phosphorous by half of the recommended doses.
- The yield was 34.5 q ha^{-1} with full dose of NPK application, while it was 32.5 q ha^{-1} with green manuring. There were about 40-45 per cent increase in rice yield was found to be 18 q ha^{-1}.

7.4.4 Crop Residues

Residues left out after the harvest of the economic portion in the field is called as crop residues. India has a vast potential of crop residues such as rice, wheat and sugarcane, which are substantially rich source of plant nutrients (Plate 7.4 and 7.5). On an average, it contains 0.5 per cent N, 0.6 per cent P_2O_5 and 1.5 per cent K_2O. They mostly constitute the staple food of and/or are mostly used as dry fodder for animals, but in certain regions where mechanical harvesting is done, the crop residues are left behind in the field which can act as a source of nutrient supply. Application of well decomposed crop residues improves the soil properties and nutrient status of the soil; they also meet a part of micro-nutrient supply and thus, help in sustaining rice- wheat crop productivity.

The results of the field experiments carried out on Inceptisol soil under subtropical irrigated conditions of Jammu (Jammu and Kashmir State) with regard to see the effect of rice residue management in wheat crop and soil properties have been described in Chapter 6 i.e., "Organic Farming for Sustainable Agriculture". Urea application with rice stubbles at the time of field preparation showed significant effect on growth and yield attributes. The effect of paddy straw on the yield of wheat crop also stands given in Chapter 6 as stated above. The effect of organic and inorganic fertilizers on sustainability of rice and wheat yields was also studied by Gupta and Gupta (1997). They found that 50 per cent NPK could be saved by continuous application of FYM or green manure during *Kharif* in rice.

7.4.5 Leguminous Crop

Legumes act as soil fertility restorers due to their ability to obtain N from the atmosphere in symbiosis with *Rhizobium*. Legumes can form an important component of INM system when grown for grain or fodder in rice-wheat cropping system, intercropped with other non-legumes or when introduced as green manure crops. The practice of rice-wheat rotation continuously has an adverse effect on physico-chemical properties and fertility status of the soil (Mahajan et al., 2002). Hence, inclusion of the legumes in rice-wheat sequences may restore the soil fertility. It has been reported that carry over of N for succeeding cereal may be 60-120 kg in *Trifolium alexandrinum* (berseem), 75 kg in *Trifolium ripens* (Indian clover), 35-60 kg in *Vigna unguiculata* (fodder cowpea), 55 kg in *Vigna mungo* (blackgram), 60 kg in *Arachis hypogaea* (groundnut), 68 kg in *Cicer arietinum* (gram) and 50 kg in lathyrus, the legume intercrop which is attributed to atmospheric nitrogen fixation by these legumes. It plays a vital role in N-fixation and economizing the resource use especially N. It has been estimated that with inclusion of legume intercrops, the extent of N addition in cereal based cropping systems would be of the order of 0.746 MT.

Plate 7.4: Crop residues of rice placed in the field

Plate 7.5: Wheat residues incorporated in the field

7.4.6 Compost

The process of decomposition of organic wastes is termed as composting and the decomposed material is called as compost (Plate 7.6). In compost making process the bulk of the material used is plant residues. On an average, it contains 1.01 per cent N, 0.5 per cent P_2O5 and 0.8-0.9 per cent K_2O. The enrichment of organic manures is of practical interest because demerits of bulky organic manures can be overcome by their enrichment. It increases the nutrient content of manures, reduces the bulk to be handled per unit of nutrients, and offers a potential for the utilization of insoluble materials such as low-grade phosphate rocks. The availability of rural compost and city refuse in India are 285 and 14 million tonnes respectively which provide 4 million tonnes of plant nutrients (NPK) per year.

7.4.6.1 Types of Compost

Compost generally prepared in India is of two types:

7.4.6.1.1 Urban or Town Compost

Urban compost is prepared from night soil, street and dustbin refuse and rich in nutrients (1.0-2.0% N, 1.0% P_2O_5 and 1.5% K_2O) as compared to FYM and rural compost.

7.4.6.1.2 Rural or Village Compost

It is prepared from farm waste products, e.g., straw, crop stubbles, crop residues such as sugarcane trash, groundnut husks and leaves, cotton stalks etc., weeds, waste fodder, urine soaked earth, litter from cowshed and hedge clippings. It contains 0.4-0.8 per cent N, 0.3-0.6 per cent P_2O_5 and 0.7-1.0 per cent K_2O.

7.4.7 Vermicompost

Vermicompost which is prepared with the technology of rearing of earthworms is an effective tool of sustainable agriculture. In India large number of vermicompost units has been established by city corporations, municipalities, NGOs besides large number of private entrepreneurs especially in the states of Madhya Pradesh, Rajasthan, Uttar Pradesh, Kerala, Karnataka etc. It has attracted the attention of not only of scientists, but farmers' worldwide, since it is a natural organic product which is eco-friendly and leaves no adverse effects either on the soil, and produces no environment effect. Thus, much interest in vermicomposting has been noticed due to the fact that earthworms play an important role in soil improvement, organic matter decomposition and in enhancing plant growth. The use of earthworms in organic waste management

Plate 7.6: Compost placed in the field

has been termed as vermicomposting and the compost is called as vermicompost. The earthworm commonly used for this purpose is *Eisenia foetida, Eudrillus eugeniae* and *Pheritima elongata* which are potential agent in breaking down animal wastes (Gupta and Bhagat, 2004). On an average, it contains 1.6 per cent N, 5.04 per cent P_2O_5 and 0.8 per cent K_2O. Apart from this, vermicompost also contains hormones like auxins and cytokines, enzymes, vitamins and useful microorganisms like bacteria, actinomycetes, protozoa fungi and others.

7.4.8 Bio-Fertilizers

Bio-fertilizers are natural fertilizers which are microbial inoculants of bacteria, algae, fungi alone or in combination and they augment the availability of nutrients to the plants. These play a very significant role in improving soil fertility by fixing atmospheric nitrogen, solubilize insoluble soil phosphates and produce plant growth substances. These micro-organisms are used for the treatment of seeds, seedlings and application to soil or in composting to increase soil fertility and for prolonged fertility thus helping in improved crop growth and high yield by quick multiplication and efficient biological activities. Preparations containing these materials can also be considered as microbial fertilizers like *Rhizobium, Azotobacter* and *Phosphobacterin* cultures, BGA and *Azolla* are the other bio-fertilizers (Mahajan et al, 2003a & b). Some of the organisms which canbe used as bio-fertilizers are listed in Table 7.3.

Table 7.3: Some beneficial organisms used as bio-fertilizers in sustainable agriculture

Name of organism	Mode of action	Crops for which used	Method of treatment	Quantity of inoculums required. (g ha^{-1})
Rhizobium	Symbiotic N fixation	Leguminous pulses, groundnut and fodder crops	Seed treatment	600
Azotobacter	Asymbiotic N fixation	Vegetables, Cotton, Cereals & millets	Seed treatment	3,400
Azospirillum	Asymbiotic N fixation	Cereals and millets Vegetable crops	Seed treatment Soil application	1,000 2,000
BGA	Asymbiotic N fixation	Paddy	Soil application	10,000
Azolla	Asymbiotic N fixation	Paddy	Soil application	One tonne (dried algae)
Phospho-bacterium	Phosphorous solubilization	Millets, wheat and paddy	Seed treatment	600

Out of the above described bio-fertilizers, BGA and Azolla are of great significance in the N economy of rice cultivation under water-logged condition. Their use may contribute 25-50 kg N ha^{-1} to rice crop. Phosphorous solubilizer also possesses the ability to bring the insoluble phosphates in soil into soluble forms by secreting organic acids (formic, acetic, propionic, lactic, and succinic acids etc.) which help in better growth and development of plants. These acids lower down the pH and bring about dissolution of the immobilized forms of phosphates and its application has shown 15-30 per cent increase in the yield of cereals such as rice and wheat. The important steps in the development of bio-fertilizer in India (Table 7.4) are as follows:

Table 7.4: Development of bio-fertilizers in India

Year	Important events
1920	First study on legume, Rhizobium symbiosis by M.V. Joshi
1956	Earliest documented production of Rhizobium inoculated by M.R. Madhok
1956	First commercial production of bio-fertilizers
1983	Setting up of National Project on Development and use of bio-fertilizer, by Ministry of Agriculture, Government of India
1986	Setting up of National Facility for BGA Collection at IARI, New Delhi by Department of Biotechnology, Government of India

Based on various kinds of micro-organisms, the bio-fertilizers can be classified in Table 7.5 as under:

Table 7.5: Classification of bio-fertilizers (Adapted from Mahajan et al. [2008a])

Sr.No.	Micro-organisms	Examples
1.	Bacterial bio-fertilizers	Rhizobium, Azospirillum, Azotobacter, Phosphobacterin
2.	Fungal bio-fertilizers	Mycorrhiza
3.	Algal bio-fertilizers	BGA and Azolla
4.	Actinomycetes bio-fertilizers	Frankia in woody plants

7.4.8.1 Rhizobium

The most widely used bio-fertilizer is *Rhizobium* culture which colonies the roots of specific legumes to form tumor like growths called as root nodules. These nodules act as factory of ammonia production. The *Rhizobium*- legume association can fix up to 100 to 300 kg N ha^{-1} in one season and can have substantial N for the following crops. The various *Rhizobium* inoculation groups suitable for different agricultural crops are listed in Table 7.5.

Table 7.5: *Rhizobium* species suitable for different agricultural crops (Adapted from Alexander [1978])

Sr.No.	Genus	Species	Crops
1.	*Rhizobium*	*R. meliloti*	Alfalfa, sweet clover, fenugreek
		R. leguminosarum bv. *vicia*	Peas, vetch, sweat pea
		R. leguminosarum bv. *phaseoli*	Beans
		R. leguminosarum bv. *trifolii*	Clovers
		R. loti	Trefoil
		R. fredii	Soybean
2.	*Azorhizobium*	*A. caulinodans*	Root and stem nodules on Sesbania (Sesbania rostrata)
3.	*Bradyrhizobium*	*B. japonicum*	*Soybean*
		B. elkanii	*Soybean*
			Cowpea, mungbean, pigeonpea, chickpea, peanut

The range of N fixed per hectare per year by different legumes is 50-55 kg for mungbean, 50-60 kg for groundnut, 60-80 kg for soybean, 80-85 kg for cowpea, 90-100 kg for lentil and 100-300 kg for alfalfa. *Rhizobium* cultivars are recommended for inoculation of seeds of various pulses and legume crops for meeting about 80 per cent of the N requirement of such crops (Gupta, 2007). In cereal based cropping systems *Rhizobium* might play an efficient role through green manuring of legumes, sequential legume or legume intercrops.

7.4.8.2 Azotobacter

Azotobacter fixes N asymbiotically and benefits the plants by growing in the rhizosphere of plants. There are six species of *Azotobacter* viz. *A. chroococcum, A. nigricans, A. paspali, A. vinelandii, A. beijerinckii* and *A. armeniacus*. Among them *A. chroococcum* is the most commonly found in arable soils. Use of *A. chroococcum* inoculants can save N to the tune of 25-50 kg ha^{-1} in cereals and millets (Gandaotra et al., 1989; Gandaotra et al., 1998; Gupta et al., 1977; Gupta et al., 2007). Apart from N fixation, this organism is also capable of producing antibacterial and antifungal compounds, vitamins and growth hormones. Similarly, *Azotobacter* is a well known for producing growth-promoting substances which favourably enhance plant growth and development (Gupta and Dogra, 1999). In general, use of *Azotobacter* reduces the inorganic fertilizer requirements to the tune of 25 to 50 per cent.

7.4.8.3 Azospirillum

It colonies the root mass and fixes N in loose association with plants. It has positive interaction with cereal crops with an average response equivalent to 15-20 kg ha^{-1} of applied N. The crops which respond to *Azospirillum* inoculation are maize, wheat, barley, sorghum, pearl millet and forage crops. Their application increases grain productivity of cereals by 5-20 per cent, millets by 30 per cent and forage crops by over 50 per cent.

7.4.8.4 Blue Green Algae (BGA)

BGA are predominant in submerged rice soils which provided an ideal condition for their growth and N fixation (Gupta and Khajuria, 1996). The common genera found in Indian rice soils include *Anabaena, Nostoc, Aulosira, Calothrix* and *Tolypothrix.* A judicious use of BGA could provide to the country's entire rice acreage as much N as is obtained from 15-17 lakh tonnes of urea. It is estimated that at farm level, it can contribute about 25-30 kg N ha^{-1}. Recent research studies have shown that BGA excrete vitamin B12, auxins and ascorbic acid which contribute to growth of rice plants.

7.4.8.5 Azolla

It is an aquatic floating water fern which harbors in its leaves a N fixing BGA. *Anabaena azollae* as symbiont. The major nutrient requirement of *Azolla* is phosphate (10-15 kg P_2O_5 ha^{-1}). The growing conditions are free floating water, sufficient sunlight, temperature around 15°C and high relative humidity (Mahajan et al, 2003a). *Azolla* multiplies at faster rate and fixes atmospheric N in flooded fields. However, it fails to grow under high temperature and it is the major barrier in using in rice fields having tropical climate. It has been estimated that application of *Azolla* can contribute at least 30 kg N ha^{-1} to rice crop. The *Azolla-anabaena* association is a live floating N factory using energy from photosynthesis to fix atmospheric N amounting to 100-150 kg ha^{-1} $year^{-1}$ from about 40-60 tonnes of biomass.

7.4.8.6 Phosphate Solubilizing Micro-Organisms (PSM'S)

Several soil bacteria particularly *Pseudomonas striata* and *Bacillus polymyxa* and fungi *Aspergillus awamorii* and *Penicillium digitatum* possess the ability to bring insoluble phosphates in-to soluble forms by secreting organic acids such as formic, acetic, oxalic, malic, gluconic, citric, propionic, lactic and succinic acids etc. These acids lower down the pH and in turn, bring about dissolution and immobilized forms of phosphates. Besides, some of the hydroxyl acids may chelate with calcium and iron resulting in effective solubilization and thereby higher utilization of soil phosphates by plants.

7.4.8.7 Vesicular Arbuscular Mycorrhizae (VAM)

VAM are the symbiotic association of fungi with roots of vesicular plants. They have a great potential especially in the acquisition of phosphorous in P deficient soils and effective tool to help in the solubilization of the unavailable P in the soil. A VAM fungus not only improves crop yields but also saves P fertilizer to the tune of 25-30 per cent and also increases the uptake of micro-nutrients such as Cu, Zn, Mn and Fe. It has also ability to withstand high soil temperatures, acid rain and biocide treatments and protecting the plants from pH extremes. About 90 per cent of plants, including the most important agricultural crops, are known to be associated with VAM-fungi.

7.5 Advantages of INM System

Finger millet-maize cropping pattern covers large area in south India. In this system of cropping pattern, application of all the three major nutrients has been found to increase the yield of these crops (Singh, 2010). It is further added that integration of FYM with NPK was found to boost the productivity of this cropping system. At Palampur (Alfisols), continuous application of N caused decline in yield in maize and wheat with time. However, use of P, K along with FYM / lime has maintained the productivity of this cropping sequence (Singh, 2010). Other advantages of INM system are briefly described below:

- To maintain soil fertility status, soil health as well as environment for attaining higher crop productivity and thereby to meet present and future food supply needs.
- To increase organic matter content by using all available organic sources and, thus, the soil physical, chemical and biological properties are ameliorated.
- It prevents soil degradation.
- There is an enrichment of soil micro-flora and micro-fauna as well as macro-organisms.
- It increases fertilizer use efficiency and generates profit to the farmers through the judicious and efficient use of chemical fertilizers, organic manures, crop residues and bio-fertilizers.
- It is an effective resourceful cycling and nutrient movement within the farm itself.
- It maintains positive nutrient balance in soil and develops eco-friendly sustainable agriculture.
- To meet the social and economic aspirations of the farmers lower the cost with high productivity and profitability.

7.6 Constraints in the Adoption of INM System

- Increasing compulsion to use cow dung as a source of fuel in rural India, thereby leaving its very little quantity for composting. Rice straw is also used as livestock bedding, feeding, thatching materials for huts and temporary dwellings in villages.
- Increasing competitive value of crop residue as animal feed, being rich in silica and oxalic acid, affecting recycling of agricultural wastes.
- Although organic residues of rice and wheat straw can be used to make up the plant nutrients need yet they require their complete decomposition due to having wide C: N ratio. If they are not fully decomposed and added as such, will cause immobilization of nutrients, especially of N. Thus, the crop yields will be affected badly.
- The organic manures like compost, FYM, poultry manure etc. have low amount of plant nutrient contents and thus large amounts are required to fulfill the nutrients requirement for the various high yielding varieties of crops, which is not possible for the country like India with a large population to feed.
- Handling problems in bulky organic manures which involve high cost of transportation.
- Organic manures release nutrients more slowly, so the plant doesn't have a flush of growth when they are applied.
- The rate of mineralization and the rate of the release of nutrients, particularly of the N and micronutrients, are slow and uniformly in organic manures.
- Problems in timely preparation of field due to incorporation and decomposition of agricultural waste and green manures.
- Growing of green manuring crops are not adopted widely because it needs additional labour/inputs, inadequate availability of water, inadequate supply of cheap and quality green manure seed locally in time and other competing uses of resources for raising other crops.
- Extra cost and time required in raising green manure crops.
- Poor quality bio-fertilizers reaching the farmers which are ineffective. Thus raising questions of their usefulness.
- Marketing of bio-fertilizers becomes difficult because the product contains living organisms in the virulent conditions.

- Unawareness and unavailability of bio-fertilizers to the farmers.
- Faulty transport and storage system raising temperature which destroy the microbial population in inoculants.

In spite of many constraints posed by the INM system as stated above, this system of crop production is of absolutely necessary, especially for the developing countries including India and requires to be followed due to following reasons:

- Adoption of INM system will ameliorate the soil health.
- It will assist in checking air, water and soil pollution caused by indiscriminate use of fertilizers and pesticides during "Green Revolution".
- INM system will assist to sustain food grain production particularly of rice-wheat cropping system.
- Manures supply a host of nutrients and improve soil physical and biological environment. Moreover, several hidden deficiencies can be prevented.

7.7 Conclusion

- The benefits of organic sources of nutrients are relevant both to the developed nations and to the developing countries.
- Application of different organic sources of nutrients alone and in combination with fertilizers increased the crop yield as well as improved the soil health.
- In country like India to meet out the ever increasing demand of huge population for food, high fertilizer doses are required, which result in deterioration of soil health and increase in cost of production. Thus, for maintenance of sustainable agriculture in our country we must go for organic sources of nutrients for some pocket and both organic as well as inorganic for others.

7.8 Future Research Strategies/Priorities

- To achieve the goal of food security, the crop production strategies should focus on increasing the crop productivity through INM without creating an adverse effect on the soil health and environment.
- The sustainable crop production is not possible through use of chemical fertilizers alone. There is an urgent need to conduct more research for adopting INM system for promoting efficient and balanced use of nutrients where the NPK chemical fertilizer be applied in the right proportion i.e. in the ratio of 4: 2: 1 and the role of organic manures and bio-fertilizers be complementary and not substitutive.

- There is need to generate more efficient integrated nutrient supply and management systems using, locally available resources like compost, FYM, farm wastes, green manures, sewage, sludge etc. Measures should be developed to quantify the effect of these resources, so as to make necessary adjustment in chemical fertilizers.
- To reduce the use of cow dung as a source of fuel, practice of cow dung burning should be totally banned as has been done in China. Agroforestry system should be popularized among the farming community to get fuel wood (Gupta, 2005). Moreover, large scale plantation of quick growing leguminous trees on road sides, wasteland and field bunds should be encouraged. Installation of biogas plants in large numbers can also solve the problem of fuel and plant nutrients.
- Quantification of nutrient losses from organic manures and devise methods to increase the efficiency of their use.
- Simple techniques be developed to handle bulky organic manures so that their processing, use and application are as simple as chemical fertilizers.
- Direct and indirect effects of crop residue incorporation on soil quality and crop productivity in rice-wheat cropping systems should be investigated in more details.
- Development of better technology for the in situ incorporation and rapid decomposition of crop residues by appropriate tillage, irrigation, fertilizer management and inoculation of suitable micro-organisms, particularly for the intensive rice-wheat system is needed to develop.
- Fertilizer recommendations on the basis of soil test-crop response correlations, locally availability and economics of different fertilizers and manures are to be formulated for cereal based cropping system.
- Estimation of the percentage contribution of organic sources to soil nutrient budget that will be available during the current growing season and residual nutrients from their applications should be conducted on priority.
- Research must be made to identify cheaper, efficient and economic sources of micro-nutrients and techniques to alleviate micro-nutrient deficiency in location specific situations for intensive rice-wheat cropping systems.
- Research efforts are needed to develop the effective use of nitrifying inhibitors and growth hormones should be included in INM technology.
- Development of technology for speedy decomposition of crop residues to be devised with the use of cellulolytic micro-organisms like *Trichoderma viridie.*

- Large scale extension education is required to popularize green manuring.
- The development of the identified soil bacterial species like *Bacillus subtilis, Bacillus cereus* var. *mycoides, Bacillus megatherium* var. *phosphaticum* based phosphate dissolution technologies for sustainable cereals production, is required to be done.
- More scientific information on supplementary use of organic, chemical and bio-fertilizers for rice-wheat system in different agro-climatic zones is needed to be generated.
- There is need to do some systematic research work to establish the role of earthworms in composting of organic wastes and the impact of vermicomposts on improving rice-wheat crop yields.
- Bio-fertilizer technology needs to be refined to make it more acceptable to farmers. Microbial strains which can compete with indigenous ones and work over a range of soil climatic conditions need to be isolated and multiplied.

Study Questions

1. Define the following:
   - i. Integrated nutrient management system
   - ii. Green manuring
   - iii. Farmyard manure
   - iv. Compost
   - v. Mineralization
2. Name the various components of INM and write the detailed accounts of any one of them.
3. What is the difference between bulky manures and concentrated manures?
4. Distinguish between the following:
   - i. Green manuring and leaf manuring
   - ii. Compost and vermicomposting
   - iii. Azotobacter *and* Rhizobium
   - iv. *BGA and* Azolla
5. Write in details about the various advantages of INM system?
6. What are the different benefits of green manuring?
7. Write a paragraph about the concept and aim of INM?

References

Agboola, A.A. (1974). Organic materials as fertilizers. FAO Bull. 27, Rome Italy, pp. 147-152.

Agboola, A.A. (1982). Organic manuring and green manuring in tropical agricultural production systems (pp. 198-222). In Symposia Papers I. Non-Symbiotic nitrogen fixation and organic matter in the tropics. Transactions in the 12[h] International Congress of Soil Science (Feb. 6-16, 1982), New Delhi.

Alexander, M. (1978). Introduction to Soil Microbiology. John Wiley and Sons Inc., New York and London.

Gandotra, V., Gupta, R.D. and Bhardwaj, K.K.R. (1989). Effect of nitrogen and seed inoculation by Azotobacter on maize in acidic soils of North Western Himalayas. Him. J. Agric. Res. 15 (2): 85-87.

Gandotra, V., Gupta, R.D. and Bhardwaj, K.K.R. (1998). Abundance of Azotobacter in great soil groups of North West Himalayas J. Indian Soc. Soil Sci. 46 (3): 379-383.

Gupta, D.K. and Gupta, J.P. (1997). Effect of organic and inorganic fertilizers on sustainability of soil fertility and grain yield in rice-wheat system. IRRI Newsletter 22 (2): 35-36.

Gupta, R.D. (2005). Environmental degradation of Jammu and Kashmir Himalayas and their control. Associated Publishing Company, New Delhi (India).

Gupta, R.D. (2007). Farmers should utilize inoculants to enhance legume production. Farmers' [F]oru ml (5): 24-27.

Gupta, R.D. and Bhagat, K.C. (2004). Earthworms do miracles in agriculture. Indian Farmers' Digest 37 (2): 41-44.

Gupta, R.D. and Dogra, S.R. (1999). Azotobacter in agriculture. The Himalayan Mail 4 (225): 6 [Jammu].

Gupta, R.D. and Khajuria, M.R. (1996). Prospects and constraints of bio-fertilizers for rice with special reference to Jammu region of Jammu & Kashmir State. Farmers and Parliament 31 (10): 21-22.

Gupta, R.D. and Sharma, A.K. (2004). Revive green manuring practice for higher food production. The Kashmir Times 61 (290): 7 [October 16, Saturday].

Gupta, R.D. and Sharma, S.P. (2006). Soil pollution and its solution through integrated nutrient management (pp. 32-38). In Balanced Fertilization for Sustainable crop Production (Eds. S.P. Sharma, S.K. Sharma, R.P. Sharma and T.S. Verma). Proceedings National Seminar on INM for Balanced Fertilization held on November 17-18, 2006.

Gupta, R.D., Jha, K.K., Sethi, A.S. and Saharan, G.S. (1977). Presence of Azotobacter in relation to physico-chemical properties of some north Indian soils. Proceedings Indian Academy Sciences 56 (6): 389-395.

Gupta, R.D., Kaistha, B.P. and Kher, D. (2007). Ecology of Azotobacter in soils of Himachal Pradesh and effects of its isolates on maize crop. Environment and Ecology 25S (4A): 1304-1308.

Jaggi, R.C., Mahajan, A., Choudhary, A.K. and Bsghla, K. (2001). Integrated plant nutrient management (IPNM) in agriculture - The need of the day. Indian Farmers' Digest 2 (6): 10-12.

Mahajan, A. and Gupta, R. D. (2009). Integrated nutrient management (INM) in a sustainable rice-wheat cropping system (pp 1-267). Springer Science + Business Media (www. Springer.com).

Mahajan, A. and Sharma, R. (2005). Integrated nutrient management (INM) system - Concept, need and future strategy. Agrobios Newsletter 4 (3): 29-32.

Mahajan, A. Gupta, R.D. and Sharma, R. (2007a). Organic farming - A need for all times. Gram Vikas Jyoti (July-September Issue): 7-8.

Mahajan, A.. Bhagat, R.M. and Gupta, R.D. (2008a). Integrated nutrient management (INM) in sustainable rice-wheat cropping system for food security in India. SAARC Journal of Agriculture 6 (2): 149-163.

Mahajan, A., Bhagat, R.M. and Trikha, A. (2003a). Fertilizing bio-fertilizers. Agriculture Today 6 (9): 52-54.

Mahajan, A., Choudhary, A.K. and Bhagat, R.M. (2002). Integrated plant nutrient management (IPNM) system for sustainability in cereal based cropping system. Indian Farmers' Digest 35 (7): 29-32.

Mahajan, A., Choudhary, A.K., Jaggi, R.C. and Dogra, R.K. (2003b). Importance of bio-fertilizers in sustainable agriculture. Farmers' Forum 3(4): 17-19.

Mahajan, A., Sharma, R., Gupta, R.D. and Trikha, A. (2007b). Organic farming and sustainable agriculture. Agriculture Today 10 (3): 48-50.

Mahajan, A., Sharma, S.K., Gupta, R.D. and Sharma, R. (2007c). Morphological, physical and chemical properties of soils from North West Himalayas. Bulg. J. Agric. Sci. 13 (5): 607-618.

Najjar, G.R. and Gupta, R.D. (1996). Effect of organic and in-organic amendments on soil properties and yield of rice. Agropedology 6 (2): 83-86.

Pandey, R.V. (1984-85). Green manure crops for soil fertility. Indian Farmers' Digest 17/18 (12/1): 33-34 and 36.

Raina, S.K., Suri, V.K., Jagi, R.C. and Mahajan, A. (2005). Get your soil tested for obtaining maximum return from the application of fertilizers. Kheii Dunia 10: 37.

Reddy. S.R. (2001). Principles of Agronomy. Kalyani Publishers Ludhiana, New Delhi Noida Hyderabad Chennai Calcutta Cuttack.

Singh, A. and Balasubramanian, V. (1980). Organic recycling in Asian agriculture. FAO Soils Bull. 43, Rome Italy, pp. 235-278.

Singh, M. (2010). Nutrient management: A key to soil health and long term sustainability. J. Indian Soc. Soil Sci. 58 (Supplement): 547- 558.

Subba Rao, A. (2010). Managing phosphorous for higher efficiency and environment quality. J. Indian Soc. Soil Sci. 58 (Supplement): 511- 514.

Subehia, S.K. and Sepehya, S. (2012). Influence of long terms N -substitution through organics in yield, uptake and available nutrients in a rice-wheat system in acidic soil. J. Indian Soc. Soil Sci. 60 (3): 213-217.

Virmani, S.M. (1994). UNCTAD Agenda-21: The new challenges for soils research. J. Indian Soc. Soil Sci. 42 (4): 516-524.

8

Use of Bio-Fertilizers in Sustainable Agriculture

Abstract

The modern agriculture has been heavily depending upon the fossil fuel-based inputs such as chemical or inorganic fertilizers, pesticides and energy intensive farm machinery. Thus, use of chemical fertilizers and pesticides on largescale has created "Environmental Pollution" i.e. Air Pollution, Water Pollution and Soil Pollution. All these kinds of pollution which have spewed due to indiscriminate use of chemical fertilizers and pesticides have recently been reviewed (Gupta and Sharma, 2006; Gupta and Singh, 2006; Gupta and Singh, 2008,; Mahajan and Gupta, 2009, Sinha et al. 2015). These writings have clearly mentioned how chemical fertilizers and pesticides have created health hazards both in human beings and animals. They have also made the soils sick in terms of deficiency of macro-nutrients including secondary ones and micronutrients and provided soil acidity, alkalinity, salinity and sodicity. Microbial activities have also been very much affected. Due to sickness of soils, their crop productivity has decreased tremendously. It is more so in case of rice and wheat crops. Hence, the use of bio-fertilizer and organic manures in combination with chemical fertilizers offers a great opportunity to increase the crop production with less cost. This is known as Integrated Nutrient Management (INM) system, details of which have been mentioned in seventh chapter of this book. In this chapter, an integrated approach of bio-fertilizer for sustainable agriculture is discussed. Use of bio-fertilizer, however, must be of right type and have to be applied at the right time to obtain maximum advantage.

Keywords Modern agriculture; crop productivity; environmental pollution; health hazards; ; INM system; chemical or inorganic fertilizers; pesticides; organic manures; bio-fertilizer; macro-nutrients; secondary nutrients; soil acidity; alkalinity; salinity; sodicity

Bio-fertilizers, as the name indicates, are the fertilizers of biological origin (Gupta, 2012). In strict sense, real bio-fertilizers are green manures and organics - materials of biological origin which are added to supply plant nutrients present in them. However, in restricted scope, the term "bio-fertilizers" or "microbial inoculants" are preparations of living organisms in alone or in combination, which may help in increasing crop productivity by way of fixing atmospheric nitrogen in soils, solubilizing of insoluble plant nutrients, stimulating plant growth or decomposing of plant residues. They have been in use since long for improving the yield of various crops as well as their quality (Mahajan et al., 2002, Mahajan et al., 2003a and b; Mahajan and Sharma, 2005; Mahajan et al., 2008a and b; Bhadauria and Singh, 2013). The comparision between chemical fertilizers and bio-fertilizers stated by Singh and Saxena (2004) are as under:

Table 8.1: Comparison between chemical fertilizers and bio-fertilizers (Adapted from Singh and Shulcla [2004])

Sr.No.	Factors	Chemical fertilizers	Bio-fertilizers
1.	Production	Industrial, centralized	Biological, small scale or decentralized
2.	Raw materials and energy budget	Fossil fuel and other imported based on non-renewable energy sources, energy intensive 40-45% for nitrogenous fertilizer for upland crops and	Atmospheric nitrogen for nitrogen fixers, unavailable P for PSM* and VAM* organic
3.	Efficiency	less than 33% for HCL, 25-33% for phosphatic fertilizers. Loss of nitrogen is more by different processes.	Efficiency is more; about 90% losses by different process are also low.
4.	Pollution effect	Generally exists due to indiscriminate use	Pollution free
5.	Accessibility	Affordable section of farmers	Small and marginal farmers
6.	Soil health	Indiscriminate use deteriorates the soil health	Improves the soil health

Phosphate Solubilizing Micro-organisms - PSM; Vasicular Arvasicular Mycorrhizae - VAM.

The term microbial inoculant or microbial culture is an appropriate term for them and is used after the name of micro-organisms they contain. *Rhizobium* inoculant or *Rhizobium* culture, *Azotobacterial* inoculant or *Azotobacterial* culture or Azotobacterin, *Azospirillum* inoculant or *Azospirillum* culture, are some of the typical examples of bio-fertilizer. BGA culture, phosphobacterin culture and *Azolla* inoculum are the other examples of bio-fertilizers.

> *"Bio-fertilizers are, infact, the carrier based products in which viable or living cells of efficient microorganisms - especially of bacteria and some fungi or algae - are mixed in maximum number to increase soil fertility or availability of nutrients to plants through seed treatments or soil applications."*

Bio-fertilizers (BF) can be classified into three main categories viz., (i) Nitrogen Fixing Bio-fertilizer, (ii) Phosphate Solubilizing Bio-fertilizer, and (iii) Cellulolytic or Organic Matter Decomposers. Nitrogen Fixing Bio-fertilizer (NBF) has further been divided into groups i.e., (i) NBF for legumes - *Rhizobium* and (ii) NBF for non-legumes - *Azotobacters, Azospirillum,* BGA and *Azolla.*

Similarly, Phosphate Mobilising Bio-fertilizer has been further grouped into two types viz., (i) Phosphate Solubilizers - Various species of Bacteria and Fungi and (ii) Phosphate Absorbers - Mycorrhizal Fungi. Cellulolytic or Organic Matter Decomposers too have been classified into two groups. These are (i) Cellulolytic Decomposers i.e., cellulose decomposer - *Cellulomonas, Cytophaga, Trichoderma* spp. etc. and (ii) Lignolytic organisms i.e., lignin decomposer - *Orthorobacier.* Briefly, the classification of the various bio- fertilizers has been diagramatically shown in Fig. 8.1.

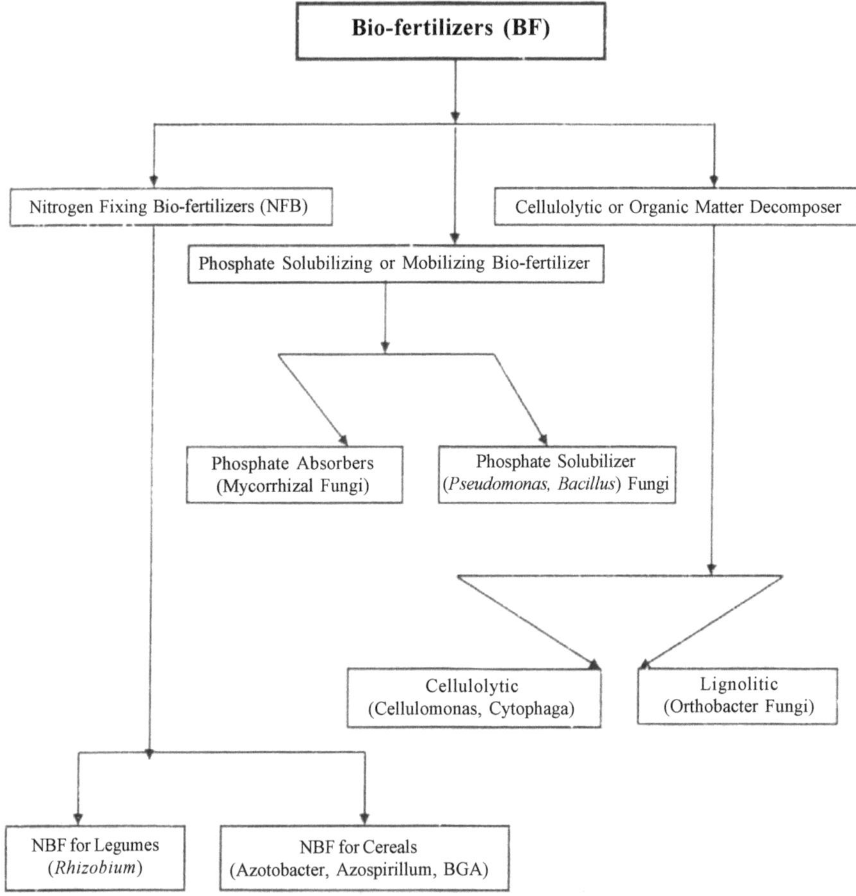

Fig. 8.1: Flow chart showing the classification of bio-fertilizers

The sky rocketing prices of chemical fertilizers in the country made it very difficult for the farmers' in the country to buy and use recommended doses of fertilizers in their field crops. This stressed the need to exploit other less expensive plant nutrient sources for obtaining maximum yields of crops. Bio-fertilizers offer a great opportunity to supplement chemical fertilizers in crop production.

Their application in crop production can help in enconomizing on the high investment requirement for chemical fertilizers (Sahai et al., 2014).

General doses of bio-fertilizers for their use in various crops viz., legumes, non-legumes and vegetables have been described briefly in Table 8.2. Details of which stand mentioned separately in each of the described bio-fertilizer in this chapter as subchapters.

Table 8.2: Doses of bio-fertilizer for various crops (Adapted from Singh et al. [2006])

Crops	Name of bio-fertilizers	Method of application	Quantity of inoculum required
A. Legumes			
Big sized seeds: Beans including soybean, chickpea, pea, groundnut	*Rhizobium* spp.	Seed treatment or soil application (as broadcast)	400 to 500 g of inoculant for 10 to 12 kg of seeds is required.
Small sized seeds: Black gram, green gram, cowpea, pigeonpea, lentil, berseem	*Rhizobium* spp.	Seed treatment or soil application (as broadcast)	200 g of inoculant for 10 kg of seed is required.
B. Non-Legumes			
Medium sized seeds: Wheat, maize, cotton	Azotobacter bacterial culture.	Seed treatment	500 g of inoculum is sufficient for hectare
Small sized seeds	Azotobacter bacterial culture	Seed treatment	250 g of inoculant should be used for hectare
Vegetables: Tomato, chilli, turnip, okra	Azotobacter, Azospirillum cultures.	Seed treatment or nursery application	2500 g of inoculant is enough for hectare
Potato	Azotobacter + Phosphobacterin cultures	Seed treatment or soil application	400 - 600 g of inoculant is enough for hectare

8.1 Role of *Rhizobium* Inoculants in Leguminous Crops

Rhizobium (singular) or *Rhizobia* (plural) - These Soil Bacteria - have the capability to fix atmospheric nitrogen (N°N) in symbiotic association with host legumes. These bacteria enter into the plants through root hairs, multiply there, form nodules and fix biological nitrogen in nodules. Leguminous plants furnish the energy to the *Rhizobia* and the nitrogen fixed in the nodules, is utilized by the leguminous plants for building- up their proteins (Plate 8.1). Some of the nitrogen fixed in the nodules of the plants comes into soil via diffusion process. Amount of nitrogen fixed, however, depends upon many factors. These factors may consist of kind of legume, type of nodules, soil conditions, presence of necessary nutrients exclusive of nitrogen etc.

It is worthwhile to state that the leguminous species are specific for a certain kind of specific nodule bacteria *(Rhizobia)*. However, there are some groups of legumes which can be nodulated by different kind of *Rhizobium*. Each group of these legumes and their specific *Rhizobia* constitute a cross inoculation group which has been mentioned in Table 8.3.

Table 8.3: Cross inoculation groups and Rhizobium legume association (Adapted from Alexander [1978])

Sr.No.	Cross inoculation group	*Rhizobium* species	Host Genera	Lequmes included
1.	Alfalfa group	*R. meliloti*	Medicago, Melilotus, Trigonella	Alfalfa, Sweet clover, Fenugreek
2.	Clover group	*R. trifolii*	*T rifolum*	Clovers
3.	Pea group	*R. leguminosarum*	Pisum, Vicia, Lathyrus, Lens	Pea, Vetch, Sweet pea, Lentil
4.	Bean group	*R. phaseoli*	Phaseolus	Beans
5.	Lupine group	*R. lupine*	Lupinus	Lupines
6.	Soybean group	*R. japonicum*	Glycina	Soybean
7.	Cowpea group	-	Vigna, Crotalaria, Archis	Cowpea, Crotalaria, Peanut

The fundamental purpose of inoculation is to add fresh inoculant of effective strains of legume bacteria to soil or seed to enhance the yields and productivity of the legumes. The need for inoculation has been felt, because all agricultural soils do not contain *Rhizobia* to promote successful growth in legumes after fixing atmospheric nitrogen.

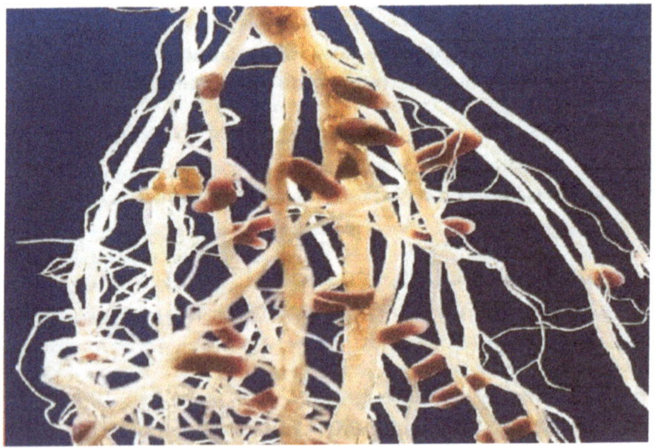

Plate 8.1: *Rhizobium*

It is now possible to produce satisfactory inoculants for legume groups. The inoculants are applied to the seed, roots of plants or to the soil. The inoculation of legumes has become an established practice in several countries like Russia,

USA, UK, Canada, Australia, New Zealand, France, Belgium, Sweden etc. The use of such culture has also become popular in India in recent years. In the present write-up, role of *Rhizobium* inoculant in leguminous crops has been described in terms of what is *Rhizobium*? How *Rhizobium* bacteria function? Kinds of nodules, contribution of legumes to soil fertility (nitrogen fixation) and response of *Rhizobium* inoculants has been elucidate.

8.1.1 What is Rhizobium?

Rhizobium is a genus of Gram negative bacteria, which are generally referred to as the root nodule bacteria. These bacteria live in the root cells of leguminous plants where their growth and metabolic activities cause a swelling to form nodules on the roots through infection thread.

8.1.2 How Rhizobium Bacteria Function?

These bacteria are capable of utilizing the free atmospheric nitrogen and synthesizing into nitrogenous compounds. The leguminous plants obtain their nitrogen from the synthesized nitrogenous compounds while the *Rhizobia* derive their energy from these plants. Prominent among the leguminous plants are such forage and green manuring plants and pulses as clover, alfa-alfa, sweet clover and vetch, *dhaincha,* sunhemp, pea, bean, lentil, gram, soybean etc. It has been already stated that one of the characteristics of these plants is their ability to live in mutual co-operation with *Rhizobia,* which form nodules on their roots Thus, the leguminous plants and the *Rhizobia* live together from the mutual benefits of each other. Such an association is known as "Symbiosis". As the *Rhizobia* fix atmospheric nitrogen in the root nodules of the leguminous plants, hence there is no need for supplying nitrogenous fertilizers to these plants, except as starter dose - which otherwise are essential for all other kinds of plants. The use of *Rhizobium* generally adds 30-40kg inorganic nitrogen ha^{-1} or evev more besides increasing crop yield by 20-30 per cent (Bhadauria and Singh, 2013)

8.1.3 Kinds of Nodules

Nodules are of two kind's viz., (i) Effective nodules; and (ii) Ineffective nodules. The fact, whether a *Rhizobium-legume* association is effective (capable of fixing nitrogen effectively) can be ascertained even in the fields by uprooting the legumes and examining the nodules by way of certain characteristics which are shown in Table 8.5.

Table 8.5: Characteristics of effective and ineffective nodules

Sr.No.	Character	Effective nodules	Ineffective nodules
1.	Colour of nodules	The nodules are pink or brown in colour with reddish pink inside.	The nodules are whitish or greenish in colour.
2.	What imparts brown or pink colour?	It is due to the presence of Leghaemoglobin.	These nodules are lacking of Leghaemoglobin.
3.	Characteristics of Leghaemoglobin	Its properties are just similar to haemoglobin, which is found in human blood.	Problems like sodicity, acidity, low amount of P, Ca, Fe, Mn and Mo are responsible in lacking of haemoglobin.
4.	Fixation of atmospheric nitrogen	They are capable of fixing atmospheric nitrogen.	They are not capable of fixing atmospheric nitrogen.

8.1.4 Contribution of Legumes to Soil Fertility - Biological Nitrogen Fixation

The important contribution of leguminous plants in agronomic practices is directly associated with their ability to fix the atmospheric nitrogen. The data given in Table 8.5, indicate that cultivation of peas in grasslands fixed the highest nitrogen varying from 400-500 kg ha^{-1} year^{-1}, whereas among forest tree species *Leucaena leucocephala* had the nitrogen fixing capacity in the rage of 410-550 kg ha^{-1} year^{-1} (Sharma and Gupta, 2003). Under Krishi Vigyan Kendra Farm of Jammu District at Ranbir Singh Pura, under SKUAST, Jammu, change of cropping pattern from rice-wheat to rice-berseem during 1998-99 not only brought a revenue of Rs. 64,000/- (Rupees sixty-four thousand only) through sale of berseem as fodder but also fixed atmospheric nitrogen in the range of 110-135 kg ha^{-1} year^{-1} (Gupta et al., 2005). According to Bhadauria and Singh (2013), *Rhizobia* can add 20-200 kg ha^{-1} year^{-1}. Apart from fixation of atmospheric nitrogen, cultivation of berseem increased the status of calcium, magnesium and potassium in soil considerably.

Table 8.5: Nitrogen fixation by various legumes

Legumes	Range of N fixation (kg ha^{-1} year^{-1})*	Legumes	Range of N fixation (kg ha^{-1} year^{-1})*	Legumes	Range of N fixation (kg ha^{-1} year^{-1})*
Soybean	40-256	Berseem	59-240	Peas	51-80
Clovers	95-212	Cowpea	62-147	Vetch	100-195
Alfa-alfa	137-375	Lucerne in grass	10-50	Peas in grass	400-500
Acacia spp.	270-300	*Leucaena* spp.	410-550	Cassia	40-188
Lentil	90-1900	Gram	85-110	Red Gram	168-200
Groundnut	50-60	Green/Black Gram	50-55		

'Range values were worked out by the author after consulting a number of books, journals of research and extension.

Thus, the cultivation of legumes assists in the maintenance of soil fertility by way of fixing atmospheric nitrogen and improving calcium, magnesium and potassium status vis-a-vis organic carbon content.

8.1.5 Factors Affecting Nitrogen Fixation

Amount of nitrogen fixed by various legumes depends upon many factors:

- Kind of legume
- Effectiveness of legumes
- Soil conditions
- Presence of essential nutrients exclusive of nitrogen

In highly fertile soils and soils supplied with available nitrogen, little or no nitrogen fixation occurs. Most noticeable results from legume inoculation are obtained on soils having average fertility.

8.1.6 Response of Rhizobial Inoculants

The fundamental purpose of inoculation is to add a fresh culture of effective strains of legume bacteria to soil. Effective inoculation of legumes has been a major factor in improving their yield and productivity. The growth of legumes contributes to the maintenance of good quality of organic matter. Organic matter in turn improves physical property of soil, increases its moisture holding capacity and helps to hold plant nutrients.

As early as 1896, the *legume-Rhizobium* symbiosis was found to be the potential source for exploitation of Biological Nitrogen Fixation (BNF). This resulted in the development of legume inoculants and artificial inoculation of various legumes at the time of sowing in order to increase their nodulation by efficient strains and in turn the yields. Now there are many reports from India and abroad which have indicated that there is substantial increase in the yield of various legumes through the use of *Rhizobial* inoculants. The response of different legumes, grown in various parts of the country to *Rhizobial* inoculants, as reviewed by Gupta (2007), has shown promising results. However, the response and performance of a particular *Rhizobial* strain is different depending upon the soil and agro-climatic conditions of the region. In general, the increase in the yield of various legumes has been observed in ranging from 4 to 75 per cent. The effect of *Rhizobium* inoculants has been detailed in the following various sub-heads (Gupta, 2007).

8.1.6.1 Effect of Rhizobial Inoculations on Quality of Legumes

Inoculation not only increases the yield of leguminous plants but also improves their quality. Numbers of reports are available which have shown that legume plants inoculated with effective strains of Rhizobia increase their protein content. Infact, the protein content is one of the indices of quality for evaluating plant for the purpose of food or green manuring. Legumes have an advantage over non-legumes in building proteins which is due to their associations with the root nodule bacteria. The protein content in grains of cowpea, groundnut and green gram were 25.3, 31.2 and 23.7 per cent under control which increased to 28.3, 32.2 and 25.2 per cent due to inoculation and increased further with phosphate application. It was found that inoculation through phosphobacteria or phosphate application alone icnreased the protein content more than the control. Protein content of black gram and horse gram were 25.33 and 23.59 per cent under uninoculated and increased to 26.30 and 24.04 per cent due to inoculation with phosphobacteria. Under inoculation conditions, an increase in nitrogen percentage of various parts of legume plants has been observed over control. Increased amount of nitrogen in shoot and root of some legumes due to inoculation or phosphatization was also observed. A positive correlation between nitrogen content of inoculated legumes grown in pot cultures and total dry weight production was also obtained at Pune and other places.

Rhizobium inoculation in combination with phosphorus alone to groundnut increased the protein and oil content of the seed but reduced the total soluble carbohydrates, free fatty acids. Inoculation alone and with fertilizers had little effect on the proportion of aminoacids in the protein, but produced variable changes in their composition.

8.1.6.2 Effect on Nitrogen Uptake

Nitrogen uptake by the legumes due to inoculation is very much dependent upon the type of the Rhizobial strain used. For example, Israel strain-6040 has been found to be far superior to all other inoculants with regard to nitrogen uptake for Bengal gram (*Cicer arientinum*) variety BEG-482. It was further reported that the highest nitrogen uptake by the plant and grain was 164.9 and 142.3 per cent, respectively, over control.

8.1.6.3 Effect on Germination

Rhizobial inoculation has been found to enhance the germination of seeds and plant growth due to the production of growth hormones or vitamins. It also recorded better germination and growth in case of sunhemp under sterilized-inoculated soil than sterilized soil alone. This indicated the necessity of symbiotic association of Rhizobium spp. with the roots to enhance the growth, which is evidently lacking in the sterilized soil. A significant increase in the dry weight of the plants grown in sterilized sand inoculated with Rhizobium spp. was also noticed. The response of Bengal gram (cv. BEG- 482) to seven Rhizobial inoculants, namely USA, Australia, IARI, Israel strains-6039, 6040, 6041 and 6042, in Black and Chaika soils (Tamil Nadu) revealed that all the inoculants contributed to good vegetative growth. The seed inoculation of gram with different strains of Rhizobium culture increased the dry weight of the grains.

8.1.6.4 Effect on Nodulation

Inoculation of legumes has been found to increase nodulation in green gram, pea, gram and methi (Fenugreek), Bengal gram, black gram and horse gram and some other legumes. Number of nodules increased with inoculation upto pod formation and decreased later on due to death of bacteria and lysis of nodule. Nodulation in all the seven legume tree species was also found to improve (Table 8.6). The dry weight of nodules is known to be influenced by the type of the Rhizobial inoculant. For instance, the dry weight of nodules per plant ranged from 324 mg with IARI culture to 1012 mg with Israel culture-6040.

Table 8.6: Effective bio-fertilizer applications on the number of nodules in seven forest species

Treatments*	Albizzia *lebbeck*	Leucaena *leucocephala*	Prosopis *juliflora*	Pongamia *glabra*	Acacia *auriculiformis*	Acacia *nilotica*	Dalbergia sissoo
Control	1.7	0.3	2.7	6.7	1.7	6.7	1.0
Rl	12.0	92.0	39.7	33.3	10.7	27.3	31.3
Rl + Al	14.7	19.3	6.0	13.7	8.3	3.0	10.7
Rl + L	14.0	23.3	3.3	73.5	8.7	10.0	8.0
Rl + Al + L	20.3	13.0	1.7	34.7	4.0	40.3	17.7
CD (5%)	9.6	26.3	8.2	17.6	3.0	6.4	3.5

'Rl - Seeds inoculated with Rhizobium. A1 - Azotobacter inoculation. L - Pelleted with lime

The data in Table 8.6 indicated better nodulation in all the legume species with Rhizobium inoculation over control. Maximum number of nodules was found in case of Leucaena leucocephala. In none of the legumes combined inoculation with Rhizobium and Azotobacter could improve nodulation.

Inoculation of seeds of soybean (cv. Bragg) with 7 days old broth cultures of SB-16, HPAU-3 and USD-110 strains resulted in an increase in nodulation, bacteroid number and leghaemoglobin content. It was also found that inoculated soybean plants with different indigenous cultures of Rhizobium japonicum had better nodule formation, higher bacteroid and leghaemoglobin content in the nodules in contrast to check plants. More number of nodules, pods and dark green colour of foliage has been noticed in case of gram (cv. G-130 and *Hare Chole*) with Rhizobium culture application. However, Rhizobium strain Arhar-

U was found to have inconsistent effects on the amount of seed, N, P and protein content in case of cowpea. Inoculation of efficient strains (cowpea) to Baisakhi moong (green gram - *Phaseolus aureus*) increased number of nodules and grains, and nitrogen uptake.

Application of Rhizobium culture to forest legume species like Albizzia lebbeck, *Leucaena leucocephala*, *Prosopis juliflora*, *Pongamia glabra*, *Dalbergia sissoo* and *Acacia nilotica* has been found to increase the height and root growth of these tree species.

8.1.7 Rhizobium-Legume Symbiosis in Problematic Soils

Soil acidity or sodicity/alkalinity affects the nodulation phenomenon adversely. The nodulation is also very poor in soils having high soluble salt content. These problems can be overcome either by using such strains of *Rhizobium* that can tolerate high concentrations of salts, high acidity and alkalinity or by pelleting inoculated seeds with buffering agent such as gypsum and calcium carbonate to reduce the inimical effect of soil in critical stages of infection of roots by *Rhizobium* at the seeding stage. It was also advocated the use of pelleting of legume seeds with lime for *Rhizobium japonicum* (SB-16) in red clayey soil of Coimbatore which was otherwise not conducive for nodulation in general and soybean in particular inspite of inoculation with *Rhizobium.*

Work done at Pantnagar has shown that some strains of *Rhizobium japonicum,* specific for soybean, can tolerate sodium chloride up to 2.5 per cent. But the degree of salinity or alkalinity optimum for better nodulation may be different from the levels of *Rhizobium* tolerance and the legume to the respective salts. Further, it has been observed that by way of treating the soil with gypsum and farm yard manure would stimulate growth and multiplication of *Rhizobia.* The use of other neutralizing agents such as ground rock phosphate, bentonite clay, milk powder, humates, basic slag etc. to inoculated seeds as pelletants, has been proved beneficial in overcoming acidic problem and thereby, increasing nodule number and their dry weight. Pelleting with gypsum and charcoal in alkaline and neutral soils respectively has also proved useful.

Gum arabic, an adhesive used in pelleting seeds, has shown to increase the number of infected root hairs, leaves of cluster bean and clover and also to promote early nodulation. The number of dry weight of nodules, and the green and dry weight of soybean plants were markedly increased by pelleting the seed with charcoal, peat, humic acid, neutralised peat and untreated peat, each having an active culture of *Rhizobium japonicum.*

8.1.8 Use of Composite Cultures

It has been found that inoculants of multistrains of *Rhizobia* perform better with some legumes like soybean, french bean over the single strain inoculants. It is believed that composite culture of *Rhizobial* inoculants is more beneficial to different agro-climatic conditions as atleast one of the strain in the inoculant can cause effective nodule formation on various legumes. It has been reported that strains of *Rhizobium japonicum* like SB-1 and SB-16 were not effective at Amaravati and Kalyani, but the composite culture was significantly effective in increasing the yield of soybean at both the places. It was also found that the composite culture of *Rhizobium* inoculant consisting of the strains from black gram, green gram and groundnut had better performance over the single strains in green gram recording maximum grain yield. Maximum dry matter production was also recorded in the treatment with composite culture.

8.1.9 Comparative Performance of Different Carriers

The survival of *Rhizobia* in peat is quite.high as compared to soil due to high organic matter content, better aeration and greater water holding capacity. The bean yield was found to be more significant in case of farm yard manure + charcoal over that of all other treatments including *nitragin.* There was no marked difference between *nitragin* and Indian peat + charcoal. *Rhizobium* species can multiply in lignite base also. The highest bacterial count obtained was 13.14 x $10^8 g^{-1}$ of the medium. The usefulness of cellulose as the base for *Rhizobia* has been indicated. Rice husk as the medium for legume inoculants has too been used. The highest counts upto 2140 million bacteria g^{-1} of medium was reached after incubation period of 4 weeks. Peat obtained from Ootacamund (Nilgiris) has been found to be an efficient carrier for *Rhizobium.*

8.1.10 Methods of Inoculations Application

Several procedures have been developed to ensure beneficial nodular associations. One of the most familiar and primitive method is the application of soil from a field previously cropped with legume. This technique, however, is not good with the reason that soil weeds, pathogens as well as other microbes are also added to the soil which as a result causes inefficiency to the nodule bacteria. On the other hand, seed inoculation of specific *Rhizobium* spp. in legumes under different Indian soil conditions, has been proved very useful. On farmer's fields in different parts of South India, seed inoculation with *Rhizobium* culture has proved essential in increasing the yield of groundnut and Bengal gram. There were, however, variations in yields from place to place due to several factors like type of soil, climate, cultural practices etc. Recently, the soil base carriers have replaced the early inoculants.

8.1.10.1 Legumes in Which the Rhizobium is Recommended

Rhizobium culture is recommended for chickpea, pea, lentil, black gram, green gram, cowpea, red gram, soybean, groundnut, berseem and lucerne.

8.1.10.2 Use of Rhizobium Culture for Effective Results

Treatment of seeds with the slury of Rhizobium culture is the most effective method of its application for obtaining maximum response. This method consists of the following steps:

- Prepare the slurry of required quantity of culture or inoculant. Generally, 400-500 ml of water is required for 200 g of culture.
- Ten per cent jaggery or gur, i.e., 40-50 g is added in the above said amount of water. The solution is then boiled and cooled.
- Pour the prepared slurry over the heap of seeds required to be treated. Mix the seeds homogeneously with hands. Thereafter, spread the treated seeds over clean floor, on plastic or on gunny bag for drying.
- Sow the treated seeds immediately.
- Never dry the seeds in the sun as some of the rays of the sun may kill the Rhizobia.

Usually, 10 kg of normal size seeds such as green gram (moong), black gram (urad), cowpea, lentil and berseem are treated with 200 g of Rhizobium culture by the above said method. However, the seeds of groundnut, chickpea, soybean and pea being bigger in size require 400 to 500 g of culture for 10 to 12 kg of seeds.

8.1.11 Conclusion

In this subchapter of bio-fertilizer for sustainable agriculture, role of *Rhizobium* inoculants in leguminous plants including certain forest tree species has been described in a number of subheads. These subheads consist of effects of *Rhizobial* inoculations on quality of legumes, nitrogen uptake, germination and nodulation as well as *Rhizobium* legume symbiosis in problematic soils, use of composite cultures, comparative performance of different carriers, methods of inoculation application and use of *Rhizobium* culture for obtaining effective results. Apart from this, what are *Rhizobium,* how *Rhizobium* bacteria function, kind of modules and contribution of legumes to soil fertility, are the other topics, which have been dealt in.

8.1.12 Future Research Strategies/Priorities

- Proper seed inoculation with efficient strains of *Rhizobium* should be used in various legumes like soybean, pigeon pea, chick pea, green gram, black gram, etc., to save considerable amount of nitrogenous fertilizers.
- Considering efficiency, proper development of *Rhizobial* cultures should be taken upon a large scale under expert supervision.
- As different strains of *Rhizobia* vary in their mode of nitrogen fixation so more research is needed in this line.
- As most of seed disinfectants are toxic to legume bacteria so legume seeds which are to be treated with disinfectant should not be inoculated in usual manner.
- To prepare viable and effective inoculants, one must have specialized training and experience. Selection should be made from most effective strains and a given number are used for production of different groups.
- Create awareness with regard to use of *Rhizobium* culture in leguminous crops, which can be created either through campaign or drive.
- To get more response of the added *Rhizobium* culture, one must add phosphatic fertilizer, especially in acidic soils which are deficient in available phosphorus. Results after conducting a number of studies have shown that inoculation and phosphate application alone increase the protein content of pulse grains significantly over control, but their combination on increased the content further.
- Advocate the use of *Rhizobium* culture in berseem to increase its production. For this purpose, *Rhizobium* culture should be prepared by the Scientists of State Agricultural Universities to provide it to the peasants.

8.2 Usefulness *of Azospirillum* for Crops

Beijerinck was the first scientist who isolated a nitrogen fixing bacterium under the name of *Spirillum.* Later on, Dobereiner et al. (1976) and Dobereiner and Day (1976) found that *Spirillum lipoferum* was very common as root and soil inhabiting nitrogen fixing bacterium in tropics. The nomenclature of this organism was'subsequently revised by Tarrand et al. (1978) as quoted by Boddey and Dobereiner (1982), and designated as *Azospirillum.* Two species of this bacterium were then proposed viz., *Azospirillum lipoferum* (Beijerinck) Comb. nov. and *Azospirillum brasilense* spp. nov. as reported by Gupta et al. (1983).

The most obvious difference between the two species are : (i) the inability of *A. brasilense* to grow on glucose and a ketoglutaric acid as carbon source, and

(ii) the giant pleomorphic cells common in old, especially alkaline cultures of *Azospirillum lipoferum.* Inoculants of both of these species have been studied in various crops under different soil situations (Lakshmi Kumari et al., 1976; Lakshmi et al., 1977; Venkataraman, 1982; Boddey and Dobereiner, 1982; HaZra, 1994; Sharma et al., 1998; Kumar and Chandra, 2006) which are detailed hereunder along with amount of biological nitrogen fixation by the *Azospirillum.* *Azospirillum* bio-fertilizers are primary inhabitants of soil, the rhizosphere and intercellular paces of root cortex of graminaceous plants. They perform the associative symbiotic relation with the graminaceous plants (Bhadauria and Singh, 2013). They are gram negative, vibrio or spirillum having abundant accumulation of polybetahydroxbutyrate (70%) in cytoplasm.

8.2.1 Biological Nitrogen Fixation by Azospirillum

Azospirillum species are group of bacteria which are found in loose association with the root system of many crop plants especially cereals and millets (Plate 8.2). They fix atmosphric nitrogen ranging from 10-25 kg ha^{-1} or 20-30 kg ha^{-1} by colonizing the root system of many vegetable plants saving 25-30 per cent nitrogen (Bisen et al., 2005). Apart from helping to save 20 to 30 kg of nitrogen per hectare and better vegetative growth, *Azospirillum* produces indole acetic acid, gibberellins and cytokinins when inoculated on crop plants which in turn result in increased root development.

The inoculation of *Azospirillum* helps in sustaining soil fertility and thereby, crop productivity of semi-tropical and tropical regions. Soils having very high percentage of organic matter favour the growth and efficiency of *Azospirillum.*

8.2.2 Effect of Azospirillum Inoculants to Crops

A number of reports with regard to the effect of *Azospirillum* inoculation on various Graminae are available Boddey and Dobereiner (1982) found extremely variable effects (Table 8.7) ranging from non-significant, positive or zero on yield parameters of various crops, to highly significant increase in dry matter production and total nitrogen content. Wani (1990) reported 4.4 to 13.8 per cent increase in wheat yield with *Azospirillum* inoculant.

Table 8.7: Crop response to inoculation with *Azospirillum* species (Adapted from Boddey end Dobereiner [1982])

Crop	Experiment	Inoculum	Parameter	% difference as compared with control
Maize	Greenhouse Oregon (USA)	Azospirillum brasilense Sp. 81 +	Plant dry weight + total nitrogen	NS
Grass Cynoden dactylon 8 cultivar	Greenhouse Floriota (USA)	Azospirillum brasilense Sp. 7+, Sp. 13t + JM $125A_2$ mixed with Azotobacter paspali	Cumulative shoot dry matter (3 harvests)	+ 17
Maize (cv. Jubilee)	Field Israel (Summer)	Azospirillum brasilense ^{2}Cd + CD-I	Grain (cobs) yield total frest weight	+8 to + r14 NS
Rice, wheat & sorghum	Field India Several trials	Various strains	Yield and dry matter	Variable

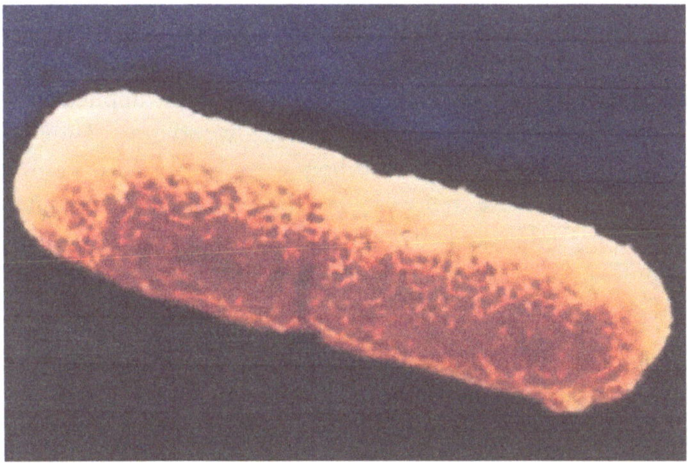

Plate 8.2: *Azotobacter*

While giving a list of benefitted and recommended crops for use of bio-fertilizer, Singh and Shukla (2004) reported that maize, paddy and ragi responded more to inoculum of *Azospirillum.* The inoculation of *Azospirillum* was found to increae rice grain yield ranging from 10 to 15 per cent depending upon the management practices and soil/environmental conditions (Kumar and Chandra, 2006). A field experiment was conducted on a clay soil at Annamalai University experimental farm to study the effect of application of organic manures, inorganic fertilizer, lignite flyash (LFA) and bio-fertilizer *(Azospirillum* and phosphobacteria) to cotton seedlings (Raghupathy et al., 1998). The results indicated that application of biodigested dung slury + diammonium phosphate + $ZnS0_4$ + *Azospirillum* + *phosphobacteria* + LFA (Treatment II) registered maximum number of bolls per plant, boll weight and seed cotton followed by combination of all except for *phosphobacteria*. This implies the importance of an integrated plant nutrient approach. The above combination of organic manures, inorganic fertilizers, LFA and bio-fertilizer enhanced the seed cotton yield by 4.64 q ha'[1] (150% increases) over control. The uptake of N by cotton plant at maturity ranged from 25.88 to 81.58 kg ha'[1]. Maximum uptake was recorded in Treatment II involving all the combination followed by the combination of all but the phosphobacteria (Raghupathy et al., 1998).

According to Hazra (1994), application of *Azospirillum* inoculant has not only helped in increasing the yield of annual crops like sorghum, maize, pearl millet and oats to the extent of 10, 10, 8 and 9 per cent, respectively as compared to control but has also enhanced the yield of perennial crops like Nepiar-pearl millet hybrid, Guinea grass, Timothi and Signal grass. Increase in the yield of these crops was to the extent of 12, 14, 17 and 13 per cent, respectively as compared to control (Hazra, 1994). According to Singh and Saxena (2005), the crops like jowar, pearl millet and lesser millets as well as barley and oat, and forage grasses responded favourably to *Azospirillum* inoculant. The yield of grain and fodder crop was equal to that attainable to the application of 20 kg N ha^{-1}. Grasses of pastures like Buffel grass, Birldwood grass, Blou buffel grass, have also showed their increase when were applied with inoculants of *Azospirillum.* The increase in the yield of these grasses was found to lie in the range of 6 to 7 per cent in comparison to control. *Azospirillum* treated sugarcane plot with application of recommended dose (5 kg ha^{-1}) over control increased the yield of sugarcane by 15 to 20 percent (Ghosh et al., 1999).

Azospirillum inoculation has been found beneficial in a wide range of vegetable crops. However, the response varied with the crop, season, location and level of native organisms (Narayan and Bhat, 2006). The use of *Azospirillum* + *Azotobacter* was found useful in increasing the yield of potato in acidic soils of Shillong. The results showed that the *Azospirillum* was slightly more effective than *Azotobacter.* Highest head weight per plant as well as per hectare yield (526 q ha^{-1}) were obtained with 60 kg N ha^{-1} + *Azospirillum* application (Sharma et al., 1998).

An experiment on the impact of inorganics and bio-fertilizer on the yield and processing quality parameters of onion cultivar, Bellary Red, was carried out during seasons of 1999-2000 and 2000-01 under irrigations in Central Zone of Karnataka. The results revealed that the plants provided with *Azospirillum* + 100 per cent NPK produced maximum bulb weight, bulb yield and dry matter yield of the bulbs (Mahanthesh et al., 2007).

The study conducted with regard to the effect of inorganic and bio-fertilizer (*Azospirillum* and *Azotobacter)* on the yield of squash (*Cucurbita* pepo) during 2003 and 2004 under Palampur, Himachal Pradesh conditions. The results indicated that treatment combinations - 80 per cent N + *Azospirillum* (soil application) in the first year and 80 per cent N + *Azotobacter* in the second year were associated with higher yields which were at par with or more than control (Sood et al., 2006). The use of *Azospirillum* culture could enhance yield by 14 to 20 per cent over control (with no inoculation of *Azospirillum)* in most of the vegetables because of their ability to fix atmospheric nitrogen to the extent of 20 to 25 kg ha^{-1} $season^{-1}$ (Tyagi et al., 2008).

8.2.3 Doses and Methods of Application

The doses and method of application are the same as described in *Azotobacter* inoculant. However, application of *Azospirillum* in rice can be done at the time of seed sowing to raise the nursery. For this purpose, the required quantity of *Azospirillum* inoculant (20 g kg^{-1} seed ha^{-1}) is applied on to the seed. The inoculant is prepared by mixing about 200 g inoculant in 300-400 ml fresh water or 10 per cent gur solution. The cooled inoculant slurry is then mixed thoroughly with seed in order to place a uniform: coating over each seed. Usually a packet of 200 g *Azospirillum* inoculant is sufficient to treat 10 kg seed. The treated seed should either be sown immediately or could be dried in shade about 10 to 15 minutes. Treated seed should not be stored.

Azospiriiium inoculant can also be applied at the time of transplanting. For this purpose, suspension of *Azospirillum* is prepared by mixing 1 kg inoculant in 15 to 20 litres water and roots of the seedlings are dipped in biofertilizer's suspension for 10 to 15 minutes before transplanting.

8.2.4 Requirements for Nitrogen Fixation

- *Azospirillum* bacteria require large amount of organic carbon such as those in crop residues with high C: N ratio.
- Bio-fertilizer pack should be crop specific and must contain the specified population count at the time of packing and expiry.
- The biofertilizer packet should be handled carefully and must be used within the specified period as the bacteria are living in nature.
- The inoculum should not be exposed to very high temperature.
- Soil moisture at time of sowing the crop should be at field capacity so that added, inoculum may get suitable condition for its growth and development (Sarangi, 2002).

- In case the seeds have been treated with seed treating chemicals, the treatment of cultures with the seed then should be done after about a week.

8.2.5 Conclusion

From the extensive field trials by various scientists under different agroclimatic conditions in the country and abroad, it is emerged as:

- Use of *Azospirillum brasilense* as inoculation, indicated a positive response to many crops, both C4 and C3 plants. C4 plants consist of bajra, sorghum, maize and C3 those of cereals - rice wheat.
- The percent increase in the crop yield was 3.1 to 22.4 for wheat, 18.4 to 26.6 for barley, 15.2 to 63.6 for sorghum and 36.5 for fodder oats.
- The extent of saving in the chemical nitrogen fertilizer input was in the range of 20 to 40 kg N ha^{-1}
- A soil + powdered FYM (1:1) carrier kept *Azospirillum brasilence* viable for a period of 6 months.
- In the Indo-Gangetic alluvium of Varanasi, the use of *Azospirillum brasilense* gave better results than an application of 30 kg N ha^{-1} to paddy, sorghum and pearl millet in the field.
- Field inoculation to number of crops like *Panicum maximum* and maize was the most successful when the plants were receiving moderate levels of N fertilizer (20 to 60 kg N ha^{-1}).
- Results have led to suggest that in view of the possible production of "Plant Hormones by *Azospirillum* spp.", yield increase due to inoculation were caused by hormonal effects on plants.
- The yield increase due to hormonal effects was evidenced from the frequent observation that acetylene reduction activity failed to account for the increase in N content in the inoculated plants.
- Thus, it might be that hormonal effects, nitrogen fixation and effects on other nitrogen transformations were all responsible to varying degrees for the response of added *Azospirillum* inoculants to the crops.

8.2.6 Future Research Strategies/Priorities

- As there is no general conclusion regarding the cause of crop yield response to *Azospirillum* i.e., whether due to hormonal effects, nitrogen fixation and effects on other transformations, so more research in this line is required to be done.

- Since there is scant information about the ecological factors with regard to the population of *Azospirillum* species, so study concerning with ecology of these bacteria becomes an imperative for various parts of India and abroad.
- After studying the ecology of *Azospirillum* species, optimum soil and other environmental conditions will be known. And, thereafter, recommended dose of plants nutrients especially phosphorus can be applied.
- The management of soil fertility using bio-fertilizer is a vital component of sustainable agriculture. Thus, the potential of this biofertilizer alongwith others need to be exploited.
- As presence of air in good amount, optimum temperature 32 to 35°C, and high percentage of organic matter are very essential to sustain the growth of *Azospirillum,* so these must be maintained to take the full advantage of the applied *Azospirillum* inoculantto crops.

8.3 Role of *Azotobacter* Inoculants in Non-leguminous Crops

Azotobacters belong to asymbiotic nitrogen fixing bacteria as they fix atmospheric nitrogen independently or freely (Plate 8.3) and do not require any partner or host like *Rhizobia.* As such, the process by which they fix atmospheric nitrogen is called "Asmybiosis" or "Azofication". Use of their culture or inoculant has been found useful in increasing the yield of variety of crops like cereals, vegetables, fruit plants, plantation crops and forest tree species.

Azotobacter inoculant is also called *Azotobacterin* or *Azotobacterial* culture. It can be prepared from various species of *Azotobacter* which grow well under aerobic soil conditions with neutral to slightly alkaline soil reactions, exception being of *Azotobacter macrocytogenes* - requires acidic soil reaction. They are heterotrophic in their nutrition (Table 8.9).

Table 8.9: Free or independent living nitrogen fixing organisms (Adapted from Gupta et al. [1983])

	Aerobes		Anaerobes
Heterotrophs (Bacteria)	Photoautotrophs (Blue green algae)	Heterotrophs (Bacteria)	Photoautotrophs (Bacteria)
Azotobacter spp.	Nostoc	*Clostridium* spp.	Chromatium
Azotomonas spp.	Calothrix	*Achromobacter* spp.	Chlorobium
Pseudomonas spp.	Anabaena Tolypothrix etc.	*Bacillus* spp.	Rhodospirillium
Spirillum spp.			
Beijerincki spp.			

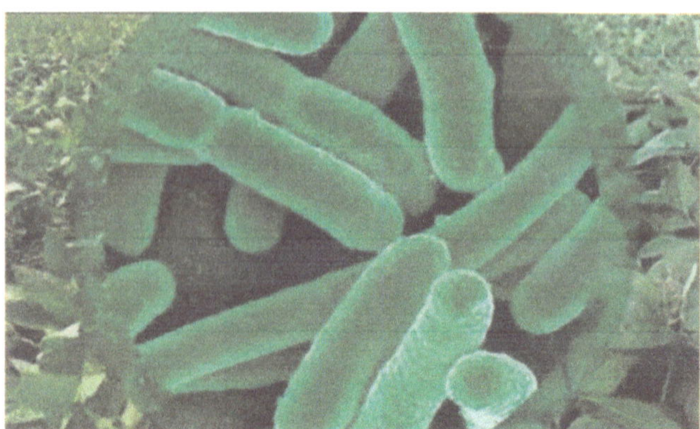

Plate 8.3: *Azospirillum*

Beijerinck (1901) was the first to isolate an aerobic micro-organism that bound molecular nitrogen and named it *Azotobacter chroococcum.* Later on, other species of *Azotobacter* like *Azotobacter agilis, Azotobacter beijerinckia, Azotobacter vinelandii, Azotobacter nigrican, Azotobacter insignis* and *Azotobacter macrocytogenes* were described.

Although a number of species of *Azotobacter* have been now isolated (as stated above) yet *Azotobacterial* culture is mostly prepared from *Azotobacter chroococcum* bacterial species after its isolation from soils. This bacterial species thrives best in well-drained soils having neutral or slightly alkaline pH and under rich organic matter content, and high available phosphorus, calcium and magnesium content.

Application of *Azotobacterial* culture ensures good seed germination, promotes root and plant growth and eventually assists in increasing the yield of various non-leguminous crops. The details of which have been mentioned in this sub-chapter of bio-fertilizers. Apart from this, nitrogen fixation by the *Azotobacter* has also been given vis-a-vis suitable conditions for using *Azotobacterial* culture and methods of its application.

8.3.1 Amount of Nitrogen Fixation by Azotobacter

Various species of *Azotobacter* have been found to fix atmospheric nitrogen ranging from 10-50 kg ha^{-1} $year^{-1}$ in comparison to symbiotic or leguminous bacteria where the amount ranges from 100-200 kg ha^{-1} $year^{-1}$ or even more. Although on an average the nitrogen fixation by *Azotobacter* is less yet they are known to produce a number of growth promoting substances like hormones, vitamins, cytokinins and many others (Singh and Purohit, 2008) which as a result enhance the germination of seeds, plant stand, vigour and yield especially of foliage crops. This has, therefore, resulted in the use of effective culture of these bacteria for various crops either through soil application or seed inoculation or other methods. *Azotobacter* when inoculated with *Rhizobium* fixed about 104 kg ha^{-1} $year^{-1}$ (Martin et al., 1995).

8.3.2 Production of Slime by Azotobacter

The several species of *Azotobacter* but *Azotobacter chroococcum* happens to be the dominant inhabitant in arable soils capable of fixing nitrogen (12-15mg N-fixed per gram of carbon source). The bacterium produces abundant slime gum which helps in soil aggregation (Bhadauria and Singh, 2013).

8.3.3 Use of Azotobacter Culture to Various Crops

Azotobacterial culture has been used as biofertilizer for several non-leguminous crops like wheat, maize, jowar, cotton and paddy sown under rainfed conditions. Its use has also shown good response to various vegetables and fruit trees both under rainfed and irrigated conditions. *Azotobacterial* inoculant has also proved useful in increasing the growth of forest tree species. Treated with *Azotobacterial* culture, wheat, maize, jowar and rainfed grown paddy have given 10 to 20 per cent more yield than control plots under various soil conditions of Madhya Pradesh (India). Besides increasing the yield, use of such a culture has brought about an early maturity in some crops too. Treatment of seed with *Azotobacter* culture has shown about 16 per cent increase in grain yield of sorghum in Udaipur, Rajasthan (India) over untreated but the differences were significant. However, the seed treatment did significantly increase in the straw yield over- untreated and the increase was to the tune of 11 per cent.

Seed treatment of maize with *Azotobacter* culture increased the plant height, dry matter production, cob length, cob girth, grain yield and stover yield as compared to control but the data were not significant under Palampur soil situations, District Kangra, Himachal Pradesh (India) as studied by Gandotra et al. (1989) and Gandotra et al. (1998).

Recent studies (Gupta et al., 2007) have indicated that seed treatment with *Azotobacter* culture along with FYM significantly increased the maize grain yield than either obtained with *Azotobacter* culture alone or FYM alone. *Azotobacter* culture alone enhanced the maize grain yield in the order of 21 to 53 per cent over the check (Table 8.10). On the other hand, with an application of FYM in conjunction with *Azotobacter* culture, the increase in maize yield was recorded to the tune of 93 to 123 per cent over control. Some Russian scientists like Mishustin and Shilnikova (1971) have reported that *Azotobacterin* when applied with well decomposed organic manure gives more response because FYM acts as mobile source of energy for the multiplication of *Azotobacter.* This might be the reason for enhancing the maize yield in both of the *Azotobacter* cultures when applied with FYM. However, the performance of *Azotobacter* culture II with application of 20 t ha^{-1} was found to be significantly superior to its application with 10 t FYM ha^{-1} (Table 8.10).

Table 8.10 Effect of FYM levels and *Azotobacter* cultures on the grain yield of maize (Adapted from Gupta et al. [2007])

Sr.No.	Treatments	Grain yield (kq ha^{-1})	% increase in qrain yield over control
1.	Absolute control	1530	-
2.	FYM @ 10 t ha^{-1}	2025	32
3.	FYM @ 20 t ha^{-1}	2428	59
4.	FYM @ 10 t ha^{-1} + Culture I	3248	112
5.	FYM @ 20 t ha^{-1} + Culture I	3267	113
6.	Culture I	2319	51
7.	FYM @ 10 t ha^{-1} + Culture II	2959	93
8.	FYM @ 101 ha^{-1} + Culture II	3415	123
9.	Culture II	1948	27
10.	Recommended NPK	3717	143
	LSD	422	-

% Treatment = [(Treatment - Absolute Control)/ Absolute Control] ^{x} 100

It is point to mention that in Russia with the use of *Azotobacterin* culture, 8 to 10 per cent increase in yield of non-leguminous field crops like wheat, cotton, maize and paddy has been seen during early 1960. The Australian experience was more or less similar in that the *Azotobacters* could colonise in nitrogen poor wheat growing fields.

Under Agronomy Research Project, several trials showed increase in yield of,wheat crop upto 24 per cent with the use of *Azotobacter* bio-fertilizer (Kumar, 2004). According to Hooda and Dahiya (1992), various strains of *Azotobacter* not only increase the grain yield of wheat crop but also the yield attributes like height of the plant, tillers, spike length etc. The inoculation of *Azotobacter* was found to increase the rice grain yield ranging from 10 to 15 per cent (Kumar and Chandra, 2006).

A field experiment of bio-fertilizer was conducted on Vertic Ustochrept of Bhubaneshwar to study the effect of integrated bio-fertilizer on yield, nutrient uptake and N-economy of rice-peanut cropping sequence. The study showed that 40 kg inorganic N integrated with bio-fertilizer (*Azotobacter, Azospirillum* and *Azolla)* and 17.5 kg of P and 32 kg of K ha^{-1} resulted in the highest grain (3.57 t ha^{-1}) and straw yield (4.32 t ha^{-1}) of rice (Pattanayak et al., 2007). On its residual fertility, peanut crop produced the highest pod yield of 2.38 t ha^{-1} with seed inoculate of *Rhizobium.*

The data with regard to crop responses to *Azotobacter* inoculant indicated that vegetables like onion, brinjal, tomato, cabbage showed an increase in their yield ranging from 2-50 per cent as compared to control. In case of wheat, sorghum, maize, cotton and rice, the increase in yield was in the range of 1 to 50 per cent than control, maximum being in maize crop (9 to 50%) followed by sorghum (8 to 38%) and wheat (10 to 30%).

The application of *Azotobacter* likewise other crops, when supported by judicious use of organic matter ensure good seed germination, promote root and plant growth, and increased in productivity in vegetables, such as tomato, potato, brinjal and cole crops (Singh et al., 2006).

8.3.4 Suitable Condition for Using Azotobacterial Culture

For the effective working or use of the *Azotobacter* culture, the following conditinos need to be fulfilled:

- Land should be well-drained to obtain sufficient aeration, otherwise *Azotobacter* bacteria would cease to work. It is because they are aerobic in nature. This is the reason, why the use of *Azotobacter* culture has not been found much useful in case of transplanted rice.
- Soil pH should be in the range of 6.6 to 7.5 or upto 8.0 in case of calcareousness of the soil. A higher or less pH than the aforesaid pH range may make the *Azotobacters* inactive, especially if the *Azotobacter* culture has been prepared from the *Azotobacter chroococcum* species.
- When the soil is acidic i.e., if its pH is less than 5.5, liming should be done. Similarly, in case of sodicity i.e., when the pH of the soil is always greater than 8.5 and exchangeable Na^+ percentage is greater than 15 per cent of the cation exchange capacity, then use of gypsum is a must to activate the *Azotobacter* bacteria.
- Sufficient moisture should be available in the soils. Normally all bacteria including *Azotobacters* function very well in soils having 40 to 60 per cent of the total moisture holding capacity (Alexander, 1978). However, *Clostridium* species being anaerobic in nature can function even in more moist soils.
- As the *Azotobacter* species are heterotrophic in nature so they require lot of organic matter for their growth and functioning in the soils. If the organic matter of any soil is less than 0.5 per cent then it should be amended by using organic manures - FYM, compost, vermicompost. Organic manure must be well decomposed and fresh one. It is because that *Azotobacterial* culture when applied with well decomposed and fresh organic manure, it provides more mobile energy source (Fulvic Acid) for the multiplication of *Azotobacter.*

- Phosphatic fertilizers should be added if the soils are low in available P.
- It is not only the phosphorus but other nutrients like K, Ca, Mg and S, are also very much required for having more response of added *Azotobacter* inoculant. For example, Singh and Pathak. (2003) found that application of 40 kg K2O + 30 kg Mg + 40 kg S + 5.25 kg Zn ha^{-1} along with *Azotobacter chroococcum* seed treatment produced highest grain and straw yield of wheat. Such treatment also showed more uptakes of nutrients and beneficial effects on quality characteristics like lysine, protein and starch contents. Similar results were also obtained by Sakai et al. (2000) and Singh (2001).

8.3.5 Methods of Application

There are number of methods of applying *Azotobacterial* inoculant to the non-leguminous crops, vegetables, fruit trees, plantation crops and forest tree species. These consist of: (i) Seed treatment; (ii) Setts/Tuber treatment; (iii) Seedlings; and (iv) Soil application/broadcasting.

8.3.5.1 Seed Treatment

The Azotobacter inoculant is coated to the seeds. 25 kg seed of medium sized such as wheat, cotton, etc., may be treated with 500 g Azotobacter inoculant, whereas 250 g ha^{-1} inoculant is enough for treatment of very small sized seeds like mustard.

Prior to treating the seeds, prepare a 10 per cent jaggery or gur solution by dissolving 50 g of gur in 500 ml water. The solution is then boiled for half an hour and cooled. Mix the solution thoroughly with the required amount of seed and add the culture gently, so that every seed gets a coating of the culture. Now sow the seed.

8.3.5.2 Setts/Tuber Treatment

This kind of application has proved beneficial for setts of sugarcane, tubers of potato, fruits and flower plants. A culture solution is prepared by mixing 2.5 kg of Azotobacteriat inoculant in 100-150 litres of water. The cut pieces of planting material required for a hectare are immersed in the suspension for 10-15 minutes. Bring out the treated tubers/setts and sow them without any delay.

8.3.5.3 Seedling Treatment

This method is useful where the seedlings are required to be transplanted. For this purpose, the required amount of inoculant is prepared in water as suspension in the ratio of 1:10 or 1:15. Dip the roots of seedlings in suspension and keep them immersed for 10-15 minutes. Treated seedlings should be then transplanted immediately.

Suspension of 2.5 kg of inoculant in 25.0 to 37.5 litres of water is enough for treatment of seedlings meant for one hectare.

8.3.5.4 Soil Treatment

Prepare the mixture of 5.0 to 7.5 kg of Azotobacterial inoculant in 100-150 kg of soil or FYM/compost. Broadcast the mixture in a hectare of land either at sowing time or 24 hours prior to sowing.

8.3.6 Azotobacter Inoculant in Composting

Composting is the time tested practice of encouraging decomposition of organic materials which is done for recycling of a variety of wastes in agriculture. Residues of various crops have great potential to contribute nearly 1.68, 2.02 and 5.04 million tonnes of N, P_2O_5 and K_2O, respectively (Gaur, 1992). However, owing to their high C:N ratio, especially of sugarcane trash, paddy straw, wheat straw, sawdust require special technology for decomposition and their application to soils for maintenance of organic matter to increase crop productivity. It has been found that the quality of compost can be ameliorated by application of 10-15 kg rock phosphate per tonne of raw material at the filling-up of the compost pit and subsequently with inoculation of *Azotobacter chroococcum* and phosphate stabilizers (Yadav et al.,1992).

Inoculation of *Azotobacter* increased the N-content of the compost. Combined inoculation of *Azotobacter* and *Bacillus megatherium* with rock phosphate yielded N-content of 1.41 per cent in compost. The municipal waste compost although showed more citrate soluble-P after composting with rock phosphate yet insignificant effect of *Azotobacter* species was recorded (Gowda et al., 1992). The effect of inoculation with *Azotobacter* and rock phosphate during composting not only increased total N but NH_4^+-cal and NO3'- forms of N also (Thakur and Sharma, 1998).

8.3.7 Precautions

- In case the seed has been treated with any fungicide or insecticide, then 2-3 packets of *Azotobacterial* culture must be used, while mixing the solution with seed.
- Always keep the culture or inoculant in cool shady place, as the temperature beyond 30°C can reduce the effectiveness of the *Azotobacter.*
- Mixing the seed with the inoculant and gur solution should never be done in the sun and dry winds.
- Always use culture before the expiry date.
- Never sow the *Azotobacterial* inoculant treated seed with fertilizer.

8.3.8 Conclusion

Various species of *Azotobacters* which fix atmospheric nitrogen in soils independently (Azofication process) are: *Azotobacter chroococum, A. agilis, A. vinelandii* etc. On an average they fix atmospheric nitrogen in the range of 10-50 kg ha^{-1} year^{-1}. Although their nitrogen fixing capacity is low yet they produce growth promoting substances which enhance seed germination growth and yield of crops. Use of *Azotobacter* culture alone or in combination with other bacterial fertilizers has been detailed along with there use in composting.

8.3.9 Future Research Strategies/Priorities

- As there is unavailability of appropriate and location specific strains of *Azotobacter* so research is required from this angle.
- Since application of local isolated *Azotobacter* culture has proved more useful under Himachal soil conditions (Gandotra et al., 1998; Gupta et al., 2007) than that of the alien, hence similar research is required to be conducted in soils of other States.
- Use of locally isolated *Azotobacter* culture along with an application of FYM has given good results as reported by (Gupta et al., 2007). Therefore, further research in this line becomes the need of the day in the light of the ill-effects/environmental pollutions caused by the chemical fertilizers.
- Exhaustive research is required to be carried out by the Agricultural Scientists, especially Soil Microbiologists for the availability of good carrier materials leading to long shelf life of the *Azotobacter* culture.
- Proper storage and distribution facilities at manufacturer, distributor and dealer levels must be provided.
- Storage of the *Azotobacter* culture should be maintained at 30° ± 2° Celsius throughout the year as under this range of temperature all species of *Azotobacter* thrive best.
- If the soils are strongly acidic then liming should be done and if the soils are highly alkaline due to salinity or sodicity gypsum should be added. Research in these lines requires to be carried out.

8.4 Use of *Azolla* for Wetland Rice Production

Azolla, a floating water fern possessing a symbiotic relationship with blue green algae (*Cyanobacteria*), is capable of fixing atmospheric nitrogen to the extent of 150 to 200 kg ha^{-1} year^{-1} or even more (Singh and Pandey, 2006). Moreover, *Azolla* has the capacity to grow profusely in shallow water system and produces large amount of biomass (Plate 8.4). Thus, owing to its fast growth rate, high nitrogen content and rapid mineralization rate to release the plant nutrients in soils, *Azolla* can be used as bio-fertilizer in rice crops in many countries of the world (Kannaiyan, 1998; Kumar and Chandra, 2006; Singh and Purohit, 2008). The use of *Azolla* as bio-fertilizer has been much realized these days to reduce the consumption of chemical nitrogenous fertilizers which is presently used in about 12 million tonnes. Nitrogenous fertilizers are also subjected to have loss through leaching and dentrification. There is also deterioration in the soil structure and environmental pollution problems. Hence, in this subchapter of the book "Use of *Azolla* for Wetland Rice Production" has been described in respect of species of *Azolla* and its phenology, amount of nitrogen fixation by *Azolla* and its response to rice crop vis-a-vis method of inoculum production.

Plate 8.4: *Azolla*

8.4.1 Species of Azolla

The genus *Azolla* is represented by six species which are mainly confined to the tropical and subtropical regions of the world. In India, the genus is represented by three species. These are: (i) *Azolla pinnata,* (ii) *Azolla filiculoides* and (iii) *Azolla imbricate.* The first two are widely distributed in India with more occurrence of *Azolla pinnata,* while the third one is mostly confined to the Himalayan region (Vaid, 2005). *Azolla nilotica, Azolla mexicana* and *Azolla microphylla* are the other species of *Azolla.*

They are small moss like floating plants or free floating water ferns, forming a dense covering on the surface of the permanent water tanks, pools, ponds, shallow ditches, channels and in the rice fields at several locations in the country. All species of *Azolla* remain in symbiotic association with the Cyanobacteria or blue green algae. *Anabaena azollae* is one of the typical examples of this kind of symbiont. *Anabaena azolla* is always found in the cavities on dorsal leaves and fixes atmospheric nitrogen efficiently which is available to *Azolla* plants. Thus, this fern algae system is unique since both photosynthesis and nitrogen fixation occur in the leaves and, therefore, fern is an attractive resource for photosynthetic production of nitrogen fertilizer.

8.4.2 Phenology of the Azolla

The stem of the fern or moss like floating plant is a horizontal rhizome. It is covered with two rows of leaves. True roots are presented in *Azolla.* The members are heterosporous and sporangia are borne within special fruiting bodies called "sporocarps".

Plant body is small which consists of horizontal rhizomes bearing leaves and roots. The rhizome is thin and delicate and floats on the water surface. It is pinnately branched; the branches are extra auxiliary On the lower side of the rhizome long adventitious roots are present.

The roots are usually close to the point of branching and are pendent in water. The rhizome and branches are covered by leaves (0.5 mm or less in size). They are borne in two alternative overlapping rows on the dorsal surface of the rhizome. Each leaf is divided into two almost equal lobes. The lower lobe (aerial) is thick, green and photosynthetic. It is somewhat obliquely placed and touches water only on the edge. Stomatas are present in both upper and lower epidermal layers. The cells of the upper epidermis give rise to a tuft of unicellular or bicelled hairs which trap considerable air. This prevents leaves from wetting.

The mesophyll is composed of columner palisade like photosynthetic cells with intercellular spaces in between them. The upper lobe has a large cavity at its base opening externally by a circular pore. The cavity contains a BGA (*Anabaena)* and spores of algae contained in next generation. It carries symbiosis relationship with *Azolla,* named as *Anabaena azollae.*

Azolla reproduces vegetatively and by spores. Vegetative propagation takes place by fragmentation. The lateral branches separate from the main axis due to formation of abscission layer at base. At maturity, lignification and deposition of tannin takes place. The sporophytic plant body of these ferns is soft and herbaceous, and the plant floats on the surface of permanent water bodies forming dense mats.

Mature plants of *Azolla* impact a characteristic reddish colour to the pond. Reddish colour is due to accumulation of anthocyanin group of pigments.

8.4.3 Survival of Azolla

As already stated, of the six species of *Azolla* distributed throughout the world, *Azolla pinnata* is commonly found in India (Singh, 1960). So, strains of *Azolla pinnata* from Thailand, Indonesia, Bangladesh, Vietnam, Nepal, Philippines, India and strains of *A. mexicana* and *A. filiculoides* from USA were examined for their growth under net house and field conditions (Singh et al., 1982). Only *A. pinnata* strains grew well. Among the various strains of *A. pinnata,* the strain from Vietnam grew better than others during summer. It remained green throughout the year. Thailand and Bangladesh strains grew well during winter but turned reddish as compared with other strains.

8.4.4 Biological Nitrogen Fixation by Azolla and Its Use as Bio-fertilizers

Algal bio-fertilizer constitutes a perpetual source of nutrients and it does not contaminate ground water. It contributes substantial amount of biologically fixed nitrogen to the rice crops. *Azolla* has been used as a biofertilizer for rice not oniy in India but also in other countries like China, Vietnam, Thailand, Philippines, Korea, Sri Lanka, Bangladesh, Pakistan, Nepal, Myanmar (also called as Burma), Indonesia, Brazil and West Africa (Kannaiyan, 1987). The *Azolla anabaena* association is a "living floating nitrogen manufacturing factory" which utilizes the energy from photosynthesis to fix atmospheric nitrogen amounting to 150 to 200 kg ha^{-1} year^{-1}. Meelu et al. (1994) reported that over the whole year, *Azolla* can fix more than 400 kg N ha^{-1}. The nitrogen is available only to the rice crop after *Azolla* has decomposed, so the ferns must be incorporated as green manure while the soil is wet.

Azolla is used as bio-fertilizer for rice crop and it could contribute 40-60 kg N ha^{-1} crop^{-1} (Kannaiyan, 1989'; Kannaiyan, 1998). The important factor is using *Azolla* as a biofertilizer for rice crop is its quick decomposition in soil and efficient availability of its nitrogen to rice plants. In tropical rice soils, the applied *Azolla* mineralizes rapidly and its nitrogen becomes available to the rice crop in very short period. In addition to contributing biological nitrogen and its rapid mineralization for the availability of rice crop, *Azolla* also adds organic matter to the soil, excretes growth promoting substances, solubilizes insoluble phosphates and amends the physical and chemical properties of the soils.

8.4.5 Nitrogen Contribution under Most Favourable Conditions

It was observed that under most favourable conditions upto 900 kg N ha^{-1} could be fixed (Singh, 1982) To get sufficient yield of *Azolla* in the paddy field, addition of phosphatic fertilizer is the most critical factor. Thailand strain performed the best by contributing 913 kg N ha^{-1} year^{-1} (Table 8.11) in Cuttack whereas Vietnam and Cuttack isolates fixed 885 and 755 kg Na ha^{-1} year^{-1}.

Table 8.11: Effect of different levels of phosphorus on the annual production of fresh and dry matter, N and P contents of three varieties of *A pinnata* (data based on continuous cultivation throughout the year) (Adapted from Singh [1982])

Azolla varieties	P levels (kg P_2O_5 ha^{-1})	Fresh wt. (t ha^{-1})	Dry wt. (t ha^{-1})	N content (kg ha^{-1})	P content (kg ha^{-1})
A. pinnata (India)	0	172.2	6.9	238.6	12.6
	4	272.5	15.0	529.4	36.9
	8	343.1	20.6	754.9	57.6
A. pinnata (Vietnam)	0	234.0	11.7	364.0	19.9
	4	345.1	18.9	725.6	49.7
	8	332.3	22.9	884.5	68.4
A. pinnata (Bangkok)	0	224.0	11.2	343.7	19.4
	4	342.9	18.9	729.7	49.3
	8	389.7	23.9	912.8	69.7

8.4.8 Application ofAzolla as Bio-Fertilizer for Rice Crops and N-Reduction Input

Use of *Azolla* as green manure could reduce nitrogen input by 25-30 per cent in lowland rice soils (Kannaiyan, 1998). Its incorporation has been found to increase the rice yield, ranging from 9 to 44 per cent. Further, it has been found that incorporation of 5 tonnes gypsum ha^{-1} along with 60 kg N ha^{-1} through urea and *Azolla* inoculation as dual crop recorded the highest grain yield of rice. Inoculation of *Azolla* coupled with urea super granule application, recorded significantly higher grain yield of rice than pure application. According to Singh and Pandey (2006), *Azolla* used for rice crop contributes 40-60 kg N ha^{-1} crop^{-1} and 10-12 per cent increase in yield. Effect of *Azolla* and BGA inoculation on grain and straw yields of rice studied by Kumar et al. (1995) indicated substantial increase in grain and straw yield (Table 8.12). Singh and Singh (1998) reported that incorporation of one crop of *Azolia* increased grain yield and N uptake as was obtained with 25-30 kg N ha^{-1} as urea, whereas 2 to 3 crops of *Azolla* increased the yield and N uptake of rice similar to 60 to 90 kg N ha^{-1} applied as urea. Use of *Azolla* + chemical N (45 kg N ha^{-1}) produced highest rice and wheat yield at Pantnagar, Uttarakhand (Mahapatra et al., 1987).

Table 8.12: Effect of *Azolla* and BGA inoculation on grain and straw yield (Adapted from Kumar et al. (1995])

Treatment	Grain yield 5^{-1} plants	Straw yield 5^{-1} plants
Unincculated	23.7	50.1
Azolla + 30 kg N ha^{-1}	28.4	57.3
Azolla + 60 kg N ha^{-1}	34.3	61.6
Azolla + 90 kg N ha^{-1}	39.7	68.8
BGA + 60 kg N ha^{-1}	30.0	66.8
BGA + 90 kg N ha^{-1}	35.5	64.6
Nitrogen 120 kg ha^{-1}	34.5	67.5
CD (0.5)	6.3	8.8

8.4.7 Content of N, P and K in Azolla

Green *Azolla filiculoides* has been reported to contain about 2 per cent nitrogen, 0.43 per cent phosphorus and 2.5 per cent potassium. Red *Azolla* contains one-third the phosphorus of that in green *Azolla* and potassium content was reduced by 20 per cent. The protein content in *Azolla* is as high as 13 per cent on dry weight basis.

8.4.8 Method of Inoculum Production

8.4.8.1 In Fields

Prepare ploughed, well leveled and bunded fields.(4 to 50 m^2). Keep 5 to 10 cm standing water in them and maintain this level of water till multiplication is continued. Inoculate fresh *Azolla* @ 0.05 to 0.4 kg m^2 depending upon the availability of inoculum. Higher inoculum density is always better for rapid spread.

Mix superphosphate (4 to 20 kg P_20_5 ha^{-1}) alongwith inoculum or broadcast after Azolla inoculation. Fresh cattle dung @ one tonne ha^{-1} can replace superphosphate. After formation of a layer in 1-3 weeks, harvest *Azolla* with the help of bamboo stick and again inoculate at the same rate for further multiplication (Singh et al., 1982).

Azolla multiplied throughout the year (31.6 to 24.6 day-night water temperature) @ 2 to 4 times $week^{-1}$. Its multiplication is affected badly when day-night water temperature exceeds 32 to 40° Celsius. Although it grows well in soils with slightly acidic (pH 6.0) to alkaline (pH 8.0) yet its optimum growth can be attained at neutral pH (6.5 to 7.5). However, very low soil pH (< 4.0) does not support its growth.

8.4.8.2 In Concrete Tanks/Trays/Earthen Pots

It is also multiplied in concrete tanks, trays, earthen pots on floated soil under field and in net house conditions. During off season, it can be maintained in the containers mentioned above besides in field, provided shade is maintained during hot summer.

8.4.8.3 Largescale Multiplication

In case the transplantation has to be done in the first week of June or July, its multiplication should be initiated in the middle of May. With the available inoculum of about 6 kg and the area to be applied in about 2 ha, its multiplication is to be started by applying 6 kg *Azolla* in an area of 4 m x 3 m after mixing with 36 g of superphosphate. After 7 days, weight of fresh Azolla will be 18 kg which can be again extended into an area of 5m × 6 m. After one week, the production of

Azolla would be 54 kg. This Azolla is further inoculated in an area of 10m ×10m and in a week the produce would be 160 kg. By extending its cultivation gradually for two more weeks as stated above, it is possible to obtain roughly about 1000 kg or Azolla which is sufficient to inoculate two hectare of paddy fields.

8.4.9 Methods of Utilization of Azolla in Paddy Fields

In paddy fields, *Azolla* can be utilized as: (i) green manuring and (iii) dual cropping (Singh and Singh, 2008).

8.4.9.1 Green Manuring

The importance of Azolla as an organic input in rice cultivation was first demonstrated in North Vietnam in the year 1957 as reported by Pareek and Maurya (1984-85). Grow Azolla before planting in ploughed, well leveled and bunded fields for which about 200 kg ha^{-1} fresh biomass of Azolla microphylla has been inoculated. It will multiply faster and cover the rice fields as a green mat in 2-3 weeks period with 15-25 tonnes biomass accumulation. Drain the field if possible and turn it into soil with plough. Then transplant rice seedlings possibly in a week.

8.4.9.2 Dual Cropping

Azolla can also be grown simultaneously with rice seedlings after transplanting. Generally, it is inoculated after one or two weeks of transplanting and the field is flooded with water. When Azolla forms a layer on water surface, water is drained off and fern is incorporated into the soil. For the second crop of Azolla. field is again flooded. Thus its spores germinate and form a mat on water surface and these again are incorporated into the soil. The Azolla layer requires repeated incorporation to avoid chocking of rice plants due to mat formation of Azolla resulting in oxygen starvation of rice crop.

8.4.10 Inoculation of Fresh Biomass of Azolla

Fresh biomass of *Azolla* is broadcast in the main field 7-10 days after transplanting rice. Inoculation of the fresh biomass of *Azolla microphylla* 200 kg ha^{-1} could multiply faster and cover the rice fields as green mat in 2-3 weeks period with 15-25 tonnes biomass accumulation. However, inoculation of even 50-100 kg biomass of *A. microphylla, A. pinnata* ha^{-1} will multiply and establish in rice fields but it takes 4-5 weeks for full coverage in the transplanted rice. Under optimum conditions with cloudy days and low temperature in day/night cycle of 25 ± 1° Celsius, the inoculated *Azolla* spread very fast in transplanted rice. Thus, the best suitable season for *Azolla* utilization with rice is wet season. It can also be used for late dry season crop. But it is very critical for its multiplication and biomass accumulation.

8.4.11 Constraints in Azolla Production

The main limitation or constraint in use of *Azolla* as bio-fertilizer is scarcity of water and high temperature in north India (Chander and Kumar, 2005). Higher temperature than 40°C is harmful for the fern. Maintenance of *Azolla* cultures in pond during winter is also a great problem. According to Singh and Singh (2008), the temperature higher than 35 ± 1° Celsius in a day time reduces the growth of *Azolla.*

8.4.12 Conclusion

In the subchapter - "Use of *Azolla* for Wetland Rice Production" - in terms of species of *Azolla,* its phenology; survival and nitrogen fixation and use as bio-fertilizer has been described. In addition to these aspects, nitrogen contribution of *Azolla* under the most favourable conditions, its application as biofertilizer for rice crop, content of N, P and in *Azolla* and methods of application have also been presented vis-a-vis suggestions and priorities for the future.

8.4.13 Future Research Strategies/Priorities

- As *Azolla* inoculation considerably increases the soil microbial and enzymic activity vis-a-vis soil nitrogen and organic matter contents, so the farmers must be advocated to use *Azolla* inoculum for increasing rice production.
- For the above said purpose, awareness among the farming community needs to be created.
- As at present, *Azolla anabaena* symbiosis is the only plant cyanobacterial symbiosis used for maintaining soil health and productivity so need now arises to investigate the use of other *Azolla* species in this regard.
- There are meager reports which indicate that an increase in nitrogen recovery when *Azolla* is incorporated into the soil rather than allowed to decompose in water or on the soil surface. This aspect requires to be studied on priority.
- In comparison to blue green algae alone, the *Azolla anabaena* association in general has been found to fix more nitrogen, and play important role in increasing rice production. Therefore, strains tolerant to high heat should also be developed for areas having high temperature.
- Looking into the usefulness of *Azolla* in rice cultivation and improvement in soil characteristics, more attention requires to be paid to make *Azolla* as a popular and successful bio-fertilizer.

8.5 Usefulness of Cyanobacterial Inoculation to Rice Crop

Blue green algae (BGA) or *Cyanobacteria* are ubiquitous. They are more widely distributed in the freshwater habitats including lakes, ponds and streams than in the marine environment (Plate 8.5). BGA are also present in the soil ecosystem. They are well known for their ability to tolerate a wide range of physical and chemical conditions of the environment. Whether it be an industry, food for human beings or feed for animals, or agriculture they play an important role in maintenance of an economy (*Nedumaran* et al., 2008).

The most important agricultural use of *Cyanobacteria* is us bio-fertilizer in paddy cultivation. The beneficial effect of *Cyanobacterial* inoculation on the grain yield of many rice varieties has been demonstrated in different parts of the world. It is because of their role in fixation of atmospheric nitrogen, addition of organic matter, acting as growth promoting substances and, thereby, improving the soil health. In this sub-chapter of the book, the usefulness of *Cyanobacterial* inoculation in terms of what are BGA or *Cyanobacterial*, their age and nature, distribution, nitrogen fixation in soils and use of BGA as bio-fertilizer, and its response to paddy under various situations.

8.5.1 What are BGA?

As stated elsewhere, most of the higher algae belonging to different classes are included in the phylum (or division) Thallophyta of the plant kingdom. The BGA, also callec Schizophyceae, Cyanophyceae or Myxophyceae (cyano - blue, phyceae = algae, myxo = slimy, due to capsule or sheath formation) have, however, been included in the "Protophyta" division along with bacteria and other primitive micro-organisms.

The BGA are the simplest forms of green plants. They resemble with bacteria in a number of characteristics. They are mainly unicellular and their cells ar usually coated by slimy capsules or sheaths

Platc 8.5: Blue green alge

They also consist of branched or unbranched. Morphologically, some of them are quite akin to the bacteria with regard to their shape of cells and the formation of chains of cells.

The BGA possess the green chlorophyll and a blue pigmnt called phycocyanin. Due to the presence of phycocyanin pigment, BGA are also known as "*Cyanobacteria*". Unlike other algae, their chlorophyll is not confined to chloroplasts but is diffused throughout the protoplasm. Like bacteria, there is no distinct nucleus present in the BGA, rather there is a diffused nucleus consisting of granules of nuclear matter diffused throughout the cell. The BGA like bacteria, multiply by binary fission and have no sexual cycle.

8.5.1.1 Age of BGA

Cyanobacteria are considered to be one of the oldest organisms on the Earth Planet. Their fossil records have shown that the BGA were evolved over three billion years ago. However, their significance in maintenance of soil fertility of rice fields was recognized and emphasized by P.K. De during late nineteen hundred forty.

8.5.1.2 Nature of BGA

Most of the nitrogen fixing BGA are free living i.e. do not require any partner for their functioning. However, many are symbiotic with plants. Many lichens are symbiotic associations of BGA and fungi and some genera like Nostoc punctiformis have been found in nodules of clover. Several bacteria like *Caulobacter spp*., *Bacillus megatherium* and *Agrobacterium radiobacter* living in the slimes'of blue green algae either double the vigour of nitrogen assimilation by Nostoc or stimulate nitrogen fixation.

8.5.2 Distribution of BGA

The population of Cyanobacteria in soils varies according to pH, organic carbon, available N and P, cation exchange capacity, exchangeable Ca and Mg as has been reported by various scientists from time to time. However, recently Das et al. (2007) from the correlation studies between Cyanobacteria and soil properties, it was observed that although population of Cyanobacteria was positively correlated with soil pH, organic carbon and negatively correlated with available N, CEC and Ca^{2+} and Mg^{2+}. But none of these relations was statistically significant. However, the population of Cyanobacteria showed a highly significant correlation with available P.

Under tropical soil conditions of rice fields of India, the occurrence and presence of a number of genera of BGA have been reported (Venkataraman, 1972; Kannaiyan, 1979; Kannaiyan, 1983; Kannaiyan, 1998). They consist mainly of

Nostoc, Anabaena, Calothrix, Aulosira, Tolypothrix and *Plectonema.* They infact are ubiquitous in all Indian rice growing soils. Contrary to these genera, *Seytonema, Cylindrospermum* and *Haplosiphon* have been found under localized soil conditions.

It is heartening to note that in the rice fields of India, the most active nitrogen fixer is *Aulosira fertilissima* while in Japan it is *Tolypothrix tenuis.* Other forms of BGA in Japanese rice soils belong to the genera *Nostoc* and *Anabaena* (Venkataraman, 1982). A number of N-fixing BGA have also been isolated from Africa and temperate soils of Russia.

Survey of four different habitats viz. paddy fields, pasture lands, ponds and ditches was performed (Shrivastava *et al.*, 2007) for two years to explore the diversity of promising species of Cyanobacteria from Kobra district of Chhattisgarh State. A total of 51 species belonging to 23 genera were identified. Maximum genera belonging to order Nostocales and minimum to Stigonematales were observed. Of the 51 species collected, 21 were N2 fixing and 30 were non-N_2 fixing form. The most abundant Cyanobacteria were *Nostoc ellipsosporum, Nostoc hate, Nostoc punctiforme, Anabaena oryzae, Calothrix juvenica, Anabaena anomla, Aulosira prolifica* and *Lyngbya alloregi.* Similar studies have also been conducted for the rice soils of Manipur (Devi'et al., 1999), Orissa (Nayak et at., 1996), Bilaspur Division of Chhattisgarh (Pandey et al., 2005) and various districts of Chhattisgarh (Shrivastava, 2000; Shrivastava et al., 2005).

8.5.3 Nitrogen Fixation by BGA and BGA as Bio-Fertilizer

8.5.3.1 Nitrogen Fixation

The ability of BGA to assimilate molecular nitrogen and its agronomic potential in rice fields in India was first recognized by De and Sarkar (1936, De and Bose 1938) and Singh (1939 and 1942) as quoted by Singh (1960). Now there is unequivocal evidence that the maintenance of soil fertility in rice soil is through nitrogen fixing BGA (Singh et al., 1982; Venkataraman, 1982, Mishra, 1993; Singh and Dixit, 2001). The amount of nitrogen fixation by different species of BGA under field conditions varies widely. In erstwhile Bihar soils, various species of BGA have been reported to fix 12 to 50 kg of nitrogen ha^{-1} year', whereas in Uttar Pradesh soils including Hill Terrain now Uttarakhand, the range of nitrogen fixation was in between 15 to 50 kg ha^{-1} $year^{-1}$ without application of phosphatic fertilizers. With application of phosphatic fertilizer, the amount of nitrogen fixation was in the range of 39 to 78 kg ha^{-1} $year^{-1}$ (Gupta et al., 1983).

8.5.3.2 Blue Green Algal Bio-fertilizer

Blue green algal inoculation or culture constitutes a perpetual source of nutrients. This culture does not contaminate ground water and depletes the resources. In addition to contributing 25-30 kg nitrogen ha^{-1} of biologically fixed nitrogen, they also add organic matter to the soil. Like Azotobacter bacteria, BGA excrete growth promoting substances, solubilize insoluble phosphates and amend the physical and chemical properties of soils (Goyal, 1982; Kannaiyan, 1998).

The blue green algal culture's composite inoculum consisting of Nostoc, Anabaena, Calothrix, Tolypothrix and Plectonema have been used for inoculation in rice which have shown better adaptability in wetland rice (Kannaiyan, 1985). Various studies conducted in the country have shown that blue green algal inoculant application @ 10 kg ha^{-1} after one week of transplantation contribute 25 to 35 kg N ha^{-1} $season^{-1}$ and grain yield increase in rice to the extent of 10 to 15 per cent (Chaubey et al., 2006). Field trials conducted in various parts of India indicated 22.2 per cent increase in rice yield due to BGA application. The benefits of BGA inoculation were also seen when applied with chemical fertilizers.

The beneficial effects of BGA on the yields of different rice varieties have been demonstrated in a number of countries and have been reviewed from time to time (Singh, 1960; Venkataraman, 1982; Kannaiyan, 1998). It is found that in erstwhile USSR, inoculation with *Amorphonostoc punctiformae* gave 13.2 per cent increase in rice crop yield, while in China with *Anabaena azotica* it was about 24 per cent. In Japan, the average rice yield's increase due to algal inoculation (*Tolypothrix tenuis*) were 2 per cent in 1st year, 8 per cent in 2nd year, 15.1 per cent in 3rd year and 19.5 per cent in the 4th year. In large number of Indian trials, complemented with recommended doses of nitrogen by BGA have been found to increase about 16 per cent higher yield of rice. On the whole, it is concluded that in many areas of India, 10 to 20 per cent increase in the grain yield of rice has been observed as a result of blue green algal application in the absence of any- added chemical nitrogenous fertilizers. There is enough evidence which evinces that blue green algal application improves the soil aggregation status, N content and organic matter content.

Recent studies have shown that use of BGA as bio-fertilizer can become an eco-friendly and economic alternative nitrogen source to increase rice productivity (Sinha et al., 2015). BGA a non cash input when applied alone contributed 20-30 kg N ha^{-1} $season^{-1}$. Besides, the integrated use of chemical fertilizer along with BGA also showed beneficial effects over chemical fertilizer application alone (Sinha et al., 2015).

8.5.4 Factors Affecting the Use Efficiency of BGA

8.5.4.1 Water

5 to 15 cm standing water is required for the proper growth or formation of blue green algal mat for efficient atmosphric biological nitrogen fixation (Chaubey et al., 2004).

8.5.4.2 Sunlight

Because of phototrophic in nature, the BGA require sufficient period of sunlight for atmospheric nitrogen fixation.

8.5.4.3 Temperature

A temperature range between 35° and 40° Celsius has been found to be the best for the growth of the BGA and nitrogen fixation. Although BGA normally prefer warmer habitats yet they have been shown to play a major role as nitrogen fixers in polar and subpolar regions as reported by Venkataraman (1982). Further, it is added that most of the BGA can withstand a higher degree of desication and regeneration on water-logging.

8.5.4.4 Soil pH and Salinity

BGA well grow under neutral to alkaline soil condition with a pH range of 6.5 to 7.5. However, they can tolerate strong acidic soil reaction as well as alkaline pH.

8.5.5 Method of Application

BGA as biofertilizer is applied at the rate of 10 kg ha^{-1}. The BGA flakes are crushed into powder and this powder is then broadcast in the standing water of rice fields about one week before transplantation. The fields are kept under water-logged conditions at least for 3 to 4 days immediately after BGA application.

8.5.6 Conclusion

Like bacteria, BGA are primitive organisms and play a significant role in rice crop production. Their usefulness as inoculants in rice culture has been mentioned in this subchapter of "Use of Bio-fertiiizer for Sustainable Agriculture". The usefulness of these Cyanobacteria in the form of their contribution in nitrogen fixation, effect on the yield of various varieties of rice as algal bio-fertilizer, factors affecting the use efficiency of BGA stands mentioned. In addition to this, what are BGA?, their age and nature and distribution have also been stated.

8.5.7 Future Research Strategies/Priorities

- Inspite of much importance of BGA in atmospheric nitrogen fixation in rice soils, production of growth promoting substances and improvement of soil physical properties, it is irony of sorts that use of BGA in rice field is still not popular among farmers. One of the reasons behind this is the unawareness about the impact of these *Cyanobacteria* which needs its wider publicity and educational drive.
- Looking at the importance of BGA, the Regional Bio-fertilizer Development Centres under Department of Agriculture and Cooperation, Ministry of Agriculture, Government of India should plan to make the publicity programmes in this respect.
- The above said planning must be done in collaboration with Colleges of Agriculture under State Agricultural Universities and State Departments of Agriculture/Horticulture/Animal Husbandry. The students of Agriculture Colleges must come forward to keep in the said publicity enthusiastically.
- Rallies should be organized in the rice cultivated villages. Some banners on BGA should be prepared. Slogans like *"Dhan Ke Adhik Paidawar, Neel Harit Kai Se"* or *"Moongi De Jaada Kheti Aastaya, Neel Harit Kai Da Istmal"* are required to be shouted.
- As the growth of BGA results in decrease in pH, so they should be used in reclamation of saline and alkaline soils.
- Use of BGA continuously for 3 years renders the rice fields purely organic, which is the need of the day to get rid of chemicalised agriculture owing to its ill-effects both for the humans and animals, and soil Hence, use of BGA in the rice fields calls for its urgency.

8.6 Role of Phosphate Solubilizing Micro-organisms in Increasing Crop Production

After nitrogen, phosphorus is a vital nutrient for the growth and development of plants and micro-organisms. In soils, it exists both in inorganic and organic forms. The inorganic forms of P are compounds of calcium, iron, aluminium and fluorine, while the organic forms are compounds of phytins, phospholipids and nucleic acids which mainly come from decaying vegetation. The soils rich in organic matter are, therefore, rich in organic forms of phosphorus also.

Rock phosphate or tricalcium phosphate is one of the basic raw materials for production of phosphatic fertilizer viz., single or triple superphosphate and diammonium phosphate. Tricalcium phosphate having insoluble form of phosphorus ($PO4^{-3}$ ions) cannot be directly applied to all soils except acidic ones. However, there are certain micro-organisms which can bring out dissolution of insoluble phosphatic compounds in soils. For example, use of cultures like *Pseudomonas striata, Bacillus* polymyxa and *Aspergillus awamori* have been found efficient in solubilization of low grade Mussoorie rock phosphate and soil fixed P (Singh and Saxena, 2005). These are called phosphorus solubilizing micro-organisms (PSM) which are described here under:

8.6.1 Why there is Necessity of PSM?

PSM contain phosphate solubilizing microbes which solubilize the phosphates from the soil and make it available to the crops. The efficiency of utilization of phosphatic fertilizers is very low due to their chemical fixation in soils. Besides native soil P is mostly unavailable to crops because of its low availability. Most of the Indian soils are characterized by poor to medium status with respect to available P. Further, there is building of insoluble phosphates in soils where phosphatic fertilizers have been applied over long periods. In view of the aforesaid facts, it becomes imperative to introduce efficient P solubilizers in the rhizosphere of crops and soils to increase the availability of P from insoluble phosphate compounds and, thereafter, its utilization efficiently. Most important phosphate solubilizing organisms are bacteria and fungi,

8.6.2 Mechanism of Phosphate Solubilization

A number of soil bacteria especially belonging to the genera of *Bacillus* and *Pseudomonas* and fungi belonging to the genera of *Penicillium* and *Aspergillus* possess the ability to bring insoluble phosphates to soluble forms by producing a number of organic acids (Bhadauria and Singh, 2013). Such acids include formic acid, acetic acid, propionic acid, lactic acid, glycolic acid, fumaric acid and succinic acid. These acids lower the soil pH and bring about the dissolution of bound forms of phosphate. Some of the hydroxy acids may chelate with calcium and iron resulting in effective solubilization and utilization of phosphate by crops (Shah et al, 2008; Gupta, 2010e).

Among the bacteria, *Pseudomonas striata, Bacillus megatherium, Bacillus mycoides, Bacillus circulan, Bacillus polymyxa* etc., are the most important species which bring about the dissolution of the insoluble phosphates. *Penicillium digitatum* and *Aspergillus awamori* are the main fungi which also help in dissolving insoluble phosphate groups in soils. These micro-organisms also produce amino acids, vitamins, growth promoting substances like IAA and GA.

8.6.2.1 Solubilization of Phosphorus from Rock Phosphate by P-Dissolving Fungi and Bacteria

Various species of fungi associated with legume root nodules have brought about the solubilization of $Ca_3(PO_4)2$ to the extent of 9.10 to 39.02 per cent as compared to control. The extent of solubilization brought about by four species of bacteria in rock phosphate has shown that soil *Bacillus megatheruim* caused gradual increase in available P upto 6 weeks and similar trend was noticed with B. cereus till the 8th week. However, *B. mycoides* and *B. circulans* showed altogether different trend, marked solubilization taking place only upo 15 days of incubation. Subsequently, it increased again at 4th, 6th and 8th week (Gupta et al,, 1986).

8.6.3 Agronomic Aspects

A commercial bio-fertilizer under the name "Phosphobacterin" was first prepared from *B. meghaterium* var *Phosphaticum* and is widely used in Russia and East European countries with yield increase in the order of 5-10 per cent over control. Field experiments conducted under various agroclimatic zones of India have indicated significant increase in yield over control. In general, 40 per cent of the superphosphate could be saved by the application of rock phosphate and phospho-micro-organisms (Singh and Saxena, 2005). Tyagi et al. (2008) reported that application of phosphate solubilizing bacteria assisted to increase the yield of vegetables like brinjal and cauliflower. Without phosphate solubilizing bacteria, their yields were 125.0 and 34.0 q ha^{-1}, while with culture of these bacteria the yields were 138.6 and 36.5 q ha^{-1}, respectively. Increased yields of wheat and potato were also found to increase due to inoculation of peat based culture of *Bacillus polymyxa* (Bhadauria and Singh, 2013)

8.6.4 Factors Responsible for Enhancing Phosphate Dissolving Micro-Organisms Efficiency

There are a number of factors which are responsible for enhancing efficiency in phosphate dissolving microorganisms. Some of them are detailed here as under:

8.6.4.1 Organic Carbon and Nitrogen Contents

As most of the phosphate dissolving micro-organisms are heterotrophic in nature, so these micro-organisms will definitely solubilize more amount of rock phosphate in presence of organic carbon and nitrogen content. The use of PSM along with FYM was effective in dissolution of insoluble forms of phosphates (Gaur, 1990). It has also been reported that response of phosphorus dissolving micro-organisms was positive in soils with high organic matter content and low available P. Prasad et al. (1998) reported that performance of *Aspergillus awamori* with or without FYM was slightly better than *Bacillus polymyxa* in increasing the yield of soybean as compared to control in moderately deep soil. However, in other two soils, *Bacillus polymyxa* did well than *Aspergillus awamori*, particularly more in soil II. Addition of FYM and low available P in this soil might be indirectly responsible for enhanced activities of Bacillus polymyxa. In calcareous soil (Vertisol) application of Bacillus polymyxa + FYM and vermiculture produced the highest yield of soybean than other treatments. Introduction of Pseudomonas striata, *Bacillus polymyxa* etc., in the riiizosphere of rice and wheat increased availability of phosphorus from insoluble phosphate (rock phosphate) and increased its utilization efficiency (Chhonker, 1994).

Vermicompost and phosphobacteria in combination with two inorganic P sources viz., single superphosphate (SSP) and Tunis rock phosphate (TRP) were tested in a calcareous black soil for their effect on the yield parameters of black gram and cotton mix (Thiyageshwari and Perumal, 1998). SSP and TRP were soil applied at 100 per cent with and without vermicompost and phosphobacteria. The application of TRP (100%) with vermicompost and phosphobacteria in black gram recorded the highest grain and haul in yield.

8.6.4.2 Soil pH

Soil pH is another factor affecting the solubilization of rock phosphate by phosphate dissolving micro-organisms. Generally, more solubilization of rock phosphate has been observed in acidic (pH < 6.5) or neutral soils (pH 6.5 to 7.5) than those of alkaline ones.

8.6.4.3 Soil Temperature

The most suitable or appropriate temperature is 30°C for bringing about more solubilization of rocks by the phosphate dissolving bacteria and otner micro-organisms.

8.6.5 *Methods of Application of PSM*

The methods recommended for application of PSM or bio-fertilizer is:

8.6.5.1 Soil Application

Usually 5.0 to 7.5 kg PSM is mixed with 100 to 150 kg compost or FYM, slightly moistened and kept over night prior to applying to the field. Organic carbon and other nutrients present in the manure or compost help the PSM to grow and hence the best results are obtained when this application method is followed.

8.6.5.2 Seedlings Treatment

This method is useful for the transplanted crops. For this purpose, prepare the suspension of the required amount of PSM inoculant in the ratio of 1:10. Dip the roots of seedlings in suspension and keep them immersed for 5 to 10 minutes. Take them out from the suspension and transplant them immediately. Suspension of 2.5 kg inoculant in 25.0 to 37.5 litres of water is sufficient for treatment of seedlings for one hectare.

8.6.5.3 Setts/tubers Treatment

This method of application is beneficial for setts of sugarcane/tubers of potato, fruits and flower plants. Prepare suspension of 2.5 kg of PSM inoculant in 100 to 150 litres of water. Immerse the setts/tubers in suspension for 5 to 10 minutes. Bring out the treated setts/tubers and sow them immediately.

8.6.5.4 Seed Treatment

Seeds to be sown are mixed with moistened PSM @ 4 to 5 kg of seed so that the inoculant adheres to the seed fully. The treated seed is spread on a gunny bag or plastic sheet and kept for a while in a shady place before applying to the soil. Care must be taken that the seeds to be treated with PSM are not pretreated with fungicide or any kind of chemicals (Choudhary and Rai, 2007).

8.6.6 Conclusion

PSM are well known to play a significant role in solubilizing the unavailable phosphates from soils into available forms for the uptake of growing plants. In this subchapter, therefore, "Role of PSM in increasing crop production" in terms of mechanism of phosphate solubilization, solubilization of P from rock phosphates by P-solubilizing bacteria and fungi, factors for enhancing PSM's efficiency and methods of application, stands incorporated.

8.6.7 Future Research Strategies/Priorities

- More research is needed regarding the phosphate solubilizing bacteria for various types of crops. These bacteria may be *Bacillus* or *Pseudomonas* species.
- There is also need to see the effect of the above said bacterial species at local conditions after their isolation.
- Similar type of work is required to be done on various species of fungi.

8.7 Potash Solubilizing Bio-Fertilizers

Some micro-organisms in soil have ability to release potassium (K) from K-bearing minerals and are known as potassium releasing micro-organismsespecially bacteria. Among the most important potassium releasing bacteria are *Bacillus mucilaginous, Bacillus circulans* and *Bacillus adaphicus* (Sahai et al., 2014). It has been proved through laboratory and field trials that these bacteria can accelerate weathering reactions of K- bearing minerals especially when in direct contact with mineral surfaces and fixed lattice K either by production of organic acids and production of metal complexing lignads.

8.8 Mycorrhizal Fungi and Their Role in Agriculture, Horticulture and Silviculture

Total amount of phosphorus (P) in soils varies from 0.02 to 0.1 per cent depending upon soil conditions (Naik, 2004). It has been already mentioned in the subchapter "Role of phosphate solubilizing microorganisms in increasing crop production" under "Use of Bio-fertilizers for Sustainable Agriculture." P occurs in soils both in inorganic and organic forms. Under acidic, alkaline and calcareous soil conditions most of the P gets fixed rendering, thereby, low availability of P to the plants. The principal soil components responsible for P fixation are clay minerals, iron, aluminium, exchangeable Ca^{2+}, Ca CO_3 and organic matter content. It is seen that upto 90 per cent of applied P-fertilizer becomes unavailable for crops uptake owing to its fixation.

The problem of low P mobility in soil can, however, is overcome nowadays by the use of PSM and mycorrhizal fungi. The role of PSM has been separately described. The present write-up, however, deals with the role of "mycorrhizal" inoculants for the sustainability of agriculture.

8.8.1 What are Mycorrhizae?

The term "Mycorrhiza" is a compound word which has been derived from Greek terms "Myco" and "Rhizo" which indicate fungus and root, respectively. The root also designates the composite organ formed by the fungus and the host plant which has a rhizone or thallus rather than root. As the organs of host plant are unharmed, although modified in structure and root growth may be stimulated, so these associations are considered to be symbiotic i.e. of mutually helpful. Thus, the word "Mycorrhiza" is used to describe a type of partnership association between a fungus and roots of the higher plants.

In 1885, the German Botanist, A.B. Frank, coined the name "Mycorrhiza" for the fungus root organs of forest trees, especially oak, beach and certain conifers.

He indicated his view that mycorrhizae constituted of much importance in plant nutrition organically. He suggested that woodland conditions in certain trees normally were infected by fungus that formed "Mycorrhiza" with their roots. Thus, it appears that there are certain fungi which are known to have played a significant role in the evolution of the "plants" as they colonized the land in the late "Silurian to early Devonian Period" - about 400 million years ago. They seemed to have become associated with such "plants" facilitating, thereby, their nutrient uptake processes. These associations were termed "Mycorrhizae". It is point to mention that the fungi make such associations with the fine roots of the plants.

Thus, the "mycorrhiza" refers to the association between fungi and roots of the plants so that the nutrients absorbed from the soil by the fungi are released to the host cells and in turn, the fungus takes its food requirements from the host (Shah et al., 2008). Because of the highly interdependent relationship established between both partners, this association is usually a mutualistic, where the host plant obtains mineral nutrients via fungal mycelium, while the heterotrophic fungus gets carbon compounds from the host's photosynthesis^Srivaslava et al., 2006).

Mycorrhiza can be found in almost any kind of soil. All except for a few vascular plant species, belonging to the families - Cruciferae, Chenopodiaceae vis-a-vis Cyperaceae, Caryophyliaceae and Juncaceae, are capable of forming mycorrhizal association.

8.8.2 Importance

Mycorrhizal fungi are of great significance both for agriculture/horticulture/ olericulture and silviculture as well as medicinal plants (Bhoon, 2008). In the first place they make a great uptake of phosphorus and free metals particularly those plant nutrients which are less soluble. Apart from increasing the uptake of P, application of Mycorrhizal inoculant enhances uptake of Zn, S, Fe, Cu and water, promotes uniform crop also. Mycorrhizal fungi improve disease resistance, soil structure and plant establishment after transplanting. They colonize roots of about 90 per cent of plants including most commercially important fruit crops viz., mango, banana, citrus, grape, papaya, litchi and pomegranate etc.

The application of phosphatic fertilizer can be reduced by 25-50 per cent of the prescribed doses depending upon the crops. The plants are colonized upto 89-90 per cent within 3-8 months. There is early flowering and fruiting of the crops, enhancing of seedling growth, increasing of crop uniformity and ameliorating of root cuttings. These beneficial effects have been noticed in several crops viz., vegetables and spices such as lettuce, onion, leek, celery, asparagus, cucumber, beans, tomato, okra, potato, musk melon, water melon, pepper etc.

Beneficial effects of the inoculated mycorrhizae have also been observed in case of foliage plants and cut flowers such as rose, lilac, primula, chrysanthemum, geranium, begonia, marigold etc. A recent study conducted by Bhoon (2008) indicated a lot of beneficial effects of mycorrhizal inoculation to some medicinal plants like *Solanum nigrum, Asparagus racemosus, Ocimum basilicum, Aloe vera* and many others.

Mycorrhizal fungi offer the protection to plants from various soil borne plant pathogenic nematodes besides enhancing the nutrient uptake by the plants. The mechanism of suppressing the pathogenic micro- organism by mycorrhizal fungi includes physical, chemical and biological means (Ramegowda et al., 2008)

8.8.3 Responses of Mycorrhizal Inoculants to Crops

Responses to mycorrhizal inoculants on some important crops under field conditions as reported by Singh and Chaubey (2007) have been given in Table 7.13. It is discerned from the data that shoot weights of maize and wheat became double with an application of mycorrhizal inoculant and the response was more than as is obtained with 50 kg P ha^{-1}. There was also increase in shoot weight of barley (30%) as compared to control.

Application of mycorrhizal inoculants showed an increase in the yield of potato, pigeon pea, groundnut and tomato (Table 8.13). Highest increase in yield was seen in tomato which was about 75 per cent more than control followed by pigeon pea where the per cent increase was 37 per cent than control (Table 8.13).

Table 8.13: Responses to mycorrhizal bio-fertilizer inoculation on some crops (Adapted from Singh [1995])

Sr. No.	Crop	Response in brief
1.	Maize and wheat	Shoot weights were doubled and were more than obtained with 60 kg P ha^{-1}.
2.	Barley	Nearly 30 per cent increase in shoot weight over control was seen.
3.	Potato	Nearly 20 per cent increase in yield was noticed.
4.	Arhar (pigeon pea)	Nearly 37 per cent increase in grain yield was observed.
5.	Tomato	About 75 per cent increase in yield was marked.

Dual inoculation with *Rhizobium* and *Globus fasciculatum* alongwith 50 per cent of P enhanced nodulation and nitrogen fixation of chick pea (Suba Rao et al., 1986). According to Konde et al. (1998) that dual inoculum of *Rhizobium* and VA-mycorrhizae across the 75 per cent of recommended dose of fertilizer phosphorus recorded maximum yields followed by VA-mycorrhizae alone and *Rhizobium* alone.

Mycorrhizal fungi inoculation increased soybean yields over the non-inoculated treatments to a certain extent depending upon the type of soil. Inoculation of tea cuttings with mycorrhizal significantly enhanced the growth of tea seedlings (Shah et al., 2008). Application of mycorrhizal inoculants has also proved useful for vegetables production (Narayan and Bhat, 2006). For this purpose mycorrhizal inoculants are added @ 50 g mixed with 500 g of soil and then broadcast in the nursery beds.

8.8.4 Types of Mycorrhizae

There are five types of mycorrhizae which have been identified and recognized worldwide. About 3 per cent of the higher plants belonging mostly to forest tree species of Fagaceae, Betulaceae, Pinaceae and some woody legumes constitute the "Ectomycorrhiza"-(Srivastava et al., 2006). Threee other types of mycorrhizal fungi which can be grouped as "Enodmycorrhizal fungi can be grouped as "Enodmycorrhiza", in which the fungus can colonize the root cortex intracellularly. One of these is restricted to some species of the family belonging to the Ericaceae (*"Ericold mycorrhiza"),* the second to the Orchidaceae (*"Orchid mycorrhiza"*) and the third, the *"Arbuscular mycorrhiza"* (AM) which the most widespread type is.

There is another kind of mycorrhizal fungi which is known as "Ectoendo mycorrhiza". It consists of plant species of the families other than Ericaceae and Monotropaceae. They have sheaths and produce intracellular penetration ("arbutoid" and "monotropoid" mycorrhiza). Briefly, various types of mycorrhizae have been shown in Table 8.14. A perusal of the table indicates that there are five kinds of mycorrhizae viz., Ectotrophic, Arbuscular, Arbutoid, Ericoid and Orchid. Main fungi involved in these various kinds of mycorrhizae are: Basidiomycota and Ascomycota, Zygomycota, Basidiomycota, Ascomycota and Deuteromycota and Basidiomycota, respectively. The host plants - Gymnosperms, Angiosperms, Bryophytes, Ptridophytes, Ericales and Orchids as reported byNaik (2004) are shown in Table 8.14.

Table 8.14: Types of mycorrhizae (Adapted from Naik [2004])

Types	Particulars				
	Main fungi involved	Penetration of host cells by fungi	Formation of sheath	Formation of Hartignet	Host plant
Ectotrophic	*Basidiomycota, Ascomycota*	No	Yes	Yes	Gymnosperms, Angiosperms
Arbuscular	*Zygomycota*	Yes	No	No	Bryophytes, Pteridophytes, Gymnosperms, Angiosperms
Arbutoid	*Basidiomycota*	Yes	Yes	Yes	Some ericales
Ericoid	*Ascomycota, Deuteromycota*	Yes	No	No	Ericale
Orchid	*Basidiomycota*	Yes	No	No	Orchids

8.8.5 Difference between Ectotrophic and Endotrophic Mycorrhizae

- Ectotrophic Mycorrhizae" are special characteristics of many forest tree species in which the fungal mycelium completely invests in tips and younger portions of the root with a sheath of compact fungal tissue. From the sheath, hyphal branches penetrate between the cortical cells to form a network but rarely penetrate into the cells themselves.
- The "Endotrophic Mycorrhizae" are more variable than the "Ectotrophic Mycorrhizae" and are characterized by the usual absence of a fungal sheath, although there may be in some cases a profuse development of extra-radical mycelium. The main variability between the two lies in the extent of penetration of the cells of the host and in the thickness of development of the sheath. In the "Endotrophic Mycorrhizae", the cells are heavily penetrated.

Out of the various kinds of mycorrhizae, the Vesicular Arbuscular Mycorrhizae (VAM) is by far the most widespread type. Their ecological and economical values can be directly inferred from the fact that they are associated without four-fifths of all land plants, including agronomically important crops. They are most common in horticultural crops also.

8.8.6 Vesicular Arbuscular Mycorrhizae (VAM) Arbuscular Mycorrhizae (AM)

Previously the mycorrhizal fungi were called vesicular arbuscular mycorrhizae but these days they have been shorted to arbuscular mycorrhize as vesicles are not formed in all the cases. Arbuscular mycorrhizae persist as large spore's upto 100 pm diameters in soils and infect roots from germinating spores by forming an appresorium like structure on the root surface. From this, the hyphae grow between the root cortical cells and penetrate individual root cells to form arbuscles, which are extremely finely branched tree like structures that occupy most of the cell volume.

Arbuscles having life span of 0-14 days, however, the supply of mineral nutrient to the host is known to occur before the arbuscles are digested. The arbuscles serve as the major sites of nutrient exchange between the fungus and the host. Through them the fungus gets carbohydrates from the host plant, and the plant in turn can get the mineral nutrients from the hyphae of the fungus which have absorbed from the soils. Mycorrhizal fungi assist in the stabilizing soil aggregates by producing a glycoprotein called glomalin. Arbuscular mycorrhizae have been observed to transfer nutrients from one plant to another via hypha connections.

After absorption of phosphate from the soils, it is changed into inorganic form in the fungal vacuoles. It is then translocated to internal fungal structures as polyphosphate granules contained inside vacuoles. These are propelled through the hyphal lumen by cytoplasmic streaming to the arbuscles. Arbuscular mycorrhizae fungi cannot be grown in pure culture. Commercial inoculants are made from infected root fragments, hyphal fragments and spores mainly of fungi in the *Glomus* genus.

8.8.7 Crops that form Arbuscular Mycorrhizae

The roots of most of the agronomic crops including wheat, maize, soybean, legumes, potato, sugarcane, groundnut and dryland rice form arbuscular mycorrhizae. Vegetables and fruit trees are also capable of forming arbuscular mycorrhizae. Among the fruits, apples, grapes and citrus are prominent. Many forest trees like maple, yellow poplar and redwood vis-a-vis plantation crops such as cacao, coffee and rubber also form arbuscular mycorrhizae. Two most important groups which are incapable of forming "mycorrhizae" are the Cruciferae (cabbage, mustard, canola and broccoli) and the Chenopodiaceae (sugarbeet, red beet and spinach).

8.8.8 Methods of Application

Various methods have been tried to introduce arbuscular mycorrhizae inoculation to field crops.

8.8.8.1 Soil Treatment

Inoculum of arbuscular mycorrhizae is added to FYM + soil + sand mixture in the ratio of 1:1:1 @ 50g inoculum 500g^{-1} soil mixture in polythene bags/pots. About 1-2 kg of bulk soil inoculum in m^2 of seed could be appropriate. Briefly, the establishment of mycorrhizae (Naik, 2004) is illustrated diagrammatically in Fig. 8.2).

Seeds of papaya or root stocks of mango, citrus, guava or sapota are planted in each bag/pot. By 45 days, root system of the plants will be colonized by arbuscular mycorrhizae and would be ready for plantation in the field pits. To the pits, 250 to 500 g of inoculum could be again added depending upon the concentration of arbuscular mycorrhizae spores in the inoculum alongwith FYM and neem cake. In case of new litchi orchards, if the culture is not available, in such situation use of soil from the tree basin of old litchi orchard is proved beneficial. It should be applied at the rate of 4-5 kg tree^{-1} (Bisen et al., 2005) as culture media. It will help in early establishment and quick growth of plants.

Alternatively, seedlings are raised in sterilized or unsterilized soil supplied with selected arbuscular mycorrhizae fungi in small nursery beds or containers and planted out mycorrhizal colonization is well established. This method has been successfully used to produce important vegetables like chilli, tomatoes, finger millet and horticultural crops like citrus, mango and forest tree species like *Leucaena leucocephala, Acacia nilotica, Tamarindus indica.*

8.8.8.2 Coating of Seeds

- Coating of seeds with mycorrhizal inoculum similar to that which is used with Rhizobium would be the easiest method of inoculating plants. Seeds are coated wittran adhesive (methyl cellulose), to which inoculum is expected to stick and subsequently infect the emerging radicles (Naik, 2004).
- Incorporating of seeds into the inoculum to form multiseeded pellets rather than coating seeds with arbuscular mycorrhizae inoculum. These pellets (about 1 cm size) are then sown.
- Placing of inoculants under or beside seed sown in a furrow, is probably an effective method of inoculating field sown crops.

8.8.8.3 Mycorrhizal Pellets

It seems technically more feasible to incorporate seeds into the inoculum to form multiseed pellets rather than coating seeds with Arbuscular mycorrhizae inoculum. These pellets (about 1 cm in diameter) consist of soil inoculum from pot cultures; stabilize with clay or peat inoculum.

8.8.8.4 Inoculation Management (Optimum Dose)

As far as the inoculum management is concerned, the doses and time of inoculation significant level, generally 1-2 kg of bulk soil inoculum having approximately about 10,000 propagules m^{-2} of seed bed is the most desirable dose'(Srivastava et al., 2006).

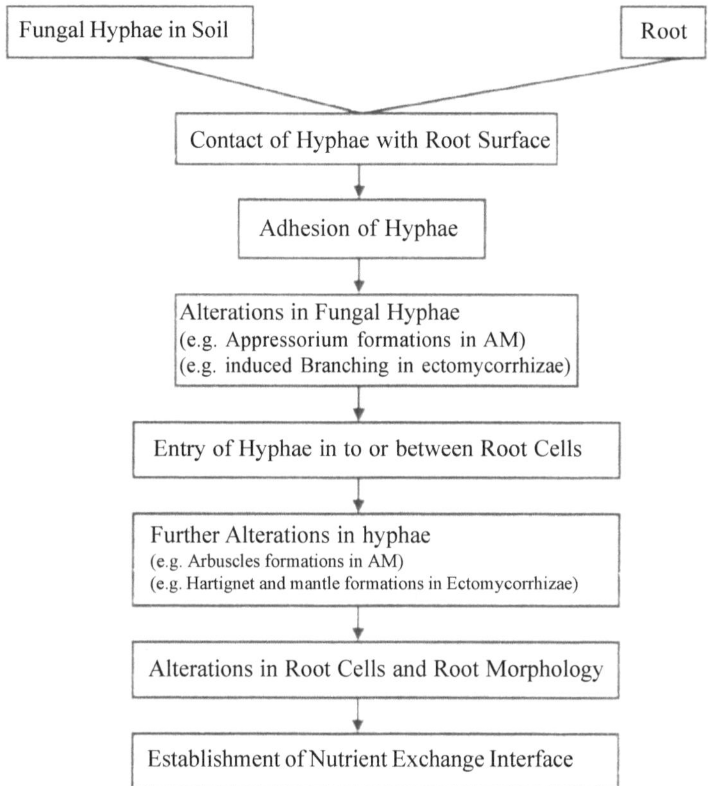

Fig. 8.2: Flow chart showing the establishment of mycorrhizae

8.8.9 Difference between Phosphate Solubilizers and Mycorrhizal Fungi

Vesicular Arbuscular Mycorrhizae (VAM) differ from phosphate solubilizing micro-organisms (bacteria and fungi - described in detail in a separate chapter) in a way that they do not solubilize the insoluble phosphate but they absorb/ assimilate phosphorus and translocate it to the host roots.

8.8.10 Factors Responsible for Arbuscular Mycorrhizae Responses

8.8.10.1 Use of Artificial Substrates

Use of artificial substrates like potting mix is a very common practice at the nursery stage, where arbuscular mycorrhiza is not present. Although this could be used successfully yet prior to its using it is necessary to see the effect of arbuscular mycorrhizae in the potting mixture. Successful colonization and beneficial effects have been achieved using sand, gravel, peat, expanded clay, saw dust, vermiculite or a mixture of these.

8.8.10.2 Different Nutrient Regimes

As the response of arbuscular mycorrhizal fungi varies according to different nutrient regimes of substrate/soil, so it is important to select appropriate nutrients containing fertilizers, their quantities and application schedules. Organic fertilizers are preferred to obtain maximum benefits out of these fungi. It is because from these fertilizer sources, the nutrients present in them are released slowly and uniformly from these manures.

8.8.10.3 Use of Pesticides

There are some pesticides which have compatibility with arbuscular mycorrhizae and can be used in an integrated manner. Captan has been found to be compatible for this purpose.

8.8.10.4 Use of Bio-Fertilizers

Micro-organisms like Pseudomonas, *Bacillus megatherium* var *Phosphaticum* have a significant role in the phosphorus solubilization and Pseudomons can be used for controlling soil borne diseases. Similarly, Trichoderma and Gliocladium can also be utilized successfully for control soil of borne diseases. Apart from this, these fungi have also proved useful in plant growth promotion and decomposition of organic wastes. Thus, combination of these kinds of organisms with arbuscular mycorrhizae would be highly beneficial to achieve the productivity in sustainable manner. However, before using these micro-organisms, they must be evaluated with arbuscular mycorrhizae for compatibility.

Worldwide, intrest in Arbuscular mycorrrhizal fungi has now reached a point where any discussion of agricultural biotechnology that does not include their role in plant productivity and functioning of natural ecosystems can hardly be considered complete. The use of Arbuscular mycorrhizal fungi in agriculture could iead to a considerable decrease in the amount of chemical pollution in soil, water as was demonstrated for maize as reported by Benjamin (2014). This clearly indicates the potential of Arbuscular mycorrhizal fungi for promoting a low chemical input agriculture.

8.8.11 Ecological and Physiological Considerations

In highly altitude forest and culturable soil, species of *Endogone* were isolated belonging to genera *Glomus, Gigaspora* and *Sclerocystis* (Jalali, 1986). Forest soils of Dehradun were dominated with *Glomus macrocarpus* in hoop pine (*Araucaria cunninghammi)* and teak (*Tectona grandis)* plantations as reported by Gupta and Banerjee (1991). Other Mycorrhizae found in new Dehradun forest soils under teak were: *Gigaspora gilmorei, Gigaspora aurigloba, Gigaspora margarita* etc. Thus, the *Glomus, Gigaspora, Endogone* and *Acauiospora* are important genera of VAM or AM.

Seasonal variations affect the density of *Ectomycorrhizal* fungi. The number of *Endogone* spores increased in rainy season and decreased in winter (Thapar and Khan, 1988). Local conditions do influence the nature of VAM. The disturbed soils owing to severely damaged ecosystem are due to intense soil erosion, rapid loss of fertility through run-off or because of shifting cultivation (Gupta and Kerni, 1991).

8.8.12 Conclusion

In subchapter, namely "Mycorrhizal fungi and their role in agriculture, horticulture and silviculture" of "Biofertilizers" has been detailed. This consists of concept of mycorrhizal fungi including their importance, response of mycorrhizal inoculation to crops, types of mycorrhizae, difference between ectotrophic and endotrophic mycorrhizae, methods of application, some ecological and physiological considerations. Suggestions and priorities for the future have also been touched briefly.

8.8.13 Future Research Strategies/Priorities

As mycorrhizal fungi serve as synergistically with N-fixing bacteria and phosphate dissolving microorganisms so it becomes the need of the day to study such effects in more details.

Since many Indian soils have now become deficient in micronutrients like Zn, Cu and Fe in soils so the use of AM must be done in crops grown in these soils due to its role in enhancing their availability. Similarly, its use should also be made in soils showing deficiency of P.

As the site quality affects the density of the mycorrhizal fungi so good fertile soil should be chosen to have very good density of living "mycorrhiza" vis-a-vis dead "mycorrhiza".

Since the disturbed soil owing to severely damaged ecosystem due to intense soil erosion, rapid loss of soil fertility through run-off loss or because of shifting cultivation greatly affects the population of mycorrhizal fungi *(Endogone* spp.), so all soil conservation measures must be adopted prior to using of "mycorrhizal fungi" inoculum.

Inoculation of mycorrhizal fungi improves crop yield both of agronomical/ horticultural and silvicultural crops which is attributed partly to solubilize phosphorus and partly by way of acting as chelating agents (Gupta and Kerni, 1991). However, less research work is available from this angle which, therefore, should be undertaken on priority.

As scanty information is available regarding the effect of various ecological factors - physico-chemical properties of soils as well as climate, soil, moisture and temperature etc., so research must be carried out in these connections.

8.9 Benefits of Bio-Fertilizers

The increase in production and use of bio-fertilizers in India is the result of special attention given by the government and interest by entrepreneurs to set up bio-fertilizer production facilities. Farmers have now also realized the benefits of the use of bio-fertilizers. The benfits of bio-fertilizers are listed as under:

- Bio-fertilizer fixes atmospheric nitrogen in soils. For example, *Rhizobium* fertilizer on an average can fix 50-200 kg of nitrogen ha^{-1} $year^{-1}$, and *Azotobacter* inoculant to an order of 10-50 kg of nitrogen ha^{-1} $year^{-1}$ or even more.
- *Rhizobium* biofertilizer increases yield by 25-30 per cent and 40-80 kg nitrogen is left over in the field useful for the subsequent crop.
- BGA can add up to about 20-25 kg nitrogen ha^{-1} $year^{-1}$.
- Bio-fertilizer like *Azotobacter,* BGA *and Azospirillum* also supply growth regulators such as Indole Acetic Acid (IAA), Gibberellic Acid (GA) and B-vitamins. Besides this, *Azotobacters* also increase the activity of beneficial *Rhizospheric* bacteria like ammonifiers, nitrifiers. phosphate solubilizers and cellulose decomposers.
- *Azotobacter* and *Azospirillum* secrete antibiotics which act as pesticides.
- *Azolla* and other bio-fertilizer not only supply nitrogen but also increase organic matter content in the form of biomass and thus enhance soil fertility. The increase in soil fertility occurs after the death and decay of the organisms present in the bio-fertilizer. The organic matter (humus) produced by these micro-organisms is of special type which enriches the soils with macro and micro-nutrients as well as humic acid, fulvic acid, hymatomelanic acid and humin fractions of humus.
- Bio-fertilizer improves physical properties of the soil like structure, water holding capacity and particle density. Actually, some of the bio-fertilizers particularly *Rhizobium* culture and *Azotobacter* inoculant produce mucilaginous substances which improve the binding capacity of the soils i.e. promote granulation in soils. In other words, we can say that soil structure is improved. They increase the cation exchange capacity of the soils also.

- They desseminate useful soil microflora and microfauna and, thereby, improve the biological properties of the soils like enzymic activity, ammonifying and nitrifying activities, etc.
- Apart from their effect on current crops, application of bio-fertilizer also leaves residual effect by way of increasing population of beneficial micro-organisms in soils. They also leave substantial amount of N in soil for use by the succeeding crops (Kumar and Chandra, 2006).
- Since bio-fertilizers are natural products so they have no harmful effect on soil and environment. They are infact, ecologically friendly, technologically feasible and socially acceptable inputs to the farmers.
- Bio-fertilizer is cheap, non-bulky and required in smaller dose.
- Application of bio-fertilizer helps in mineralization of plant nutrients.
- There is increase in the yield of the various crops with the use of bio-fertilizer. Increase in yield of crops, however, depends on the type of bio-fertilizer used and soil conditions.

8.10 Constraints Relating to Bio-Fertilizer Use

As the bio-fertilizers are live or viable preparations of bacteria/fungi, so many constraints may exist in use of bio-fertilizer leading, thereby, to their inconsistence response to crops and poor adoption by the farmers. The main constraint is their poor quality products available in the market. Actually, the constraints associated with bio-fertilizer can be classified into the following five categories:

8.10.1 Constraints Pertaining to Production and Distribution

This group itself consists of three constraints viz., unavailability of appropriate and location specific strains, unavailability of good carrier material and poor inoculant quality. The details of these constraints are as follows:

8.10.1.1 Unavailability of Appropriate and Location Specific Strains

There is always lack of appropriate and location specific strains with high nitrogen fixing ability under various soil conditions. High nitrogen fixing ability here means to survive the micro-organisms of the bio-fertilizer in moisture deficient soils, soils showing high temperature and other adverse soil conditions.

8.10.1.2 Unavailability of Good Carrier Material

Although good carrier materials like peat, charcoal, lignite, mixture of FYM and soil are currently available yet they have short shelf life. Good alternatives are called for in this respect.

8.10.1.3 Poor Inoculant Quality

Poor inoculant quality is not only due to poor production facilities of bio-fertilizer but can be caused owing to poor standards, transport, inadequate distribution and storage facilities.

8.10.2 Marketing Level Constraints

Because of short shelf life of bio-fertilizer, the retailers of fertilizers do not keep them for sale. Moreover, there is limited demand of the bio-fertilizer and lack of storage facility. Shelf life of carrier based inoculants which are presently being produced is 3 months or at the most 6 months. Further, due to poor awareness among the farming community, the demand of bio-fertilizer has not increased much so far.

8.10.3 Field Level Constraints

The benefits of bio-fertilizer application depend upon a number of soil and crop management aspects under field conditions. Soil and climatic problems such as high nitrate content, deficiency of available P, Mo, Fe, Cu, Co, presence of toxic elements, competition from native strains and compatibility with host genotypes, use of substandard inoculants etc., are the important field level constraints for getting poor response of the added bio-fertilizer.

8.10.4 Quality Control Constraints

One of the major reasons of failure with bio-fertilizer is the supply of spurious quality of bio-fertilizer. Therefore, quality control of bio-fertilizers is very essential to gain farmer's belief because it is a natural product carrying living micro-organism having very short shelf life. At present there is no control order of bio-fertilizers and the whole quality control activity is confined at voluntary basis in the country which leads to poor popularity among the farmers. So, it is the need of the sustainable agriculture that some quality control standards should be set up for bio-fertilizers to prove its effectiveness in farmer's field.

8.10.5 Pricing Policy

There is no appropriate price mechanism of bio-fertilizer either at National or International level. Consumers have no idea on Ex-factory price of retail price. It has been also observed that some organizations are selling the same product at different prices which is quite illegal. This is the high time to formulate the strategy on pricing policy.

8.11 Guidance and Precautions for the Use of Bio-Fertilizer

Following are the principal precautions/guidances which are required to be taken up prior to using bio- fertilizers:

- It is necessary that bio-fertilizer to be supplied should be of good quality containing minimum 10^7/g viable cells and purchased from the reputed sources/manufacturers only.
- Use a bio-fertilizer only for the specified crop written on the packet, especially in case of *Rhizobium* culture,
- The name of the crop should be mentioned on culture bag for which it has to be used.
- Excess culture should be used or left over can be put in furrows of the field so that micro-organisms live in rhizosphere.
- Store them before use in cool and shady place at room temperature of 25-28°C for high shelf life.
- Avoid direct contact of bio-fertilizer with pesticides/weedicides/chemical fertilizers during storage or application.
- A packet of 200 g bio-fertilizer is sufficient for 10 kg seed treatment.
- It is better to use bio-fertilizer just before sowing/planting of the crop in field because micro-organisms survive well in soil instead seed/seedling surface after treatment.
- If soil conditions are not favourable, then it would be desirable to use some soil amendments like lime or rock phosphate in strongly acidic soils or gypsum or phosphorus-gypsum in saline-alkali soils along with bio-inoculants to produce higher crop yield.
- Avoid making seeds too wet or mixed in excess water while using bio-fertilizer and rough handling of seed treatment may also be avoided.
- If seeds have to be treated with any fungicide or insecticide or any toxic chemicals, apply *"FIR* formula for treatment of bio-fertilizers where *F* denotes first fungicide, the I denotes insecticide and in last *R* denotes bio-fertilizer.
- Do not put culture in warm or hot water, which could destroy the living bacteria contained in the biofertilizer.
- Avoid together use of bio-fertilizer and full recommended dose of N chemical fertilizers. There should be 15-20 days gap in their application for better N fixation.

- The response of bio-fertilizers depends greatly on the quality of the inoculants and therefore, it is advocated that farmers should purchase it from reliable source. Moreover, the farmers should check the expiry date on the bio-fertilizer packets and beyond that it should not be used in crop.
- The pH of the soil should be near neutral pH (6.5-7.5); otherwise the efficiency of bio-fertilizers would be adversely affected.
- Adequate nutrition, for instance, with P, K and Ca etc. for the normal growth and activity of *Rhizobium* is a must which should be restored. To meet the requirement of Ca and P choice should be in favour of fertilizers like CAN and SSP.
- It is necessary to follow up all the instructions prior to using bio-fertilizers as mentioned on the packets.

8.12 How to Get Bio-Fertilizer

Realizing the importance of development of sustainable agriculture, Government of India has launched a "National Project on Development and Use of Bio-fertilizers" in 1983 and established seven Bio-fertilizer Development Centers in different parts of the country namely at Hisar, Jabalpur, Bangalore, Bhubaneswar, Nagpur, Imphal and with headquarter at Ghaziabad. Bio-fertilizers of *Azotobacter* and BGA can be obtained from the following address:

The Director

National Bio-fertilizer Development Centre,

C.G.O. Complex 11204-B, Kamla Nehru Nagar,

Ghaziabad (Uttar Pradesh) - 201 001 India

The farmers must plan in advance and orders should be placed, at least three months prior to planting/sowing date, so that bio-fertilizers may reach well in time. Information regarding the crop and area under the crop should be furnished with clearly.

8.13 Conclusion

Some beneficial organisms used as bio-fertilizers have been recognized as important inputs in INM system and their use is of recent origin. They are apparently eco-friendly, low cost and non-bulky agricultural inputs, renewable and pollution free. They also play a significant role in plant nutrition which act as supplementary and complementary sources of plant nutrients and have no deleterious effect on environment. Various biofertilizers above mentioned can be used according to needs of the crops.

8.14 Future Research Strategies/Priorities

- The management of soil fertility using bio-fertilizers is now-a-days a vital component for sustainable agriculture, so potential of various bio-fertilizers needs to be exploited.
- Although production and use of bio-fertilizer has increased in India in various states yet it is confined mostly to Tamil Nadu, Karnataka, Gujarat, Maharashtra and Madhya Pradesh. Their production and use should also be popularized in other states.
- Their use requires to be tested under various soil and agroclimatic conditions.

Study Questions

1. Define the following:
    i. Microbial fertilizers
    ii. Micro-organisms
    iii. Cyanobacteria
    iv. Mycorrhizae
2. Differentiate the following:
    i. Ectomycorrhizae and endomycorrhizae
    ii. Chemical fertilizers and bio-fertilizers
    iii. PSM and VAM
    iv. Bacterial bio-fertilizers and algal bio-fertilizers
    v. Effective and ineffective nodules
    vi. Symbiotic N fixing bio-fertilizers and asymbiotic N fixing bio-fertilizers
3. How can bio-fertilizer play a dominant role in enhancing productivity?
4. Explain the role of bio-fertilizers in context with sustainable agriculture?
5. Define bio-fertilizers and from where to order the bio-fertilizers?
6. Explain prospects and constraints for the use of bio-fertilizers?
7. Draw up a flow chart diagram showing the establishment of mycorrhizae?
8. What are the climatic factors affecting the growth and production of *Azolla*?

References

Alexander, M. (1978). Introduction to Soil Microbiology. John Wiley and Sons Inc., New York and London.

Benjamin, U.E. (2014). Functional diversity of Arbuscular mycorrhizal fungi in management of agro-ecosystem. Indian Farmers' Digest 47 (5): 17-19.

Bhadauria, B.P. and Singh, Y. (2013). Exploration of bio-fertilizers and bio-pesticides in IPM for sustainable aariculture. Indian Farmers'- Digest 46 (7): 14-17.

Bhoon, Y.K. (2008). Some medical plants field trials mycorrhizal inoculation. Gram Vikas Jyoti(October-December Issue): 74-78.

Bisen, B., Tiwari, J.P. and Tripathi. S. (2005). Bio-fertilizers in fruit crops - a way towards organic farming. Indian Farmers' Digest 38 (7): 35-37.

Boddey, R.M. and Dobereiner, J. (1982). Association of Azospirillum and other diazotrophs with tropical gramineae (pp. 28-47). In Symposia Papers I. Non-Symbiotic nitrogen fixation and organic matter in the tropics. Transactions in the 12th International Congress of Soil Science (Feb. 6-16, 1SS2), New Delhi.

'Chander, R. and Kumar, S. (2005). Bio-fertilizer: An ecofriendly source of plant nutrients. Indian Farmers' Digest 38 (4): 9-14.

Chaubey, A.K., Gangwar, S.P., Sharma, M. and Kumar, Y. (2006). Bio-inoculants: cost effective source of plant nutrients for sustainable agriculture. Indian Farmers' Digest 39 (2): 25-28.

Chaubey, A.K., Kumar, R., Rawat, S. and Chandra, R. (2004). Blue green algae - A potential bio-fertilizer for rice crop. Indian Farmers' Digest 37 (9): 15-16.

Chhonker, P.K. (1994). Crop response to phosphatic fertilizers. Fert. News 39 (4): 39-41.

Choudhary, B.R. and Rai, M. (2007). A bio-fertilizers: A boon to vegetable production. Agrobios Newsletter 5 (10): 17-18.

Das, S., Mandal, N., Bose, P., Saha, K.C. and Kole, S.C. (2007). Occurrence of diazotrophic Cyanobacteria in relation to available phosphorus in some rice soils of Purba Midnapore. Environment and Ecology 25S (4A): 1249-1251.

De, P.K. and Bose, N.M. (1938). A study of the microbiological conditions existing in rice soils. Indian J. Agric. Sci. 8 (4): 487-498.

De, P.K. and Sarkar, S.N. (1936). Transformation of nitrate in water-logged soils. Soil Sci. 42 (1): 143-155.

Devi, G.A., Dorykanta H. and Singh, N.L. (1999). Cyanobacterial flora of rice field soils of Manipur. Phykos 38 (1&2): 13-18.

Dobereiner, J. and Day, J.M. (1976). Associative symbiosis in tropical grasses: Characterization of micro-organisms and dinitrogen fixing sites (pp. 518-538). In Proc. First International Symposium on N_2 Fixation (Eds. W.E Newton and C.J. Nyman), Washington University Press, Pullman WA.

Dobereiner, J., Nery, M. and Marriel. I.E. (1976). Ecological distribution of Spirillum lipoferum Beijerinck. Can. J. Microbol. 22 (10): 1464- 1473.

Dutta, A.K. (1970). A Textbook of Bacteriology. Bharat Bhawan (Distributors), Govind Mitra Road, Patna-4, Bihar (India).

Gandotra, V., Gupta, R.D. and Bhardwaj, K.K.R. (1989). Effect of nitrogen and seed inoculation by Azotobacter on maize (Zea mays) in acidic soil of North West Himalayas. Him. J. Agric. Res. 15 (2): 85-87.

Gandotra, V., Gupta, R.D. and Bhardwaj, K.K.R. (1998). Abundance of Azotobacter in great soil groups of North West Himalayas. J. Indian Soc. Soil Sci. 46 (3): 379-383.

Gaur, A.C. (1990). Phosphate Solubilizing Micro-organisms as Bio-fertilizers (pp. 176). Omega Scientific Publications, New Delhi (India).

Gaur, A.C. (1992). Bulky organic manures and crop residues (pp. 36-51). In Fertilizers, Organic Manures. Recyclable Wastes and Biofertilizers (Ed. H.L.S.Tondon). Fertilizer Development and Consultation Organization, New Delhi (India)

Ghosh, T.K., Tyagi, M.K. and Duhan, J.S. (1999). Azospirillum - A nature's boon to boost sustainable cane production. Farmers and Parliament 36 (11): 18.

Goyal, S.K. (1982). Blue green algae and rice cultivation (pp. 346-376). Proceedings of National Symposium on Biological Nitrogen Fixation, Indian Agriculture Research Institute, New Delhi (India).

Gowda, T.K.K., Radhakrishna, D., Balkrishna, A.N. and Sreenivas, K.N. (1992). Studies on the manurial value and nutrient enrichment of municipal solid waste compost produced from Bangalore city garbage (pp. 39-41). Proceedings of National Seminar on "Organic Farming", MPKV Pune.

Gupta, R.D. (2007). Farmers should utilize inoculants to enhance legume production. Farmers' Forum 7 (5): 24-27.

Gupta, R.D. (2010e). Role of phosphate solubilizing micro-organisms in increasing crop production. Indian Farmers' Digest 43 (2): 20-22.

Gupta, R.D. (2012d). Bio-fertilizers for sustainable agriculture in north western Himalayan soils (pp. 142-154). In Model Training course on agroforestry in sustainable farming and natural resources management in hills of north western Himalayas (Eds. S. K. Gupta, M.D. Saleem, K.K. Sood, N.S. Raina, L.M. Gupta, S. Sehgal and M. Gupta). Sponsored by Department of Agriculture and Cooperation, Ministry of Agriculture, Government of India, Division of Agroforestry FOA, SKUAST-J, Chatha, Jammu, J&K (India).

Gupta, R.D. and Banerjee, S.K. (1991). Problems and management of soil and forest resources of North West Himalayas. Mahajan Book Centre, Last Morh Gandhi Nagar, Jammu (Jammu and Kashmir State), India.

Gupta, R.D. and Kerni, P.N. (1991). Mycorrhizal fungal flora in forest species with reference to North Himalayan conditions (pp.157-162). In Problems and Management of Soil and Forest Resources of North West Himalayas (Eds. R.D. Gupta and S.K. Banerjee). Mahajan Book Centre. Last Mcrh Gandhi Nagar, Jammu (Jammu and Kashmir State), India.

Gupta, R.D. and Sharma, S.P. (2006). Soil pollution and its solution through integrated nutrient management (pp. 32-38). In Balanced Fertilization for Sustainable crop Production (Eds. S.P. Sharma, S.K. R.P. Sharma Sharma and T.S. Verma). Proceedings National Seminar on INM for Balanced Fertilization held on November 17-18, 2006.

Gupta, R.D. and Singh, H. (2006). Indiscriminate use of fertilizers poses health hazards. Farmers' Forum 6 (9): 20-24.

Gupta, R.D. and Singh, H. (2008). Indiscriminate use of pesticides in agriculture: Public health issues and their control. Indian Farmers'Digest 41 (1): 8-13.

Gupta, R.D., Bhardwaj, K.K.R., Marwah, B.C. and Tripath, B.R. (1986). Occurrence of phosphate dissolving bacteria in some soils of North West Himalayas under varying biosequences and climosequences. J. Indian Soc. Soil Sci. 34 (3): 498-504.

Gupta, R.D., Dev, S.P. and Paul, Y.S. (1983). Non-symbiotic nitrogen fixation in crop cultivation. Paper Presented in Seminar held at Kafota (Sirmour) on 30th May, organized by National Fertilizer Limited, (Nangal), India.

Gupta, R.D., Kaistha, B.P. and Kher, D. (2007). Ecology of Azotobacter in soils of Himachal Pradesh and effects of its isolates on maize crop. Environment and Ecology 25S (4A): 1304-1308.

Gupta, R.D., Kher, D. and Jalali, V.K. (2005). Organic farming: Concept and its prospective in Jammu and Kashmir. J. Res., SKUAST-J 4 (1): 25-37.

Gupta, R.D. 2012. Biofertilizers for sustainable agriculture in north western Himalayan soils. Model Training course on Agroforestry in sustainable farming and natural resource management in Hills of North Western Himalayas. Division of Agroforestry, Facultry of Agriculture, SKUAST, Chatha, Jammu

Hazra, C.R. (1994). Response of bio-fertilizers in forage and fodder crops. Fert. News 39 (4): 43-53. Hooda, I.S. and Dahiya, D.R. (1992). Effect of bio-fertilizer on wheat production (pp. 67-69). In Biofertilizer Technology Transfer (Ed. V.L. Gangawarne). Associative Publishing Co., New Delhi (India)

Hooda, I.S. and Dahiya, D.R. (1992). Effect of bio-fertilizer on wheat production (pp. 67.96). In Biofertilizer Technology Transfer (Ed. V.L. Gangawarne). Associate Publishing Co. New Delhi, (India)

Jalali, B.L. (1986). Vesicular arbuscular mycorrhiza: current status (pp. 437-449). In Vistas in Plant Pathology (Eds. A Verma and J.P. Verma) MPH Publisher, New Delhi, (India)

Kannaniyan, S. (1976). Blue green algal bio-fertilizer for rice crops. Formers and Parliament 14 (11): 19-25.

Kannaiyan, S. (1979). Blue green algal bio-fertilizer for rice crops. Farmers and Parliament 14 (11): 19-25.

Kannaiyan, S. (1983). Studies on the algal bio-fertilizer inoculation of rice crop. National Conference, Society Basic and Applied Microbiologist, Dungar College, Bikaner (Rajasthan), India.

Kannaiyan, S. (1985). Studies on the algal application for low land rice crop. Tamil Nadu Agric. Univ. Bull., Coimbatore (Tamil Nadu), India.

Kannaiyan, S. (1987). Azolla biotechnology. Tech. Bull. Tamil Nadu Agric. Univ. Coimbatore, Tamil Nadu (India), pp. 1-10.

Kannaiyan, S. (1989). Azolla fertilizer for rice (pp. 15). In Proceedings National Seminar on Biotech Molecular Biology, CMS College, Kottayam (Kerala), India.

Kannaiyan, S. (1998). Bio-fertilizers for sustainable rice production (pp. 14-45). In Integrated Plant Nutrient Supply System for Sustainable Productivity (Eds. C.L. Acharya, P.K. Ghosh and A. Subba Rao). Bull, of Indian Institute of Soil Science No. 2, Bhopal (Madhya ^ Pradesh), India.

Konde, B.K. Managave, P.M. and Sonawane, R.B. (1998). Effects of Rhizobium loti, vesicular arbuscular mycorrhizae and P levels on * nodulation, yield and nutrient uptake by chickpea (Cicer arientinum). J. Indian Soc. Soil Sci. 46 (3): 465-467.

Kumar, N. (2004). Role of bio-fertilizer in rice-wheat cropping sequence. Indian Farmers' Digest 39 (2): 7-9.

Kumar, N., Pareek, R.P., Chandra, R., Singh, D., Katiyar, A.K. and Singh, D.P. (1995). Response of cereals to microbial inoculants (pp. 151-157). In Organic Farming for Sustainable Crop Production and Environmental Protection (Eds. Y. Singh et al.).

Kumar, R. and Chandra, R. (2005). Bio-fertilizer: An ecofriendly source of plant nutrients. Indian Farmers' Digest 38 (4): 9-13.

Kumar, R. and Chandra, R. (2006). Bio-fertilizers for saving fertilizer nitrogen in rice crop. Indian Farmers' Digest 39 (5): 7-10.

Lakshmi Kumari, M., Kavimandan, S.K. and Subba Rao, N.S. (1976). Occurrence of nitrogen fixing Spirillum in roots of rice, sorghum, maize and other plants. Indian J. Expl. Biol. 14 (5): 638-639.

Lakshmi, V., Rao, A.S., Vijaylakshmi, K, Lakshmi Kumari, M., Tilak, K.V.B.R. and Rao, N.S. (1977). Establishment and survival of Sprillum lipoferum. Proc. Indian Acad. Sci. 86 (6): 397-404.

Mahajan, A. and Gupta, R. D. (2009). Integrated nutrient management (INM) in a sustainable rice-wheat cropping system (pp.1-267). Springer Science + Business Media (www. Springer.com).

Mahajan, A. and Sharma, R. (2005). Integrated nutrient management (INM) system - Concept, need and future strategy. Agrobios Newsletter 4 (3): 29-32.

Mahajan, A., Bhagat, R.M. and Gupta, R.D. (2008a). Integrated nutrient management (INM) in sustainable rice-wheat cropping system for food security in India. SAARC Journal of Agriculture 6 (2): 149-163.

Mahajan, A., Bhagat, R.M. and Trikha, A. (2003a). Fertilizing bio-fertilizers. Agriculture Today 6 (9): 52-54.

Mahajan, A., Choudhary, A.K. and Bhagat, R.M. (2002). Integrated plant nutrient management (IPNM) system for sustainability in cereal based cropping system. Indian Farmers' Digest 35 (7): 29-32.

Mahajan, A., Choudhary, A.K., Jaggi, R.C. and Dogra, R.K. (2003b). Importance of bio-fertilizers in sustainable agriculture. Farmers' Forum 3(4): 17-19.

Mahajan, A., Gupta, R.D. and Sharma, R. (2008b). Bio-fertilizers - A way to sustainable agriculture. Agrobios Newsletter 6 (9): 36-37.

Mahanthesh, B., Harshavardhan, M., Srinavasa, V., Sajjan, R.P. and Thippeshappa, G.N. (2007). Impact of inorganic and bio-fertilizer and on the yield and processing quality parameters of onion bulbs (cv. Bellary Red) in Kharif season under irrigated condition. Environment and Ecology 255 (4A): 1244-1248.

Mahapatra, B.S., Sharma, K.C. and Sharma, G.L. (1987). Effect of bio, organic and chemical N on yield and N uptake in rice and their effect on succeeding wheat crop. Indian J. Agron. 32 (1): 7-11.

Martin, D., Pareek, R.P. and Chandra, R. (1995). Estimation of nitrogen fixation by inoculated rice-chickpea cropping sequence by 15 N dilution method (Abstract). Indian Society of Soil Science, Division of Agricultural Chemistry and Soil Science. New Delhi (India), pp. 74-75.

Meelu, O.P., Yadvinder, S. and Singh, B. (1994). Green manuring for soil productivity improvement. FAO World Soil Resources Report No. 76, pp. 123.

Mishra, U.C. (1993). Bio-fertilizer: IPNS Approach. Employment News 18 (7): 1-2.

Mishustin, E.N. and Shilnikova, V.K. (1971). Biologial Fixation of Atmospheric Nitrogen. The Macmillan Press Limited, London, Great Britain.

Naik, S.K. (2004). Mycorrhizal fungi - A boon for agriculture. Indian Farmers' Digest 37 (9): 17-19.

Narayan, S. and Bhat, S.A. (2006). Bio-fertilizers and vermicompost for sustainable vegetable prodcution. Indian Farmers' Digest 39 (10):32-36.

Nayak, H., Sahu, J.K. and Adhikari, S.P. (1996). Blue green algae of rice fields of Orissa State II. Growth and nitrogen fixing potential. Phykos 35 (182): 111-118.

Nedumaran, T., Rajasekar, K.T. and Perumal, P. (2008). Alternative uses of Cyanobacteria. Agrobios Newsletter 7 (1): 24-25.

Sandey, F.A., Shrivstava, A.K. and Shrivastava, D.K. (2005). Cyanobacterial flora of rice fields in Bilaspur Division of Chhattisgarh. Chh. J. Sci. Tech. 2: 203-208.

Pareek, R.P. and Maurya, B.R. (1984-85). Azolla as a bio-fertilizer. Indian Farmers' Digest 17/18 (12/1): 25-26.

Pattanayak, S.K., Rao, D.L.N. and Mishra, K.N. (2007). Effect of bio-fertilizer on yield, nutrient uptake and nitrogen economy of rice-peanut cropping sequence. J. Indian Soc. Soil Sci. 55 (2): 184-189.

Prasad, J., Hajare, T.N. and Mandal, D.K. (1998). Effect of P solubilizers on soybean grain yield in swelling clay soils. J. Indian Soc. Soil Sci. 46 (3): 468-469.

Raghupathy, B., Ananthalaksmi, K. and Poonkodi, P. (1998). Effect of organic manures, inorganic fertilizers, bio-fertilizer and lignite flyash on transplant rice-fallow cotton cv. Pach 2 in a Typic Chromustert (pp. 108-114). In Integrated Plant Nutrient Supply System for Sustainable Productivity (Eds. C.L. Acharya, P.K. Ghosh and A. Subba Rao), Bull, of Institute of Soil Science No. 2, Bhopal (Madhya Pradesh), India.

Ramegowda, G., Naik, M.K. and Devika Rani, G.S. (2008). Mechanism of diseases suppression by mycorrhizal fungi. Agrobios Newsletter. 6 (12): 43-44.

Sahai, P., Singh, A.P. and Chandra, R. (2014). Bio-fertilizers: A renewable source of plant nutrients. Indian Farmers' Digest 47 (6): 11-14.

Sakal, R., Sinha, R.B., Singh, A.P., Bhogal, N.S. and Ismail, M.D. (2000). Influence of Sulphur on yield and mineral nutrition of crops in maize-wheat sequence. J. Indian Soc. Soil Sci. 48 (2): 325-329.

Sarangi, S.K. (2002). Bio-fertilizer for cereal fodder. Intensive Agriculture 40 (1-2): 18-19. v-

Sfiah, R., Devi, S and Rana, B. (2008). Efficient ecofarmer friendly fertilizers forfarmrs. Indian Farmers' Digest 41 (12): 10-12.

Sharma, A.K. and Gupta, R.D. (2003). Only effective legume association fixes atmospheric nitrogen. Co-operative Bulletin 46 (4): 13-14.

Sharma, R.C., Sood, M.C. and-Khan, M.A. (1998). Integrated Nutrient Management for sustaining crop productivity in the hills (pp. 59-72). In Integrated Plant Nutrient Supply System for Sustainable Productivity (Eds. C.L. Acharya, P.K. Ghosh and A. Subba Rao). Bull, of Institute of Soil Science No. 2, Bhopal (Madhya Pradesh), India.

Shrivastava, A.K., Pandey, F.K. and Shrivastava, D.K. (2005). Nitrogen fixin Cyanobacterial flora and properties of paddy field soils of four districts of Chhattisgarh State. Nat. J. Life Sciencb 2: 501-506.

Shrivastava, D.K. (2000). Cyanobacterial flora of paddy fields of Durg district, Chhattisgarh State. Phykos 39 (1&2): 125-128.

Shrivastava, D.K., Pandey, F.K. and Shivastava. A.K. (2007). Distribution pattern of Cyanobacteria in different habitats of Korba district of Chhattisgarh. Environment and Ecology 25: 5 (3A): 842-847.

Siingh, A., Singh, Y.V. and Kamal, S. (2006). Bio-fertilizer for sustainable vegetable production. Indian Farmers' Digest 39 (11): 14-16.

Singh, B. and Singh, Y. (1998). Integrated nutrient management for a sustainable rice-wheat systems (pp. 46-53). In Integrated Plant Nutrient Supply System for Sustainable Productivity (Eds. C.L. Acharya, P.K. Ghosh and A. Subba Rao). Bull, of Indian Institute of Soil Science No. 2, Bhopal (Madhya Pradesh), India.

Singh, B.V. and Saxena, S. (2005). Sustainable organic farming. Farmers Forum 5 (9): 17-21.

Singh, G.R. and Pandey, D.S. (2006). Grow Azolla for wetland rice production. Indian Farmers' Digest 39 (6). 22-24.

Singh, H.P. (1995). Summer Institute Bulletin, June 22-July 11, 1995, Department of Agronomy, GBPUA & T-263145 Nainital (India), pp. 152-171.

Singh, H.P. and Chaubey, A.K. (2007). Know about a new mycorrhizal bio-fertilizer. Indian Farmers' Digest A0 (7): 35-38.

Singh, P.K. (1982). Azolla as an organic nitrogen fertilizer for medium and lowland rice. Review of soil research in India. 12th International Congress of Soil Science, New Delhi, India (February 14-16).

Singh, R.N. (1939). An investigation into the algal-flora of paddy field soils of the united provinces. Indian J. Agric. Sci. 9 (1): 55-57.

Singh, R.N. (1942). The fixation of elementary nitrogen by some of the commonest blue geen algae from the paddy field soils of the united provinces and Bihar. Indian J. Agric. Sci. 7 (5): 743-756.

Singh, R.N. (1960). Role of blue green algae in nitrogen. Economy of Indian Agriculture, Indian Council of Agricultural Research, New Delhi (India).

Singh, R.N. (2001). Ph.D. Thesis submitted to Chandershekhar Azad University of Agricultural and Technology, Kanpur (Uttar Pradesh), India.

Singh, R.N. and Pathak, R.K. (2003). Response of wheat (Triticum aestivum) to integrated nutrition of K, Mg, Zn, S and bio-fertilization. J. Indian Soc. Soil Sci. 51 (1): 56-60.

Singh, R.P. and Singh, Ram Pratap (2008). Production technology of blue green algae and Azolla for maintaining the fertility of rice field. Agrobios Newsletter 6 (8): 13-14.

Singh, T. and Dixit, S.K. (2001). Use of bio-fertilizers for higher crop yield. Farm Information Unit, Directorate of Extension, MOAGOI, Krishi N/istar Bhavan, Pusa (New Delhi), India.

Singh, T. and Purohit, S.S. (2008). Bio-fertilizer Technology. Agrobios (India), Behind Nasrani Cinema, Chopasani Road, Jodhpur (Rajasthan).

Singh, T.R. and Shukla, V. (2004). Bio-fertilizers: An introduction. Indian Farmers' Digest 37 (1): 28-31.

Stngh, V.P., Tomar, I.S. and Singh, R.P. (1982). BGA and Azolla - Low cost bio-fertilizer for rice culture. Krishi Agradoot (October-December Issue): 33-38.

Sinha, R., Verma, O., Gautam, P. and Nainwali, K. (2015). Use of blue green algae in rice culture. Indian Farmers' Digest 48 (3): 9 & 20.

Sood, R., Vidyasagar and Sharma, C.M. (2006). Effect of inorganic and bio-fertilizers on yield of summer squash (Cucurbita pepo) [pp. 153-154]. In Balanced Fertilization for Sustainable crop Production (Eds. S.P. Sharma, S.K. Sharma, R.P. Sharma and T.S. Verma).

Srivastava, R., Tewari, V. and Sharma, A.K. (2006). Arbuscular mycorrhizae: A potential bio-fertilizers for enhancing plant productivity in horticulture crops. J. Indian Soc. Soil Sci. 39 (1): 23-25.

Subba Rao, N.S., Tilak, K.V.B.R. and Singh, C.S. (1986). Dual inoculation with Rhizobium sp. And Glomus fasciculatum enhances nodulation yield and nitrogen fixation in chickpea (Cicer arietinum Linn.). Plant and Soil 95 (3): 351-359.

Tarrand, J.J., Krieg, N.R. and Dobereiner, J. (1978). A taxonomic study of the Spirillum lipoferum group, with descriptions of a new genus, Azospirillum gen. nov. and two species, Azospirillum lipoferum (Beijerinck) comb. nov. and Azospirillum brasilense sp. nov. Can. J. Microbiol. 24 (8): 967-980.

Takur, S.K. and Sharma, C.R. (1998). Effect of rock phosphate enrichment and Azotobacter inoculation on the transformation of N and N during composting. J. Indian Soc. Soil Sci. 46 (2): 228-231.

Thapar, H.S. and Khan, S.N. (1988). Seasonal frequency of Endogene spores in new-forest soils (pp. 161-162). In Trends in Tree Sciences (Eds. P.K. Khosla and R.N. Sehgal). Indian Soc. Tree Scientists, University of Horticulture and Forestry, Solan (Himachal. Pradesh) India.

Thiyageshwari, S. and Perumal, R. (1998). Integrated plant nutrient supply with vermicompost and phosphobacteria and inorganic phosphorus in black gram (pp. 191-195). In Integrated Plant Nutrient Supply System for Sustainable Productivity (Eds. C.L. Acharya, P.K. Ghosh and A. Subba Rao). Bull, of Indian Institute of Soil Science No. 2, Bhopal (Madhya Pradesh), India, Tyagi, A.K. Umrao, V.K. and Kumar, S. (2008). Use of bio-fertilizer in vegetable production. Agrobios. Newsletter 6 (12): 24-25.

Tyagi, A.K., Umrao, V.K. and Kumar, S. 2008. Use of biofertilizers in vegetables production. Agrobios Newsletters, 6(12):25-25.

Vaid, T.K. (2005). Azolla - a nutritious fodder fern. The Kashmir Times 62 (87): 13 [March 29, Tuesday].

Venkataraman, G.S. (1972). Algal Bio-fertilizer and Rice Cultivation. Today and Tomorrow's Printers, New Delhi, pp. 75.

Venkataraman, G.S. (1982). Nitrogen fixation by blue green algae and its economic importance (pp. 69-82). In Symposia Papers I. Non-Symbiotic nitrogen fixation and organic matter in the tropics. Transactions in the 12th International Congress of Soil Science. (Feb. 6-16, 1982), New Delhi.

Wani, S.P. (1990). Inoculation with associative nitrogen fixing bacteria: Role in cereal grain production improvement. Indian J. Microbiol. 30(4) : 363-393.

Yadav, K., Devendra, P., Prasad, C.R. and Mandal, K. (1992). Effect of enriched compost and Rhizobium culture on the yield of green gram. J. Indian Soc. Soil Sci. 40 (1): 71-75.

9

Vermicompost/Earthworm's Role in Sustainable Agriculture

Abstract

As soil degradation is a major threat to the long term sustainability of agricultural production worldwide due to indiscriminate use of chemical fertilizers and pesticides accompanied with soil erosion/denudation etc., so application of organic manures is one means to prevent its degradation including green manure and vermicompost. In the present chapter of the book, the role of vermicompost in sustenance of agriculture has been given. A brief about earthworms, their ecology, vermicomposting, composition of vermicompost, vermin-wash, effects on physical, chemical and biological properties of soils and crop production have also been highlighted.

Keywords: Vermicompost, Its preparation, uses and effects

Ever increasing use of chemical fertilizers without an adequate organics has deteriorated the fertility and productivity of soil and also aggravated multi-nutrient deficiencies in soil and plant system (Gupta and Singh, 2006). Further, change in the cropping pattern, intensive cultivation and indiscriminate use of pesticides with an objective to obtain more and more crop yield, has not only made agriculture a profitable commercial venture but has also brought in several problems of pests and diseases, and ecological crises. As the nitrogenous fertilizers are soluble and as such they may leach down into ground water as for example presence of nitrate in underground waters of some parts of UP, Haryana, Punjab, Gujarat, has become a potent source of blue baby disease (Gupta, 2005). Contamination of metals like arsenic, nickel, chromium, cadmium, iron, etc., in soils through organic wastes or industrial effluents and their leaching in the underground water and flowing into the various water bodies, have caused a number of health problems in West Bengal, Orissa, Assam, Gujarat, Andhra Pradesh, Jammu and Kashmir (Gupta, 2016).

Use of fertilizers and pesticides may enhance cost of agriculture production due to their increased prices, year after year which bring more poverty to the farmers. It has, therefore, become necessary to find an alternative system, which can sustain agricultural production. Organic farming is one such an alternative. Organic farming refers to farming in the spirit of organic relationship. It is defined as, "Production system which shuns or excludes the excessive use of chemical fertilizers, pesticides, growth regulators and livestock additions, and mainly relies on biological aspects for pest control, bio fertilizers, crop rotation, crop residues, green manuring, organic manure and vermicompost.

However, complete organic farming is not immediately feasible in order to meet the challenges of increasing demand of food production for the huge population. With increase in population, our compulsion will not only to stabilize agricultural production but also keep it increasing, on a sustainable basis. The way to improve and sustain soil productivity in order to get maximum crop production for more profit is "Integrated Nutrient Management". Integrated Nutrient Management (INM) involves use of manure including composts, bio-fertilizers, top rich forest soil and some quantity of chemical fertilizer (Mahajan and Gupta, 2009). In this context, an application of vermicompost has proved the best way to ameliorate the soil strength and also to reduce the fertilizer requirements. Vermicompost in fact serves as an eco-friendly fertilizer, which is made of organic wastes such as farm and organic residues with the help of earthworms. It assists the soil to retain its natural value in respect of texture, structure, soil organic matter and plant nutrients, bacterial and other microbial population, which in turn enhance crop production on sustainable basis. In this chapter of the book, therefore, the role of vermicompost/earthworms in sustainable agriculture has been detailed.

9.1 A Brief About Earthworms

Earthworms, belong to Phyllum - Annelida and Class - Chaetopoda and order Oligochaeta, are long, narrow, tubular, brown-coloured animals measuring 2-7 cm in length having 100-200 segments. They are hermaphrodite and constitute the most important category of macro-animals of soils. On the whole, the earthworms of the world are classified into 10 families, comprising 240 genera and about 3320 species (Gupta and Bhagat, 2004). However, many areas concerning with earthworms still remain to be exploited.

As far as Indian earthworm fauna is concerned, it is reported to consist of more than 500 species which refer to 67 genera. Even there are reports indicating that India possesses about 3,000 species of earthworms which are adapted to wide range of environment and vermiculture needs. They have been classified into two categories based on their ecological niche (Perel, 1977) as:

9.1.1 Surface Living (Epigeic)

They are surface dwellers which feed predominantly on litter and excreta of other organisms, used extensively in composting process. The common species available in different soils are: *Eisenia submontana, E. foetida* and *Eudrilus eugeniae.*

9.1.2 Endogeic

These are also called burrowing worms or soil formers. They feed extensively on organically rich soils. The popular species is *Octahaetona thurstoni.* They ingest organic and mineral fractions in soil, thus promoting the formation of organo-mineral complexes.

9.2 Origin of Earthworms

The origin of earthworms is believed to be about 120 million years ago. It is propounded that the earthworms evolved after the origin of dicotyledonous plants. However, the other school of thought believes in that these worms originated prior to the appearance of dicotyledonous plants. Therefore, annelids appeared to be originated in Precambrian and Cambrian Periods that is around 570 million years ago.

9.3 Ecology of Earthworms

The population of earthworms is variable, depending upon number of ecological factors such as organic matter, soil reaction, soil moisture content, type of the soil. Their details stand discussed elsewhere (Gupta and Bhagat, 2004). Cocoon morphology hatching and emergence pattern in tropical earthworms have also been appeared in literature (Dash and Senapati, 1980). The effect of pesticides on the densities of Collembola in the layer of decomposing plants was not similar in all cases (Atlavinyte et al., 1977). The least number of these organisms were found in the layer of Agropyron in the experiment.

9.4 Origin of Vermicompost

Once again there has been evidence that fundamental research, which is a priority of a number of economic interest, can lead to important applications. This is the case with the study of earthworms' family, which has opened up new vistas for setting an industry.

Marcel Bouche, a leading agricultural scientist of French, was fascinated by earthworms on which he has been studying since 1963 at the French National Agronomic Research Institute. Although his subject matter did not appear to be very inspiring at the beginning yet it seemed to act as companion for gardeners, grave diggers and fishermen. It all began about some where during 1973, when Marcel Bouche after continuous research noticed that an enormous mass of organic wastes could be digested very actively by a tonne of earth worms ha^{-1} yet discretely, participated in the ecology of soil and subsoil. One species of earthworm in particular, *Eisenia Andrei* attracted his attention because it had the capacity to yield about 400 Kg of nitrogen rich matter (compost) ha^{-1} $year^{-1}$, which was far higher than in good manure.

Atleast, Marcel Bouche, devoted 20 years to draw attention to his research ideology, however, these creatures were got neglected to deal with garbage dumps. As, at that time, most of the people were concerned only with collecting rubbish and creating more or less unofficial garbage dumps, so his publication only aroused polite interest. The idea of having earthworms to transform rubbish into compost was premature. And the garbage dumps made by the "Waste Consumer Society", continued to grow abundantly without mentioning the smell/ odour and damage to the countryside. The birds, which feasted upon them and the larvae which, swarmed there contributed to the breeding of all kinds of disease causing germs and parasites in the environment.

9.5 Patents

Once the laboratory and experimental stages had been completed successfully, it was another quite separate matter to move on to an industrial level. It was disheartening to go around knocking at the doors and he turned away by firms specialized in waste processing and management. At last, when a garbage dump at La Voulie in the Ardeche region of France had to be closed, owing to complaints from the locals, Marcel Bouche seized the opportunity. With the agreement of Mayor and small firm from the town of Montelimar, he went into "earthworm compost making". It took 2 months for the complete cycle to carry out this process, which was protected by "International Patents", which is called "Naturba". The Naturba process had lower cost than other processes because it was not carried out in the open but in specially constructed buildings which were at the same time had an advantage.

The advantages were that since the operation was carried out in a closed building, hence there was no smell or odour. For the same reason, the garbage occupied up far less space than in a garbage dump. Moreover, since the earthworms worked in an airtight space, so the carbon dioxide gas they released helped more fermentation. Every day 10 to 12 tonnes of garbage were brought to the plant in closed plastic bags. After opening the plastic bags, the garbage used to calibrate. Using magnetic field, iron used to be separated from the garbage and fermentation usually used to take month's time. During the fermentation process, which took about a month the heat produced (70°C which could eventually had other constructive uses) was enough to kill the larvae of insects, parasites and bad seeds. During the 2nd month, the *Eiseiniandrei* went into action. Tonnes of fermented waste entered a kind of grated column, where the worms started their work of digestion, which took a month's time. The nitrogen rich compost gradually came out mixed with debris that cannot be absorbed (7% glass, plastic, batteries, aluminum tins), is sorted out or recycled. About 300 tonnes of compost is produced from one tonne of earthworms. Now, the compost prepared from the garbage with the help of earthworms is called "vermicompost".

9.6 Why There is Necessity of Using Vermicompost?

Although the use of modern agricultural practices has succeeded in giving higher production of various crops yet due to rapid use of urea, diammonium phosphate and pesticides has deteriorated in the quality of the land and soil, eventually loss of entire ecosystem and health of human beings, turned to be a matter of serious concern. The chemical fertilizers and pesticides are not only deteriorating the land/soil but also lead to a barren and infertile land. As the fertilizers become soluble they may leach out into the soil and force the earthworms and other macro organisms vis-a-vis microorganisms to seek refuge else where, there by turning the soil into inert material. Hence, greater efforts are required to produce living soil by using vermicompost to increase bacterial, microbial population and organic contents in soil. The living soil can in turn, provides healthy and balanced nutrition to the plants which result in healthy plant growth and increase in crop yield. Further it is added that vermicompost is the limelight in this direction to save human and animal life and to maintain the balance of the ecosystem. It is an eco-friendly fertilizer which is made of organic waste (plant and animal residues) and earth worms, helps the soils to retain their natural value and structure/texture without harming the production level.

9.7 Vermicomposting - Simple but Unique Technique

Vermicomposting technique is totally different from the traditional composting methods. In traditional methods composts are prepared through decomposition process, while in vermicomposting the decomposable organic wastes are eaten by the earthworms, digested in the body and materials not required for their nutrition are excreted. The excreta of these earthworms is called vermicompost (Sahu, 2004).

Briefly, vermicomposting is a faster, eco-friendly very efficient and cost effective natural biodegradation process which turns wastes and garbage into gold - the nutrient rich and very good quality of organic manure.

9.8 What is Vermicomposting?

Vermicomposting is the technique of using earthworms systematically to turn farm wastes and other organic wastes into valuable compost or organic manure or vermicompost. Vermicomposting, infact, is one of the best methods of recycling of organic wastes to improve soil fertility as well as disposal, of these wastes to check environmental pollution and health hazards. The partial replacement of mineral or inorganic fertilizers and sustainability of agriculture are the two most important objectives of Vermicomposting. The Vermicomposting is also called vermi- culture technology. The earthworms have given to man the almost unbelievable power of turning his "garbage into gold". The Vermicomposting or vermi-culture is an innovative type of biotechnology that does not call for extensive laboratories or sophisticated industrial equipment.

A recent study conducted by Chauhan (2007) to prepare vermicompost from poplar *(Populus deltoids),* bamboo *(Dendrocalamus strictus)* and *safeda (Eucalyptus hybrid),* showed that bamboo leaves provided highest amount of vermicompost both in terms of wet and dry weight. The net dry weight of the vermicompost formed by bamboo leaves was 27.5 kg, while in case of poplar and eucalyptus leaves it was 25.6 and 26.6 kg. For bamboo leaves, the net dry weight of vermicompost was 14.9 kg as compared to poplar and eucalyptus leaves where the figures were 12.9 and 13.7 kg.

9.9 Composition of Vermicompost

Vermicompost is a mixture of earthworm casts, which are rich source both of macro and micronutrients, vitamins, growth hormones and microflora. Vermicomposting technology also has multidirectional impact on human life as it produces organic fertilizer which assists in safe disposal of organic wastes, recovery of energy rich resources. It helps to combat environmental pollution and creates job opportunities.

Vermicompost holds promise to play a significant role in building up of soil fertility and improving soil health for sustainable agriculture. It contains 0.5 to 2.5 per cent nitrogen, 0.1 to 2.0 per cent P_2O_5 and 0.15 to 1.5 per cent K_2O (Tiwari, 2002; Sahu, 2004). Besides these major nutrients, it also possesses secondary nutrients like calcium and magnesium to the extent of 0.44 and 0.15 % respectively. Apart from N, P, K, Ca and Mg, vermicompost also contains Fe (175.3 ppm), Mn (96.5 ppm), Cu (4.8 ppm), Zn (24.4 ppm) as reported by Sahu (2004).

9.10 Economic Importance of Earthworms

Because of numerous beneficial effects of earthworms in nature, they are covering perhaps every important aspect towards "Sustainable Organic Farming" except control of pests, particularly insects causing damage to crops. Indirectly, they are beneficial to reduce the effects of soil-borne pathogens.

In view of the economic importance or usefulness of earthworms, one must appreciate beneficial activities performed by these organisms in soils. Such beneficial activities may comprise tillage of soil, to promotion or hastening of humification process, sustenance of fertility of soils, improvement in the physico-chemical properties, of soils preventing soil erosion, reclamation of waterlogged soils, enhancement of biological nitrogen-fixing activities etc. The detail of each of these beneficial activities is mentioned here under:

9.10.1 Tillage of Soil

Burrowing activity of the earthworms brings about tillage of soil. Interestingly in most conventional forms of tillage implements, it is possible to plough the soil upto about 30 cm depth while earthworms can do this upto 3m or more without adversely affecting plants in any manner. In addition to this, the earthworms feed on soil organic matter thus break down its various particles and mixing of soil nutrients and bacteria in digestive process as well with deposit of excretory pellets known as vermicasts. These casts act as "bio-fertilizers" having upto 100 times more microbes in the surrounding soil. The locomotory and burrowing habits of earthworms are essential for the recycling of organic matter and other nutrients in the soil. They improve soil aeration.

9.10.2 Humification

Earthworms are capable of breaking down the complex organic compounds of organic matter into monomers like glucose, amino acids and number of other organic acids (benzoic acid, phenols, catechuic acid etc.) which, thereafter, polymerize to form humus or humic substances. The whole of the process by which humic substances like humic acid, fulvic acid, humin, etc., are formed in conjunction with the help of other macro and micro-organisms is called "humification".

Humic substances so formed have multipurpose uses. Firstly they act as store house of plant nutrients both macro and micro. Secondly they regulate the ionic exchange phenomenon like clay particles, which is of great significance in this universe after photosynthesis. Thirdly, various fractions of humic substances are amenable to maintain C: N ratio in soils. Fourthly, they serve as chelating agents which are responsible for rendering the availability of micronutrient cations slowly and uniformly.

9.10.3 Maintenance of Soil Fertility

Earthworms mostly consume organic materials like leaves, weeds, and peels of vegetables, animal excreta and partially humified matter. They also derive their nutrition from the decaying microorganisms based on the food preference. Earthworms can be categorized into two classes: (1) Those feeding on the soil are known as geophagous; and (2) That feeding on vegetation is termed as phytogeophagous.

Soon after consuming various organic materials, the earthworms worked soils are always rich in plant nutrients than the soils without earthworm activities. It has been found that soils with worm casts contain more nutrients i.e. 5 times nitrogen, 7 times phosphorus, 11 times potassium and 2 times each of magnesium and calcium than the soils without earthworms. Similarly the soils having wormcasts harbour about 7-8 times more of bacteria and actinomycetes than those having no worm casts.

Of various roles of earthworms, benefits accrue even after death of these organisms in natural conditions. Nitrogen content in the soil is increased with the decay of the dead earthworms. It has been estimated that dead earthworm species can produce upto 10-12 mg of nitrate individually.

9.10.4 Change of Physio-Chemical Properties

Soil material passing through the bodies of earthworms undergoes physical and chemical changes. Physically the earthworms improve the soil structure, aeration capacity and water holding capacity, and volume of expansion.

9.10.4.1 Amelioration of Soil Structure

In certain areas of England earthworms work over and bring to the surface more than 25 tonnes of fine (dry mould) soil rich with manure ha^{-1} $year^{-1}$. This results into the improvement of the soil structure and texture. The whole superficial bed of vegetative mould passes through their bodies in the course of a few years. Usually the soil horizons having earthworms show a granular to sub- angular blocky structure. The amount of soil that these creatures pass through there bodies may account to as much as 37.5 tonnes of dry earth ha^{-1} $year^{-1}$.

Soil aggregates having earthworms are consistently more water stable than those containing no earthworms. A study made at Shilarou, district Shimla, Himachal Pradesh (India) regarding the effect of snowbound earthworms on soil aggregation indicated that there was significant improvement of water stable aggregates (Sharma et al., 1976). The improvement was more in the top (0-15 cm) soil (Table 9.1) than in the lower soil (15-30 cm). This refuted the complaints that earthworms in the soils of studied area deteriorated soil structure and made it hard to till.

Table 9.1: Cumulative weight percentage of aggregates (expressed on oven dry basis)[3]

Treatments	Cumulative weight percentage of aggregates retained on sieves of diameter (mm)				
	1.00	0.50	(L25)	O10	0.10
A	6.46	12.96	20.06	32.84	100.00
B	12.82	19.10	30.18	52.62	100.00
C	34.06	38.16	42.50	63.42	100.00
D	35.60	49.56	54.24	65.40	100.00
E	38.90	56.90	66.56	82.34	100.00
F	39.40	57.30	73.30	84.32	100.00
LSD (0.05)	0.934	0.641	0.650	0.692	
			Sub Soil (15-30 cm)		
A	6.46	12.96	20.06	32.84	100.00
B	12.96	18.94	26.08	32.84	100.00
C	28.06	31.16	34.98	38.55	100.00
D	32.94	38.04	43.96	51.06	100.00
E	34.90	45.58	52.40	58.46	100.00
F	38.64	48.72	58.20	68.20	100.00
LSD (0.05)	0.469	0.438	0.263	0.343	

Than in the lower soil (15-30 cm). This refuted the complaints that earthworms in the soils of studied area deteriorated soil structure and made it hard to till.

"Adapted from Sharma et al. (1976)

9.10.4.2 Other Physical Properties

Generally with increase in number of earthworms the bulk density is reduced while the total pore space and water holding capacity is increased significantly (Sharma et al., 976) both in surface and subsurface soils (Table 9.2).

Table 9.2: Single value physical constants (on oven dry basis)[3]

Treatments	Bulk density (g cc^{-1})	Pore space (%)	Water holding capacity (%)	Sticky point (%)
		Surface Soil (0-15 cm)		
A	1.31	40.48	60.03	41.05
B	1.28	42.13	62.83	42.60
C	1.25	44.56	64.57	44.95
D	1.20	46.20	66.68	49.07
E	1.15	49.14	71.17	53.92
F	1.12	51.11	76.59	57.69
LSD (0.05)	0.034	0.098	0.563	0.702
		Sub Soil (15-30 cm)		
A	1.31	40.48	60.03	41.05
B	1.29	41.96	61.62	41.51

C	1.27	43.71	63.07	42.26
D	1.23	44.88	63.88	43.32
E	1.19	47.64	66.42	45.61
F	1.16	49.18	68.68	51.24
LSD (0.05)	0.030	2.161	0.435	0.375

"Adapted from Sharma et al. (1976).

Earthworms indeed improve soil aeration by making pores with their burrowing, feeding and casting activities, which in turn improve soil water drainage and help better root penetration. Large pores (2-11 mm size) receive appropriate oxygen supply essential for the growth of aerobic micro organisms and plant growth. The volume of expansion in soil is also known to be increased by the earthworms. The data given below indicate that there was about 27 per cent increase in volume of expansion in soil treated with earthworms than that without earthworms

Treatment	Volume of expansion (CC)	Per cent increase over control
Soil without earthworms	240	—
Soil pot with earthworms	340	26.66

9.10.4.3 Chemical Properties

The chemical constituents of soils are too altered by passing bodies of earthworms through them. Usually there is an improvement in the chemical properties. For example, soil containing 28.5 mg of citric acid soluble P_2O_5, after passing through the bodies of earthworms had 34.6 mg of P_2O_5, showing thereby 21.4 per cent increase in this essential plant nutrient in soil treated with earthworm as compared to control (Gupta and Bhagat, 2001). Higher amount of CaO and MgO in A3/B1 and B2, and B1/B12 horizons of soil profiles from Holta and Bir areas (Himachal Pradesh, India) have been attributed to the presence of earthworms (10 to 20% on volume basis) in these soil horizons. This has further been corroborated by higher pH than surface soil horizons where its value was 6.1 to 6.2. The amount of organic carbon also increased considerably. In soil horizons having no-earthworm its content was 0.96 to 1.13 per cent. On the other hand, the soils showing the presence of earthworms had organic carbon in the range of 1.16 to 1.34 (Gupta and Tripathi, 1992).

9.10.5 Soil Mixing

Another important effect of the earthworms is the mixing and transportation of the lower soil to the surface or vice versa due to their burrowing habits. The earthworms when are present in large number, tend to interfere with the weathering and soil forming process, preventing thereby, the distinct horizon differentiation. In contrast to the processes of horizon differentiation like calcification, podzolisation, laterisation etc., other processes also operate which tend to make soil profile more uniform vertically. The mixing process has been grouped together under the term pedoturbation and can be subdivided according to the agent causing the mixing. Faunal pedoturbation is mixing by animals. Floral pedoturbation is mixing by plants. As the earthworms create constant mixing within the soil profiles and as such textural or structural horizon never develops. This has been supported by studying the soil of the first three layers (up to 40 cm depth) in case of R.S. Pura soil profile, which appear to have been mixed up owing to the presence of earthworms and their activities because there was an abrupt change in the soil characteristics like bulk and particle densities, pore space, water holding capacity after the third layer (Gupta, 1992). In the light of the soil mixing process as stated above, the mixing of these three layers of R.S. Pura (Jammu region, Jammu and Kashmir) soil profile could be designated as faunal pedoturbation. However, now a days the population of earthworms has considerably decreased in lower soil horizons of this area.

9.10.6 Soil Erosion

Use of vermicompost has proved an asset in preventing the soil erosion. In fact, the channels formed by the activity of earthworms can assist in the easy and quick percolation of water in the subsoil. Moreover, the humic acids in vermicompost act as binding agents, thereby, reducing the soil erosion. Similarly, the vermicasts promote soil granulation being rich in calcium.

9.10.7 Reclamation of Waterlogged Soil

Like Jammu and Kashmir, the soils of many parts of India, are waterlogged where none of the crop is grown except rice crop which is cultivated during Kharif. Such soils can be reclaimed by the introduction of the earthworms. These organisms with their burrowing habit will produce several channels in waterlogged soils with the result that superficial water moves into the deeper layers of profile. It has been found that earthworms burrow normally to a depth of 1.3 m in soils, sometime up to 3 m or even more. It is point to mention that the high organic matter in the marshy soils helps earthworms to thrive best under these conditions.

9.10.8 An Ideal Breeding Abode for Micro-Organisms

The gizzard of the earthworms serves as a novel colloidal mill in which the feed is ground to a single micron giving an enormous surface area for the microbial processing. The gut of an earthworm is a perfect natural bioreactor, which provides an ideal breeding home for aerobic bacteria as well as actinomycetes and fungi. In this gut, the feed or raw material enters from one end and the processed soil (vermicastings) comes out from the other end.

9.10.9 Utilization of Wastes

Vermicomposting and/or presence of earthworms in the field is an appropriate technique for disposal of biologically degradable waste material like chaff of wheat, rice, soybean etc., and also organic wastes along with partially decomposed dung, and in turn it produces high quality compost. Earthworms are considered as the "*best friends of farmers*" since time immemorial and are referred to as, "*intestines of the earth*". Thus wastes can be converted into a potential source of organic manure with the help of earthworms.

In soils, earthworms are the major secondary decomposers in the soil ecosystem. They utilize organic matter as food and release a part of carbon as carbon dioxide during respiration, which is eventually utilized by crop, roots in the rhizosphere. Beside affecting the physicochemical properties (as stated above), application of vermicompost aqueous extract reduces the level of Al^{3+} in acid soils due to chelation combined with pH-induced precipitation of Al^{3+}'

9.10.10 Enhancement of Biological Nitrogen-Fixing Activities

Vermicompost, when added to the soil along with half recommended dose of chemical fertilizer in rice, selectively enhances the activities of nitrogen fixers (blue green algae). Symbiotic nitrogen fixation by mycorrhizae which receive recommended dose of chemical fertilizers and vermicompost or farmyard manure is also known to have been enhanced (Tiwari, 2002).

9.10.11 Plant Growth

A handful of earthworms can be added in the pits or in the soil in which the trees or plants are growing. The earthworm then burrows into the soil and aerates the root zone. The worm castings together with their cocoon and undigested feed play a significant role in improving the soil and thereby, enhancing plant growth. Earthworms help to provide excellent conditions for a variety of useful micro-organisms. Soil teeming with other life is ideal for supporting luxuriant plant growth. The worm castings also known as vermicastings are not only rich in plant nutrients but are also beneficial for bacterial and actinomycetes population.

9.10.12 Vermi-Wash and Crop Growth

Vermi-wash serves as an excellent growth promoting substance when it is used on crops. It is used as foliar spray when plants exhibit yellowing symptoms

(Ramesh, 1995). Two to three days after sowing, all the leaves turn dense green. Weak seedlings grow and yield well after spraying of vermi-wash. Vermi-wash, when added to compost pits, helps in hastening of decomposition process of compost. Having seen all these advantages, a simple and economical technique has been devised by the section of Entomology in Regional Agricultural Research Station, Andhra Pradesh Agricultural University, Jagtial - 505327. The detail of the technique or system for preparation of vermi-wash is as follows:

"*It consists of 30 cm high earthen pot, which is filled with pieces of stone upto a depth of 10 cm from the bottom. On the surface of the stones, a filter paper or muslin cloth is spread. Then a layer of humus (well-decomposed organic matter), along with 150-200 earthworm is laid down. The hole at the bottom of the pot is fixed with a rubber hose, through which vermi-wash is collected. Hose should be twisted back and tied with twine to prevent leakage of wash. The whole assembly is kept inside a thatched shed or any cool corner of farm shed. Everyday, 2 to 3 spoonful slurry of fresh dung is added on the humus as a feed for the earthworms. About 100 ml of water is also added every day from the top. The water slowly percolates down and collects at the bottom of the pot. Since water passes through earthworm burrows, so a portion of it washes down the gut of the worms as well as castings. Every day 100 ml of wash can be obtained by this method*". To get good results, the vermi-wash should be applied at the rate of 100 ml per plant at the root zone or diluted four times and applied as foliar spray (IA).

9.10.13 Nutritional Status of Vermi-Liquids

Vermiliquids contain on an average of 200 ppm N, 70 ppm P and 1000 ppm K as well as organic carbon (0.93%). Growth promoting substances like Indole Acetic Acid and Gibrelic Acid are also present in the order of 15.2 and 3.1 ppm, respectively.

Vermiliquids can be applied as spray and soil application as well soaking of different type of cuttings. It was reported that vermiliquids increase the germination per cent, growth parameter and growth indices of many agricultural crops and fruit crops (Vijayanathan and Ganeshkumar, 2006).

9.11 Use of Vermicompost on Crops

Now in India, many of the farmers are using vermicompost in various crops. An experience of Sh. R. Madhavan, Farm Manager, of one of the farms in Chingleput district, Chennai is cited here as one of the example for using vermicompost in various crops. He used 2 tonnest (t) of vermicompost on dry weight basis in J-18 paddy variety and obtained a yield of 18 q acre^{-1} (45 q ha^{-1}). After this experience, now a number of farmers of Chingleput are going on in for this method of cultivation (Nagarajan, 1997).

Vermicompost and phosphobacteria in combination with two inorganic-P sources *viz.*, single superphosphate (SSP) and trirock phosphate (TRP) were tested in calcareous black soil (Coimbatore, Tamil Nadu) for their influence on yield parameters of black gram (CO5) and cotton (LRA 5166) mix (Thiyageshwari and Perumal, 1998). SSP and TRP were applied in the soil at 100 per cent with and without vemicompost and phosphobacteria. Application of TRP (100%) with vermicompost and phosphobacteria gave the highest grain and haulm yield (Table 9.3). In cotton the effect of SSP and TRP on cotton yield and stover yield were at par. Its highest yield was recorded due to application of SSP in combination with vermicompost (2405 kg ha^{-1}), a 68 per cent increase over the control (Table 9.4).

Table 9.3: Effect of integrated nutrient supply of SSP and TRP with V and PB on the grain and haulm yield of blackgram - Co 5 in a calcareous black spoil*

Treatments	Grain yield (kg ha^{-1})	Haulm yield (kg ha^{-1})
Control	530	1096
SSP (100%)	800 (51)	1212
SSP (100%)+ V	983 (85)	1356
SSP (100%) + V + PB	986 (86)	1337
TRP (100%)	730 (38)	1132
TRP (100%) +V	968 (83)	1228
TRP (100%) + V + PB	998 (88)	1390
LSD (0.05)	12.0	6.64

V = Vermicompost; PB = Phosphobacteria (Figures in parenthesis indicate percentage increase over control) "Adapted from Thiyageshwari and Perumal (1998).

Table 9.4: Effect of integrated nutrient supply of SSP and TRP with V and PB on the yield parameters of cotton (LRA 5166)[a]

Treatments	Kapas yield (kq ha^{-1})	Stover yield (kq ha^{-1})
Control	1427 (-)	2040
SSP (100%)	2311 (62)	2292
SSP (100%)+ V	2405 (68)	2515
SSP (100%) + V + PB	2374 (66)	2393
TRP (100%)	2141 (50)	2284
TRP (100%)+ V	2340 (64)	2353
TRP (100%) + V + PB	2383 (67)	2477
LSD (0.05)	22	38

"Adapted from Thiyageshwari and Perumal (1998).

After mixing mothbean seed with 66 kg vermicompost some farmers have obtained wonderful results in the form of 14.7 q grains of mothbean (Rajasthan Mothbean 40 variety) in 4 *bighas* of land (Chaudhary et al., 2007). This yield was more than double of wheat the farmers have harvested in the same area with local moth without vermicompost. Influence of cojoint use of vermicompost and chemical fertilizers on the yield of okra and onion was studied. Results indicated highest marketable yield of okra in the treatment of 100 per cent recommended NPK + vermicompost @ 10 t ha^{-1}. Similarly maximum, marketable yield of onion was noticed in plots receiving 100 per cent recommended NPK plus 25 tonnes vermicompost ha^{-1} during both the years *i.e.,* 2003 and 2004. Application of vermicompost increased N uptake over 100 per cent NPK alone by 17.5 to 33.7 kg ha^{-1} in okra and by 9.2 to 30.3 kg ha^{-1} in onion. The increase in P uptake ranged from 4.1 to 7.2 kg in okra and 2.4 to 4.2 in onion. Similarly, K uptake increased by 7.1 to 13.7 kg ha^{-1} in okra and by 7.6 to 16.3 kg ha'[1] in onion. It was concluded that cojoint use of organic manure including vermicompost is better over the sole use of chemical fertilizers (Sharma et al,, 2006).

A field experiment was conducted during winter season of 2002-2003 and 2003-2004 on rapeseed-mustard at Bharatpur (Rajasthan) to evaluate the feasibility of vermicompost and FYM and their method of application along with inorganic fertilizers in mustard (Sinsinwar et al., 2004). The results showed that application of FYM @ 5 t ha^{-1} or vermicompost 2.5 @ t ha^{-1} along with 75 per cent recommended doses of fertilizers in furrows enhanced the productivity of mustard in semi-arid region of Rajasthan.

9.12 Preparation of Vermicompost

Vermicompost can be prepared by digging 3 pits of convenient size (1.20 x 1.2 m) under shade in a somewhat moist site but not complete dry. At the bottom of the pit broken bricks (3-4 cm) are laid followed by a layer of top rich soil from the upper portion of the land where earthworms have lived earlier to a thickness of about 15 cm. This is followed by a layer of small lumps of cattle dung. In this bed are then inoculated about 100 earthworms. Over the earthworms, nitrogen rich green leaves especially of leguminous crop *vis-a-vis* vegetable wastes obtained from kitchen are applied every day for a period of one week. The pits, thereafter, are left as such but covered with broad leaves for 10 days. Watering the unit is done once in a week. Now tiny earthworms would be seen and addition of vegetable waste is continued until pit is full. After 45-50 days, the contents from the pits are taken out and heaped on ground in open sunlight. Due to the heat of sun, the earthworms move to the bottom of the heap and settle there. The top portion of the heap is collected and sieved through sand.

The soft dark brown material obtained through the sieve is called the vermicompost.

The unsieved material along with the earthworms obtained from the bottom of the heap is again used for preparing vermicompost. About 2.5 tonnes wet vermicompost is obtained which on drying attains the weight of 1 tonne. Two tonnes of dry compost can be obtained in two operations from 3 pits. Farmers of South India use 2 tonnes such vermicompost acre^{-1} or 5 t ha^{-1} for paddy and other crops like sugarcane and ragi.

9.13 Use of Earthworms as Poultry Feed

Earthworm meal is processed from earthworms, which acts as an excellent material for preparation of poultry, cattle, pig and fish feed concentrate (Pal et al., 2008).

9.14 Conclusion

It is concluded from the foregoing account concerning vermicompost/ earthworms' role in sustainable agriculture, that the importance of earthworms as a crucial component of soil fertility and productivity, was known to farmers long ago. But their significance faded away with the advent of modern agriculture. Many farmers are, however, now realizing their worth. It is because "earthworms can do all that a plough or tractor can render though in a small way". They aerate the soil by building tunnels; while burrowing increase the water holding capacity of the soil. The conventional ploughing upturns the soils to a maximum depth of 30 cm whereas the ploughing by earthworms through burrowing goes down as much as 3 m or even more. In the burrowing process the soil's porosity increases to allow more rain water to percolate downwards and recharge the ground water. The earth worms also deposit worm's castings on the surface of the soils at night, while they bring the subsurface soil to the top and expose to bacterial action.

Bacteria help in breaking down of cellulose, hemicellulose, starch, proteins and fats. Earthworms carry the rich humus from the surface to the roots and help to maintain the pH of the soils. Unlike chemical fertilizers, vermicompost is economical as besides increasing soil fertility and water holding capacity as well as aeration, it also produces humic substances. Humic substances or humus, in addition to serving as storehouse of plant nutrients, also modifies an ion exchange phenomenon which is next in significance after photosynthesis. So vermicompost is important potential input for sustainable agriculture. It can become an important component of Integrated Nutrient Management and can play a significant role in areas where yields of crops are declining despite application of chemical fertilizers. Therefore, the judicious use of chemical fertilizers along with vermicompost can go a long way to increase yields of crops without deteriorating soil quality.

The earthworms have infact, given to mankind the almost unbelievable power of turning his garbage into gold. The vermicomposting or vermiculture technology is an innovative type of biotechnology that does not call for extensive laboratories or sophisticated industrial equipment. But alas! Inspite of so many benefits provided by the earthworms, their population in soils is perishing due to number of reasons. The most important among them is use of pesticides/chemical fertilizers. We must take necessary steps to increase their population as has been explained in the following quote: "*Enhance the dwindling earthworms population in soils. They will inturn, increase fertility and productivity of soils and, thereby; assist in maintaining the sustainability of agriculture" as said by Dr. R.D. Gupta, Former Associate Dean, Sher-e-Kashmir University of Agricultural Sciences and Technology, Jammu (Jammu and Kashmir State)".*

9.15 Future Research Strategies/Priorities

- As in India, about 700 million tonnes of organic wastes or even more are generated every year in the form of farm residues, rural and urban wastes, so wherever it is possible they must be utilized for preparing vermicompost using vermiculture.
- Unlike chemical fertilizers, vermicompost is more cheaper and increases fertility, water holding capacity and nutrient value of soil, so it would be very potential input for sustainable agriculture.
- Vermicompost can also become an important component of Integrated Nutrient Management and can play a significant role in areas where yields especially of rice and wheat are declining inspite of application of chemical fertilizers. Hence, judicious application of chemical fertilizers alongwith vermicompost can go a long way in increasing yields without deteriorating soil quality.
- The Central and State Governments need to introduce scientific and sustainable research to the agriculture failing which it would not be able to face consequences of the increasing population.
- More research is needed to see the effect of vermi-wash on the yield and other parameters of various crops vis-a-vis vermicompost on various crops.

Study Questions

1. What is vermicompost? How it is prepared?
2. Write a few advantages of vermicompost than chemical fertilizers.
3. Write down the chemical composition of the vermicompost.
4. What is meant by epigeic and enogeic earthworms? Explain with example.
5. How earthworms assist in making sustainable agriculture?
6. Write a paragraph on earthworm ecology.
7. What is vermin-wash? Write few advantages of vermin-wash.
8. Vermicomposting - Simple but Unique Technique, give coments.

References

Atlavinyte, O., Daciulyte, J. and Lugauskas. (1977). The effect of Lumbricidae on plant humification and soil organism biocenoses under application of pesticides. In Soil Organisms as Components of Ecosystems. Ecol. Bull. 25: 222-228. [Stockholm],

Chaudhary, M.K., Rathore, S. and Singh, Y.V. (2007). Success story: Vermicomposting can become boon for farmers. Indian Farmers Digest 40 (2): 30-31.

Chauhan, S. (2007). Earthworms for earthy solutions - preparation of vermicompost from popular (Populus deltoids), bamboo (Dendrocalamus strictus) and eucalyptus Eucalyptus hybrid). Wastelands News 23 (1): 49-51.

Dash, M.C. and Senapati. (1980). Cocoon morphology hatching and emergence pattern in tropical earthworms. Pedobiology 20 ( ): 317-324.

Gupta, R.D. (1992). Soil Genesis in the Jammu and Kashmir Himalayas. In Himalayan Environment Man and the Economic Activities (Ed. J.L. Raina) Pointer Publishers, Jaipur.

Gupta, R.D. (2005). Environmental degradation of Jammu and Kashmir Himalayas and their control. Associated Publishing Company, New Delhi (India).

Gupta, R.D. and Bhagat, K.C. (2001). Humble Earth worms: Tillers of the soils. The State Times 6 (14): 1-4 (Sunday Magazine, 14th January).

Gupta, R.D. and Bhagat, K.C. (2004). Earthworms do miracles in agriculture. Indian Farmers Digest 37 (2): 41-44.

Gupta, R.D. and Singh, H. (2006). Indiscriminate use of fertilizer poses health hazards. Farmers' Forum 6 (9): 20-24.

Gupta, R.D. and Tripathi, B.R. (1992). Genesis of soils in wet temperate and subalpine/moist climatic zones of the north-west Himalayas. J. Indian Soc. Soc. Sci. 40 ( ): 507 and 508.

Gupta, R.D.2016. Wildlifea and wetland Ecosystem of Jammu and Kashmir Himalayas. N.R. Books Internation al. Pacca Danga, Jammu, Tawi, J&K.

Mahajan, A. and Gupta, R.D. (2009). Integrated nutrient management (INM) in a sustainable rice-wheat cropping system. Springer Science + Business Media B.V. , pp. 1-267.

Nagarajan, S.S. (1997). Super Poni (J-18) and Vermicompost in organic farming. Farm Progress 35 (1): 1-3.

Pal, D.P., Deo, A.K. and Tripathi, A. (2008). Earthworms is effective poultry feed. Agrobios Newsletters (11): 64.

Perel, T.S. (1977). Differences in Lumbricid organisation connected with ecological properties. Soil organisms as components of ecosystems. Ecol. Bull. 25: 56-63. [Stockholm],

Ramesh, P. (1995). Vermi-Wash Promotes crop growth. Intensive Agriculture 33 (1-2): 39.

Sahu, N.C. (2004). Vermicomposting - A simple but unique technique. Indian Farmers Digest 37 (9): 30-34.

Sharma, R.P., Datt, N., Chander, G. and Sharma, A. (2006). Influence of cojoint use of vermicompost and chemical fertilizers, on productivity, nutrient uptake and available nutrient status in okra-onion sequence. In Balanced Fertilization for Sustainable crop Production (Eds. S.P. Sharma, S.K. Sharma, R.P. Sharma and T.S. Verma). Proceedings National Seminar on INM for Balanced Fertilization November 17-18, 2006.

Sharma, S.K., Agarwal, G.C., Misra, V.K. and Thakur, P.C. (1976). Effect of snowbound earth worms on soil structure. Him. J. Agric. Res. 4 (1-2): 9-14.

Sinsinwar, B.S., Kumar, M., Premi, O.P. and Kumar, A. (2004). Effect of FYM and vermicompost methods of application on mustard productivity in semi-arid region of Rajasthan. In Abstracts of National Conference on Resource Conserving Technologies for Social Upliftment. Indian Association of Soil and Water Conservationists, CSWCRTI, Dehradun (Uttranchal), India, pp 121-122.

Thiyageshwari, S. and Perumal, R. (1998). Integrated plant nutrient supply with vermicompost and phosphobacteria and inorganic phosphorus sources in black gram. In Integrated Plant Nutrient Supply System for Sustainable Productivity (Edited by Acharya et al.). Bulletin Indian Institute of Soil Science (No. 2), Nabi Bagh, Berasia Road, Bhopal (MP), India.

Tiwari, M.K. (2002). Vermicomposting for sustainable Agriculture. Intensive Agriculture 40 (1-2): 28-29.

Vijayanathan, K. and Ganeshkumar, M. (2006). Nutritional status of vermi-liquids. Agrobioas Newsletter 4 (10): 8-9.

10

Checking the Crop Pests Naturally A Novel Way to Sustain Agriculture

Abstract

The traditional farming system though was less productive yet it was sustainable for soil environment and economic point of view as it depended upon local farm inputs. With increasing demand of food consumption with human population increase, the so called modern scientific agriculture came to fore, ushering in what is known as Green Revolution. The main pillars of the Green. Revolution technology used were: use of high yielding varieties of crops, application of high doses of chemical fertilizers and utilization of pesticides for controlling insects, pests and diseases. In India, Green Revolution occurred in 1967-1968 with spectacular increase in rice and wheat production. Although the period from 1968-90 (Green Revolution Period) has witnessed quantum jump in total production but the last 24-25 years had been disappointing. The annual rate of increase in production has drastically come down which could be attributed to indiscriminate use of chemical fertilizers and pesticides. Not only this, many health effects/toxicity of pesticides have been noticed in human beings and animals. Various water bodies have also been affected (Gupta, 2016). To overcome, these cropped up issues; there is imperative need to check the crop pests naturally. This chapter of the book deals with the use of bio-pesticides, cultural controls, agro-ecosystem analysis, integrated pest management etc. for checking insect pests.

Keywords: Green Revolution; traditional farming system: modern scientific agriculture; human population; soil environment; high yielding varieties; chemical fertilizers; pesticides; rice production; wheat production; cultural controls; agro-ecosystem analysis; integrated pest management

The term pesticide includes fungicides, bactericides, nematicides, rodenticides, soil fumigants and herbicides. They cover a wide range of formulations: chlorinated hydrocarbons, organophosphates, metallic salts, acetic acid derivatives, thiocarbamates etc. They all have deleterious effects on humans, domestic animals, wildlife and plants.

An estimate made by International Labour Organization (ILO) has evinced that sale of pesticides has peaked worldwide worth 2.7 billion US Dollars. Use of potent broad spectrum insecticides has formed an integral part of industrialized agriculture in the west, and in the crops in which Green Revolution had brought about in India, and other third world countries. These insecticides though are supposed to solve pest problems yet it is evident that their continuous use has led to the senseless saturation of the environment, and is responsible for severe black lash in the form of pest outbreak induced by sprayings.

There is no doubt that the industrialized countries, use 80 per cent of the world's insecticides and other agrochemicals but probably suffer only one per cent deaths from pesticide poisoning. On the other hand, the developing countries that are using merely 20 per cent of the world's pesticides are afflicted with 99 per cent of all such deaths. This disparity stems not only from climatic conditions but also from indiscriminate use, and ignorance and lack of adequate awareness of the hazards involved in handling the pesticides.

The problem of pesticide usage in India has assumed an alarming proportion due to its erratic practising in various crop ecological zones, cropping systems, crop rotations, and their indiscriminate use, misuse and overuse. Indians have been reported to have the highest amount of pesticide residues in their bodies as compared to other population groups in the world (Gupta, 2005c). Indiscriminate use of hazardous pesticides such as DDT and BHC - though both have banned - for preventive public health and agriculture farms has resulted in an intolerably high level of residues of these pesticides in food grains, fruits, vegetables, and milk. According to Neurologists of the Lucknow based King George Medical College, these residues cf pesticides get accumulated or stored in the tiny fat globules in the body of humans and that of animals.

Continuous use of pesticides with extensive amount, especially non-degradable and persistent nature, will not only have a deleterious effect on human and animal life but also create environmental pollution i.e., air, water and soil pollutions. Hence, the only solution to these non-degradable and persistent poisons is to phase out many of them using Nature's own ways including eco-friendly bio-pesticides. Use of these alternate methods will not only check insect-pests, diseases and weeds, which bring about 20-30 per cent loss in agricultural production (Chander and Munshi, 2000; Mainac et al., 2011; Deshmukh and Sabalpara, 2011), but will also prevent contamination of food by toxic residues and render the environment free from pollution.

In this chapter of the book, various methods to check the crop pests naturally have been described along with various public issues relating to indiscriminate use of pesticides.

10.1 Pesticide Usage Trend

Pollution free environment constitutes an imperative for sustaining any kind of life on the Earth, which is already in doldrums these days due to over exploitation of its various resources. On the other hand, technologically sound farming is the need of the hour for feeding the burgeoning population which is more than 7 billion. And on the other hand, the indiscriminate use of chemical fertilizers and pesticides has caused environmental degradation and environmental problems to all the living beings on the planet Earth, Like consumption of chemical fertilizers, the consumption of pesticides in some countries of the world is also much higher than India. Japan uses about 12 kg of pesticides ha^{-1}, while Holland and France utilized 10.5 kg ha^{-1} and 5.9 kg ha^{-1}, respectively. In some European countries average consumption of pesticides is, however, 5 kg ha^{-1}.

In India, the use of pesticides is approximately 56,000 tonnes which works to be hardly in the range of 0.03 to 0.56 kg ha^{-1} with an average figure of 0.261 kg ha'¹ (Gupta, 2006a) or 0.288 kg ha^{-1} (Singh, 2006) which is as far less than the foreign countries as stated above. Yet, the problem of pesticide use has assumed an alarming proportion due to its erratic use in various regions, cropping patterns, crop and their discriminate use, misuse and overuse.

Recent estimate has indicated that in India, pesticide use accounts for about 39 per cent in northern states like Uttar Pradesh, Punjab, Haryana and Rajasthan followed by southern states like Andhra Pradesh, Tamil Nadu, Karnataka, Kerala, which account for 29 per cent. In central states like Madhya Pradesh, Gujarat and Maharashtra, its consumption is 18 per cent, whereas in eastern states like Bihar, Orissa and West Bengal the use of pesticides is only 6 per cent of its total consumption. It is evident from the foregoing paragraph that 14 states of the country utilize about 92 per cent of the total pesticide use. The remaining states including Jammu and Kashmir use about 8 per cent of the pesticides. In Jammu and Kashmir, and Himachal Pradesh, apple fruit crop utilizes the major share of the pesticides consumption.

The number of pesticides used in the United States is 775; in European Union its number is 600 and Thailand 280, as compared with India where this number is found to be 179. A recent study by the Centre for International Trade in Agriculture and Agro-based Industries found 87 pests in apples imported from the US, which were not present in India earlier. So, time has now come to take the stock of certain environmental problems, which hitherto have been mustard at the desired level of attention by the Indian health policy makers. It is, therefore, essential to allude certain health issues which have been emerged out from the indiscriminate use of various pesticides during “Green Revolution”.

Public issues relating to pesticides use are many - economic, social, cultural, environmental and public health (Gupta and Singh, 2008). However, for the present oration main endeavour is to highlight the matter of pesticides of the environment occurred through "Green Revolution Technology" and its cumulative effect - Post Green Revolution.

10.2 Public Health Issues

Public health issues related to pesticides may be operational hazards during manufacture, formulations and distribution of pesticides, operational hazards during application in agriculture or public health and post application hazards due to pesticide residues through food chain. There are both acute and chronic effects. Acute effects are relatively easily recognizable and can be managed. It is the chronic toxicity which is of more concern from public health point of view. Such effects manifest after unpredictable duration and influence unsuspecting vulnerable groups like foetus, children malnourished millions with already compromised immune mechanisms. Based on published literature, it can be summarized that pesticide exposure leads to concerns, teratogenicity, mutagenicity, endocrine disruption and related syndromes, asthma and allergies etc. (Singh, 2006).

10.2.1 Pesticides Residues in Human Blood and Occurrence of Diseases

A study by the Centre for Science and Environment, the Delhi based premier research and advocacy organization, has found "very high levels" of pesticide residue in human blood samples taken from some Punjab villages (Anonymous, 2006). The study consisted of 20 randomly selected blood samples from four villages - Mahi, Nangal, Jajjal in Bathinda district and Drier in that of Ropar district. Each of the samples was tested using an internationally accepted methods and was found to contain 6 to 13 different pesticides. The levels of organo-chlorines were found astounding: 15 to 605 times higher than those found in blood samples of the people in United States as tested by the US Centre for Disease Control Prevention in a report of 2003. The levels of lindane were shockingly high at about 605 times than found in US population. The levels of DDT were 188 times higher.

The study detected hexachlorocyclohexane (HCN) in all the blood samples and DDT in about 95 per cent of the samples. Organophosphorous pesticides were found equally in high levels. For example, average level of monocrotophos in the blood samples of Punjab people were four times higher as per the limits fixed by the WHO. The presence of above said pesticides can suppress the immune system, which in turn creates a number of diseases. Although there is no conclusive proof regarding the cancer incidents in some parts of Punjab yet it is suspected due to indiscriminate use of pesticides by the farmers in Punjab.

Pesticide exposure besides leading to cancer - prostrate, breast, pancreas, soft tissue sarcomas, leukemia's, brain tumors and so on - enhances loss of female fetuses, endocrine disruption and related syndromes also. Asthma and allergies, cerebral damage and dysfunction, damaged personality scores, male sterility, infertility, premature ageing, anaemia, metabolic aberrations are the other diseases/disorders which have been produced by indiscriminate use of pesticides (Singh, 2006).

The above findings are based on the study conducted by Agriculture Health Study, National Cancer Institute, USA and Pesticide Resource Data - US Environment Protection agency and Pesticide Literature Review, April 2004, Ontario College of Family Physicians, Canada.

10.2.2 Contamination of Food Stuffs

In Haryana, about 70-80 per cent of the cereals, milk and milk products, feed and fodders, fruits and vegetables vis-a-vis water bodies were contaminated with pesticide residues (Mahajan et al., 2004). Out of these food stuffs and water bodies, about 30-40 per cent contained toxic residues above the prescribed limits. Though consumers could save themselves from hazards of organophosphates and carbonates by applying simple processes like washing, peeling and cooking yet contamination of milk with pesticides like BHC and DDT is hard to reduce. The residues of these pesticides being fat soluble get concentrated in butter and ghee, and their amount usually goes beyond the prescribed safe limits.

Pesticides like DDT, BHC, aldrin, toxaphene, chlordane etc., had been banned in the country and their residue accumulated so far in the environment would dissipate with the passage of time. Infact, Indians have the highest levels of DDT in their bodies, more than half of the vegetables we consume are laced with deadly pesticides of which about 10 per cent are dangerously loaded to cause a number of diseases. On worldwide scale, pesticides poisoning is to the extent of 30 million people and in India about 2 lakh (0.2 million) people are poisoned by pesticides annually.

It is worthwhile to mention that children and elders or aged persons are quickly attacked and subject to diverse type of diseases. Most of the chemical pesticides cause skin troubles, allergy, hypertension, high bloods pressure, baldness, eye disorders, scabies.

10.2.3 Pollution of Soil, Water and Air

Indiscriminate use of pesticides in some of the areas of different States of Indian Union has led to pollution of soil, water and air which have proved to be health hazards. The main causes of all these types of pollution have been detailed elsewhere (Gupta, 2005c; Gupta, 2006a).

Reckless use of chemical pesticides in agriculture, animal husbandry and public health operations for an apparently successful control of plant pests and vector borne communicable diseases have been reported as an emerging and a long term irreparable damage to humans, predators, parasites, soil macro and micro flora. Useful animals, fish, birds, food and fodder crops are also of great concern. Studies at Cornell University (USA) have established that 99.9 per cent of the pesticides sprayed irrespective of developed or developing countries go into the environment, with only 0.01 per cent of pesticides reaching the target pests.

The continuous use of DDT and HCN in India, which was more than 70 per cent of the pesticides for the last few years, has shown their residues in soil, water and air and rain water in many areas including Delhi. Out of the 50 samples each of soil and earthworm collected, possessed about 0-2.61 and 0-37.74 mg kg^{-1} of total DDT residues, respectively. The area near the vicinity of DDT factory showed high level of DDT residues. Two years survey of the Yamuna River in Delhi showed that water contained on an average of 0.24 mg kg^{-1} DDT and the bottom sediments 0.20 mg kg^{-1} of total residues. The air and rain water samples monitored from three different areas for two years indicated that DDT was more near the DDT factory area, while HCH was more near commercial complex. The toxicological effects can cause serious metabolic and physiological disorders. In certain cases reproductive potential is lost and some compounds cause neurological disorders, behavioural disorders, paralysis etc.

Even in the age of 21st century more than 50 per cent of the women in the various parts of the world haul water for domestic use (Gupta, 2010c). In India, Himalayan states like Uttarakhand, Himachal Pradesh, Jammu & Kashmir, women in the rural areas walk along distance to transport water for domestic use. Rachel Carson, Maria Cherkasova and Vandana, are some of the women environmentalists. Rachel Carson believed that pesticides and chemicals from industries could damage the earth environment. Maria Cherkasova is a Journalist, Ecologist and Director of Centre for Independent Ecological Programme. Vandana Shiva is an internationally famous Indian Physicist Eco-feminist and Activist.

The toxic residues in food, air and water damage the ecosystems of plants and animals. The pesticides reach the aquatic life by way cf surface run-off, industrial wastes and accidental spillage (Gupta, 2016).

10.2.4 Illness of the People Dwelling in Cashew Plantation Areas

The cashew plantation areas in Kerala have been badly affected due to a pesticide, endosulfan that was sprayed on cashew plantation for more than 20 years (Anonymous, 2006). A study by the Ahmadabad based National Institute of Occupational Health had clearly pointed out the close link between endosulfan spraying and the severe illness of the people in the area. Subsequently, the Union Agriculture Ministry had banned the use of endosulfan in the entire State.

Nature Lovers Movement Ernakulum and other had filed a writ petition in this connection. An affidavit hitherto submitted by the Kerala Government stated it had set up five expert committees since 2000. The committee found residues of endosulfan in samples of blood, water, vegetables, soil and milk collected from the area. Apart from this, they also identified health problems of various natures in these areas. The experts though did not directly attribute these health problems to endosulfan spraying of pesticide yet they have recommended permanent ban on spraying of the pesticide in the area.

10.2.5 Suicide of the Farmers

In its recent study called "Insecticides/Pesticides - 2005", conducted by ICAR and has noted many problems cropped up after the introduction of the "Green Revolution". It was primarily due to intensive indiscriminate use of chemical fertilizers and pesticides. The ICAR reports say, "The higher dose of pesticides coupled with repeated uses increased the production costs of cotton cultivated manifold". Combined with lower cotton prices in the late 1990's, many farmers even could not repay their loans. As a result, a large number of them committed suicide during 2000 to 2002 in Andhra Pradesh, Karnataka and Punjab (Sharma, 2005). It has been found that 80 to 90 per cent of the cotton growers buy the pesticides on credit from dealers and sometimes due to failure of crops; they can not pay back the amount. As such they commit suicide.

Punjab state happens to represent only 1.5 per cent of the geographical area of the country and 2.5 per cent agricultural area, but consumers are over 18 per cent of total pesticides in the country (Singh, 2006). Cotton growing districts utilize pesticide at the rate of 17 litres capita^{-1} year^{-1} as compared to other districts, where its use is less than 5 litres capita^{-1} year^{-1}. Accordingly, cotton cultivators of Punjab use about 54 per cent of total pesticide consumed in India (Singh, 2006).

10.3 Pesticide Resurgence

Indiscriminate use of pesticides besides causing contamination of soil, air, water, crop plants and produce, has also resulted in pest resurgence. For example, increased use of pesticides has been accompanied by the massive resurgence of brown plant hopper, the worst enemy of paddy fields. In 1989-90, Thailand had the largest brown hopper outbreak, causing enormous loss in rice. Likewise, Indonesia lost 9 million tonnes of rice due to brown plant hopper attack (Gupta, 2006a).

In India, reckless use of pesticides has also given rise to brown hopper, a common pest of paddy crop. This was more so in rich rice fields of Punjab and Haryana. Although the pesticide consumption in Jammu and Kashmir is very low as compared to other states of the country, yet the situation seems to be alarming, regarding the ecological backlash. For the past few years severe outbreak of mite pests has been observed on brinjal and marigold in Jammu region and vale of Kashmir. It is pertinent to mention that mites never appeared before as major pest of above said crops. However, they have assumed a serious status in the present situation.

It is worthwhile to mention that many of the noxious pest problems, which we are facing today in Jammu and Kashmir, are the invasions of certain pests. For instance, sanjose scale on apple, coddling moth on apricot and leaf folders on paddy, are the result of pest outbreak in the absence of their natural enemies - predators, parasites etc. Infact, pests and their natural enemies co-exist in Nature, creating thereby a natural balance between them. And the pest population normally remains within an equilibrium level and does not show outbreak due to the action of their natural agents. Ecological changes are the other main causes of pest outbreak.

10.4 Pesticide Poisoning

The World Health Organization estimate that 3 million cases of pesticide poisoning occur worldwide annually with 20,000 deaths of which 96 per cent take place in the developing countries where regulatory health laws are lax. In addition many thousands of cases that result in minor respiratory ailments skin irritation and nausea, go unreported because of the victims do not even connect them to pesticide use.

Pesticides and other chemical contaminants that enter water ways through agricultural run-off, storm water drains and industrial discharges may persist in the environment for longer periods. They may also be transported by water or air over long distances (Gupta, 2016).

As early as 1962, Rachel Carson brought into the public notice about the poisoning of entire “Silent Soring”. Uncontrolled use of agricultural pesticides has led to the contamination of entire watersheds in country like USA, where deleterious effects of chlorinated hydrocarbons on the nervous system and sex hormones on vertebrates including man have recently come to light. The United

Nations Population Fund's report (UNPF) indicates that in Delhi and Agra, the drinking water coming from rivers is heavily polluted by a number of pesticides, and women show the highest levels of exposure. Apart from this, there are many reports about the population of water bodies with Hg (mercury). One such area is 20 km stretch close to new Mumbai in Maharashtra. There are about 1200 industrial units in the Thane-Balapur industrial area, which are polluting the sediments of the river Ulhas. It has registered high levels of Hg and As (Arsenic) and is a potential source disaster (Gupta, 2013d). Although not even a single report is so far available regarding the pollution of Hg in various aquatic ecosystems of Jammu & Kashmir yet much more use of Hg containing fungicides, in apple production might have polluted the Dal lake and Wular lake. It is worthwhile to mention that excess of Hg causes Minimata Disease which first affected the fish and later on human beings after taking affected fish by them.

10.5 Falling Prey of Children to Pesticides

India being the second largest cotton producing country in the world after Peru, having only 5 per cent of the total arable land, consumes 55 per cent of pesticides used in the country. The farmers growing cotton spend Rs. 246 crores every year on pesticides to save the crop from the prey of various insects. Out of the various types of pesticides used, the most of them are ineffective on the 40 types of pests and insects which feed on cotton. However, whenever the pesticides are exposed to children, have a number of serious ailments.

The study having title, "Arrested Development" based on field research was conducted in six cotton producing states by the International Non-Government Organization Green peace in November 2004. These states were - Gujarat, Karnataka, Tamil Nadu, Maharashtra, Andhra Pradesh and Uttar Pradesh. The purpose behind this study was to disseminate the results of the study to the farming community on the impact of pesticides on their children's mental health and physical development. The results of this study showed the hellish effects of pesticides on the most vulnerable section of our population. It made comparison between pesticides exposed children and those who did not expose such poison. The children exposed to pesticides had slow physical and mental growth - poor reflexes and low stamina. They had also poor tactile perception, non verbal cognition, eye hand co-ordination and their sense of balance, concentration and weak memory. Girl children reached puberty abnormally late. According to pediatricians the children who used to process cotton seeds treated with dangerous pesticides, were suffered with a number of serious ailments. These mostly consisted of concerning with reproductive health both in men and women, including irregular menstrual cycles in girls.

10.6 Control Measures

In the years to come, competition of the countries in the production of various agricultural commodities is likely to play a big role in the International Trade. Their cost of production has to be reduced considerably and profitability of the farmer's maximized. This can be possible if the agricultural products become free from the hazards of chemical fertilizers and pesticides. So, use of pest avoidance tactics, enhancement of biological pest suppression and adoption of other non chemical methods of pest management are required to take-up to tackle farmers above said problems. The detail of these methods is given here under:

10.6.1 Use of Bio-Pesticides

Bio-pesticides are eco-friendly bio-degradable, non-toxic and cost effective alternatives for chemical pesticides. These have been developed from the sources of higher plants, lower plants - bacteria, viruses. Accordingly, they are named as botanic or botanical pesticides, bacterial insecticides and viral insecticides, respectively. Besides being eco-friendly these pesticides can also be used against the pest strains (insects, fungi, weeds) which have resistance to chemical pesticides. Some of the entomopathogenic viruses which infect insect have been used as safe and effective pesticides. Other specific pest species that have no adverse affects on useful insect pollinators, insects yielding useful products, insect, predator, and warm blooded mammals including man. They are safe even in their large scale use (Gupta and Tikoo, 2005).

10.6.1.1 Botanical Pesticides

Botanical pesticides are bioactive products of plant origin. Being less persistent in environment and safe to mammals and other non-target organism, they have become of an imperative need of the day. Many plant based formulations are now available in the market at many places in the form of botanical pesticides. For example, neem (*Azadirachta indica*), is the most conspicuous source of bio-pesticides of botanical origin. Margocide (OK 80% EC), neem bacidene, bioneem, nimbasol, neemazal F. are the main products of neem which can be used to control a number of pests of different crops. Even freshly prepared neem seed kernel can be recommended for checking different harmful pests. Neem products are easy to handle. The toxic effects of neem are attributed to the occurrence of large number of bitter compounds called meliacins i.e. azadirachtin, nimbin, etc. (Chander and Munshi, 2000). Among them, the effect of azadirachtin is the most potent one.

It is worthwhile to mention that leaf, fruit, seed, seed coat, bark and root of neem tree have multipurpose uses (Gautam and Chauhan, 2005). Flower contains sterols, thinomyl alcohol, benzyl alcohol, benzyl acetate and other essential oils, while leaf possesses nimbin, nimbinine and querctin. Similarly, nimbin is found in its bark. Nimbidol present in seed is antipyretic in nature. Neem is also used to reduce mental tension, cure cancer at early stages, anaemia, jaundice, fever and skin diseases.

Safe storage of grains is very important as insects alone can themselves cause a lot of damage either in the fields or in storage. Research conducted either in this line indicates that insects create loss about 25-30 per cent in total grain production. Use of pesticides for safe storage has produced a number of hazards like occurrence of pesticide residues in the form of DDT, BHC etc., in various food stuffs. But application of dried leaves of neem after wrapping in the papers and placing them in the upper, middle and bottom of the granaries, has found effective in making safe storage of grains (Gupta and Sharma, 2005). Use of dried leaves of drek (*Melia azedarach*) and vana (*Vitex negundo*) are also used in the similar manner as has been explained above incase of leaves of neem (Gupta, 2010a and Gupta, 2010c).

In recent years, apart from neem a large number of other plants have been found to possess toxic compounds, which are effective against an array of obnoxious pests, A list of such plants, their parts to be used for control of insect pests is presented in Table 10.1. The data indicate that the leaf extracts of lantana (*Lantana camara*), Tulsi (*Ocimum sanctum*), and Vetiver (*Vetivera zizanoides*) are found useful for controlling leaf minors infesting potato, beans, brinjal, tomato, chilies, cucurbits and onion. Botanical product RD-repelin, a combination of extracts from seed of neem, pongamia, mahua and sesame can be effectively used to control the incidence of aphids in pea and white fly in lady finger (Chander and Munshi, 2000). Plant species like garlic, turmeric and sweet flag are the other alternatives, whose products have given promising results in controlling various pests (Table 10.1).

Table 10.1: List of plants, whose parts are used for controlling insect pests (Adapted from Chander and Munshi [2000])

Name of the plant	Parts to be used	Name of the pest to controlled
Neem (*Azadirachta indica*)	Various products	These are effective against 200 pests
Champac (*Michelia champaka*)	Flower extract	Mosquito larvae
Lantana (*Lantana camara*)	Leaf extract	Leaf miners infesting potato, tomato, brinjal, chilies, onion
Tulsi (*Ocimum sanctum*)	Leaf extract	Leaf miners infesting potato, tomato, brinjal, chilies, onion
Vetiver (*Vetivera zizanoides*)	Leaf extract	Leaf miners infesting potato, tomato, brinjal, chilies. onion

Contd.

Marigold (*Tagetes erecta*)	Crushed roots when applied to soil	Root knot nematodes
Marigold (*Tagetes erecta*)	Extract from roots	Yellow fever
Grapefruit (*Citrus paradist*)	Fruit extract	Colorado potato beetle
Drek (*Melia azedarach*)	Bark extract	Potent anti-feedants against tobacco caterpillar
Garlic (*Allium sativum*)	Various parts	Acts as a repellent for various pests
Drumstick (*Moringa oleifera*)	Root extract	Inhibits the growth of many bacteria
Bael (*Aegle marmelos*)	Leaf and fruit extracts	Various pests
Basuti (*Adhatoda vasica*)	Leaf extract	Various pests
Sweetfly (*Acorus calamus*)	Leaf extract	Various pests

* Adapted from, Chander and Munshi (2000)

10.6.1.2 Use of Herbal Pest Repellents

A karikali based group in Tamil Nadu, Vazhviyal Multiversity which called itself a university with multifarious ecological roles, has produced an herbal pest repellent. This repellent is prepared from the leaves of five plant species. These species vary from place to place, but ideally the repellent will have neem, tulsi and datura. The leaves are collected, cut into pieces and pounded. Keeping it in a compost pit for 10 days ferments the biomass which is then put in an earthen pot filled with cow urine. The fermented solution is filtered with muslin cloth and water is added. This solution can be used as an herbal spray. The only catch is that it should be used before the insects appear ance. In Tamil Nadu more than 8,000 farmers in some about 10 districts have been using this herbal pest repellent.

10.6.2 Use of Cow's Urine and Dung

Many of the traditional farmers still use the urine of cows to combat crop pests and diseases. When cow's urine is sprayed at 10 per cent concentration to diseased plants, they soon recover and become healthy. Cow's urine is also served as foliar application of nitrogen to the withering plants (Gupta and Sharma, 2005)

Although recently after neem, turmeric and jamun, cows urine traditionally used for medicinal purposes as well in India has been patented in USA as distilled bio-enhancer yet its role in containing the insects on crops has to be realized.

10.6.3 Use of Neem for Crop Production

The estimated market amount of neem based pesticide in the country is Rs 1000 million (100 crore) and is growing by 7-9 per cent annually, according to EXIM Bank study (Anonymous, 2015). There are about 20 million neem trees in India and over 60 per cent of the entire global neem trees population is found in India. Neem bears 3.5 million tonnes of Kernels every year and from this around 0.7 million tonne of neem oil can be obtained. Total neem oil produced in India about 0.25 million tonne, which is only 30 per cent of the total potential. Neem based products exports by India in 2012 were at about 5.73 US million dollars. USA and Italy are the leading importers of neem extracts.

10.6.4 Biological Control Technique

Most of the insects, plants and animals are friends of farmers. For example, out of 100 insects found in farmers fields, only one kind of insect will be a pest and others are friendly insects, so he can reduce number of pests by putting these friends. This technique called "biological control" does not kill all the pests, but some crop damage may still occur. The yield, however, may not be affected. For instance, insects such as leaf caterpillars chew on the leaves of corn plants, making them to seem sick, but are not seriously injured and continue to grow. Thus biological pest control works better when no pesticides are applied.

10.6.5 Bacterial Insecticide or Microbial Control

Use of pathogenic bacteria, fungi, nematodes and viruses for curtailing crop pests and diseases is called Microbial Control. For example, a spray of Bacillus thuringiensis (Bt) which infects digestive tract of certain larvae, is now available in its several strains. Many commercial formulations of Bt have now been registered in the country for controlling pests of cotton, pulses, tobacco and vegetables (Sood et al., 2008).

10.6.6 Use of Other Microbes

Of the several fungi known to infect insect-pest, green muscardine fungus, has shown promise against pests like pyrilla on sugarcane. Fusarium oxysporum and Verticillium lecanii control the green bug on coffee. Metarhizium anisopilae, Bavuria or Beauveria bassiana and Nomurea relyi, are the most popular and available in different parts of the country. Hirsutella thompsoni, is of the latest which showed a great promise for managing the coconut mite and is commercially marketed in southern India as Mycohit. This fungus is the only solution to this pest as all the pesticides failed to manage this pest on coconut. Trichoderma species are used to control soil borne disease and offers protection throughout the crop growth period (Rajeshwari et al., 2002).

10.6.7 Bio-Agents

Bio-agents like Trichogrammatids, Chrysopids and Nuclear Polyhedrosis Viruses (NPV) are very common these days. Trichogrammatids are used against caterpillars damaging sugarcane, cotton, paddy, oilseed, pulses and vegetables. NPVs are specific to their respective hosts and are used for controlling two insects viz., Helicoverpa armigera and Spodoptera litura. All these bio-agents are being produced on large scale and commercially available. However, of the various types of viruses, the NPV has shown the greatest potential because they are more virulent (Sood et al., 2008).

10.6.8 Cultural Controls

It is one of the oldest and cheapest methods of pest suppression. Rotation of crops, mixed cropping, intercropping, tillage and sanitation are some of the cultural control techniques.

10.6.8.1 Crop Rotation

Crop rotation can be one of the most effective means of controlling insect-pests, diseases and weeds in low input agricultural systems. In fact, crop rotation makes life difficult for insect-pests, as they are discouraged by growing different crops each season. For example, nematode population will be reduced in fields infested with Meloidgyne graminicola if mustard is grown prior to sowing deep water rice mixed with millet or sesame. Similarly a change of crop rotation in Krishi Vigyan Kendra Farm Jammu, J&K State from rice-wheat to rice-berseem during 1997-98 has not only proved profitable in enhancing the yield of rice and wheat, and soil fertility, but also control obnoxious Phalaris minor weed in wheat during 1998-99 (Gupta and Sharma, 2002).

10.6.8.2 Mixed/Intercropping

Mixed cropping and intercropping systems are traditional practices and in a way cover the principle of risk distribution in rain-fed agriculture as well as dis-favouring of insect-pests. Mixed cropping refers to the system of growing of two or more different crops in the same area with no row arrangement. Mixed cropping imitates nature and does not favour one pest species over another. Intercropping refers to the system of growing two or more different crops together in the same area in separate rows. It is in fact one form of mixed farming which helps in alleviating certain pests by making microclimate less favourable for them. It also hinders the free movement of pests along the plants of some species. For example, intercropping of cabbage with tomato reduces the infestation of diamond back moth in cabbage. Cabbage grown with white clover or French beans mitigated the population of aphids and root fly compared to monoculture.

In china, farmers intercrop sorghum between cotton plants because sorghum attacks the natural enemies of cotton. In the Philippines, farmers control corn stalk borer by mixing corn crops with those of groundnut. The groundnut crops provide hiding places for predator spiders which attack stalk borer. If we plant strong smelling herbs among crops and around crops, we will find that pests are less likely to eat in these areas. For instance, if we grow coriander among crops we will attract the parasites of pea aphids. When some rows of leguminous crops are grown in this system after every five to six rows of main crop, the farmers are helped to prevent soil erosion to some extent and increase the soil fertility by fixing the atmospheric nitrogen. Besides attracting natural enemies, intercropping controls weeds because they cannot find room to grow and spread.

10.6.8.3 Ploughing and Tillage

Ploughing and tillage expose inhabiting insects and destroy off plants or weeds which provide food and breeding sites for many pests. Summer ploughing resulted in high mortality of cut worms, white grubs, grasshoppers and soil inhibiting caterpillars and beetles. For example, heavy ploughing has been found to kill grubs of red pumpkin beetle in cucurbits and cut worms in various other vegetables (Chander and Munshi, 2000).

Deep tillage not only helps in reducing the insect-pest and weed problem but also breaks the hard pan in subsoil layers, allowing thereby the roots to penetrate the sub-soil and extracting more moisture as well as plant nutrients from the deeper layers for the growing plants. Deep tillage is usually carried out with the help of mould board plough or tractor drawn mould board or disc plough.

10.6.8.4 Sanitation

Sanitation involves destruction of sites where pests breed or hibernate. Destruction of crop refuse, disposal of garbage, fallen fruits from orchards, removal of diseased plants and pruning's of infested plant part offers an economical and effective way to check plant pests and diseases. For example hibernating pink boll worm larvae in double seeds of cotton are easily killed if cotton seeds are crushed or fumigated in ginneries and crop refuse burnt before the emergence of adults in the next season. De-trashing and destruction of infested lower leaves of sugarcane is proven practice for suppressing scale insects and red rot disease (Saxena, 1996). To safeguard farmers' interest, there is urgent need to supply pure water to rural communities (Pandit, 2011).

10.6.8.5 Resistant Varieties

The first and the foremost endeavour should be to look for resistant varieties of the crops because if resistant varieties are available then there remains little or no need for other controls. Plant resistance provides an ideal adjuvant with other methods when resistance alone can not maintain a past problem below economic level. But even varieties with low levels of resistance may require less insecticides than susceptible varieties of different vegetables are shown in Table 10.1 (Chander and Munshi, 2000).

10.6.8.6 Clean Cultivation

Clean cultivation refers to the use of healthy seed and destruction of crop residues and weeds assist in reducing pest population and preventing its carry over to meet the next crop. For exampie, destruction of cucurbits vines after the cropping season overcomes the carry over of pumpkin beetle to the next season.

10.6.8.7 Planting Time

Planting time for growing different crops and particularly vegetables should be adjusted in such a way that susceptible stage of the crop synchronizes with the most inactive period of pest number. For example, early sowing of the cucurbits escaped from the attack of the red beetle. Similarly, early planting of chilies suffer from leaf curl than late sown crop. Likewise spring summer crop of lady finger inflicts less with yellow vein mosaic virus than rainy season crop. October to November planting of tomato afflicts less with white fly attack and leaf curl disease.

10.6.8.8 Mulching

It can be used as an important tool to reduce the incidence of pests in crops, especially vegetables. Black plastic mulch can decrease the incidence of aphids in summer squash; aluminium painted plastic mulch is quite effective against onion thrips at seedling stage (Chander and Munshi, 2000).

10.6.8.9 Judicious Use of Fertilizers

Plant growth is dependent upon the nutritional status of the soil, which in turn has indirect effect on pests. Generally higher doses of nitrogen favour the pest build up, while high doses of phosphorous and potassium have depressed effect on the pests. High dose of nitrogen increases the incidence of fruit and shoot borer in brinjal, aphids in cauliflowers and fruit borer in okra. Application of potassium and phosphorous can help to reduce the injury caused by fruit and shoot borer in brinjal.

The incidence of stem borer in lowland rice crop grown in clay loam and sandy loam soil was studied to find out the efficiency of potash and its time of application. Addition of 50 kg of potash through MOP reduced stem borer incidence by 4.2 and 55.9 per cent in these soils, respectively. Similarly, its split application at basal, active tillering, panicle initiation and boot leaf stage reduced incidence by 2.81 and 28.9 per cent in clay loam and sandy loam soils, respectively as compared to basal application.

10.6.8.10 Trap Crops

Trap crops provides protection by preventing the pest from reaching the main crop. The trap crop must be more alluring to the pest and less economical than main crop. Mustard along with cabbage as trap crop can reduce the incidence of diamond black moth, leaf webworm and aphids. Marigold planted with tomato as trap crop is highly effective against fruit borer Heliothis. Similarly putting trap rows of tobacco or marigold in cotton field is used for diverting the egg laying females of Helicoverpa on these crops.

Putting few lines of pearl millet or sorghum in between fields of okra or tomato have proved very effective in controlling spread of leaf curl virus disease as they act as barrier for the movement of white fly which is the vector of the disease.

10.6.8.11 Flooding

Flooding is considered as an effective control method for checking leaf minor in Cole crops. Flooding is also effective against cutworms.

10.6.9 Mechanical Control

Mechanical control refers to the use of mechanical devices or manual operations for pest control. Hand picking of inactive adults, conspicuous egg masses and large sized larvae can be done during early stage of infestation when the population is low. Dipping in kerosene water can kill these pests. The examples include collection and destruction of adults, larvae and eggs of hadda beetle infesting brinjal, tomato and potato, adults of red pumpkin beetle in cucurbits melon fruit flies and okra fruits infested with fruit borer.

10.6.10 Organic Soil Amendments

The use of well decomposed organic matter as soil additives is well established and ancient agricultural practice used by small and large scale farmers alike. Amendments alone improve the nutrients and water holding capacity of the soil, thereby, improving the plant growth and hence increasing tolerance to nematodes. Higher organic content in soils also stimulates microbial activity of beneficial soil microorganisms antagonistic nematodes sustainable agriculture systems would be expected to have soils with the high organic matter content.

10.7 Agro-Ecosystem Analysis

Regular pest monitoring survey and surveillance is essential to increase the activities of the peasants for meeting the challenges. The analysis of the agro-ecosystem under which the pests and their natural enemies thrive is vital for Integrated Pest Management (IPM). The crops, other plants, pests, their natural enemies, pollinators etc., changes in land use pattern, agronomical practices followed and ever fluctuating weather factors have to be monitored under an improved surveillance programme. Now a days, computer aided pest prediction are available for main crop, vegetable and fruit pests in many of the developed countries.

10.7.1 Integrated Pest Management (IPM)

Challenges and opportunities of sustainable food supply for the next 35 years are critical in terms of biotic stresses, environment, and water energy for the Nation and across the globe. Keeping in view the losses in food production due to biotic stresses, the role of IPM is well cut out to ensure sustained rise in food supply while minimizing use of chemical pesticides an ensuring safety to water and environment (Chattopadhyay, 2015). Injudicious use of chemicals for pest management in certain crops have resulted in the development of pest resistance to pesticides, pest resurgence, development of secondary pests into major ones, an increase in production costs and uneconomic crop production. IPM provides a long term solution to these problems by employing as many pest management techniques as possible for using chemical pesticides rationally.

Practically, it is not possible to get rid of all the pests that feed or live on plants because the natural ecological balance is lost if all the insects are destroyed. Infact, in the biological cycle all the insects depend upon each other and if a particular stock of insect is perturbed, the others will be deprived of the food. In other words, we can say that every category of insect species have to exist for the maintenance of ecological balance. Thus, it is on this principle, the notion of IPM has been devised (Gupta et al., 2005, Singh, 2012).

The IPM system is a broad ecological approach for managing pest problems rather than to have complete pest eradication. As a matter of fact, |PM system is considered to be the best in containing the insect pests of various crops, fruits and vegetables. It is because the IPM system is a mixture of all the available methods and techniques for total crop management in order to check the pest population below the economic threshold level (ETL). Insecticides are used only when they can be justified economically and ecologically.

In IPM approach, much more stress is given on the use of resistant/tolerant varieties of crops vis-a- vis varieties having transgenic genes, surveillance and monitoring top find out ETL for using need based pesticides judiciously. Apart from the above, conservation and augmentation of natural enemy resources like Nuclear Polyhedrosis Virus, parasites, predators, fungal antagonists and use of non-chemical methods like botanic pesticides and pheromones, are the other methods included in the IPM system.

10.7.2 Aim and Objective of IPM

The main aim and objective of the IPM system is to optimize pest control in respect of overall economic, social and environmental gains. Another objective of IPM is to produce maximum crop production using minimum cost, with minimum damage to the environment. Further, it is added that all the necessary precautions are taken up to protect the natural enemies of insect pests.

10.7.3 Basic Premise of IPM

The basic premise of IPM is that no single method can be successful over a long period. Therefore, a mixture of biological, physical and chemical methods is integrated to sustain a pest management system.

The ultimate goal of IPM is having sustainable agricultural systems with minimal use of pesticides (Singh, 2012).

10.8 Conclusion

In the present chapter of the book natural ways/methods to check the crop pests have been deliberated, which is the need of the hour. These mostly consist of use of bio-pesticides in the form of botanicals and herbal pest repellents. Use of cow's urine, biological control techniques, and microbial and cultural control measures have also been discussed. A brief account with regard to Integrated Pest Management stands discussed. Apart from the natural methods to check the pests, pesticide usage trend, public health issues occurred due to misuse or overuse of pesticide in various crops have also given.

10.9 Future Research Strategies/Priorities

- The Environment Awareness Societies must launch an Environmental Awareness Campaign in various areas regarding the impact of indiscriminate use of pesticides.
- The M.Sc. and M. Tech. Students of Environmental Sciences can be chosen to act as a unique means of disseminating environmental awareness at the grass root level. For this purpose; they can present street plays on various themes. Involvement of such students has performed a very good job in gathering villagers - men, women and children - in large numbers, in parts of Haryana and Uttar Pradesh to create environmental awareness.
- The village folk can be persuaded to enter into a live interaction with the students and faculty members of the Agricultural/Traditional Universities, showing their keenness to know more about the environment which belongs to all of them. As an insensitive, the students involved in this connection must be given prizes. While distributing such prizes the *sarpanches,*

panches of villages and chairmen block samities must be invited to enhance an impact.

- The plays like *"Nai Subah", "Paryavaran (Environment) Hamara Jeevan Data Hai. "Paryavaran Bachao Desh Bachao"*, must be centered on the themes of importance of trees. Ill-effects of deforestation, the problems of polythene bags and waste disposal and adverse effects of environmental pollution - soil pollution, water pollution, air pollution etc., should also be emphasized to focus among the rural masses.
- A multi-residue technique which can detect about 30 pesticides in a single analysis must be popularized in the various Agricultural Universities, ICAR Institutes and other National Institution as has been done in Haryana Agricultural University.
- The farmers must wash their hands after spraying the pesticides to various crops, fruit plants and vegetables. They should put on masks on their mouth while spraying.
- The peasants should always use high quality chemical pesticides and must purchase them from right and authentic sources.
- It is very essential that pesticide be used judiciously failing which they will have adverse effects on humanity and the ecosystem.
- It is desirable that all the fruits be washed thoroughly prior to their consuming. Similarly before cooking the vegetables, these must be washed thoroughly.
- Timely weeding helps to protect the crop from pests of all kinds, so it must be done in time.
- Varieties of crops, which are sown should be recommended ones for the region and must be resistant against insects or diseases.
- There is a dire need to monitor the use of the pesticides and creation of awareness among the masses about the right use of the pesticides.
- Advise the farmers to use balanced doses of NPK fertilizers in the ratio of 4: 2: 1.

Study Questions

1. What is meant by bio-pesticides? Write its importance in brief.
2. What do you understand by Integrated Pest Management?
3. Name various cultural methods used for controlling insect pests and write the details of any one of them

4. Write a paragraph on the importance of mechanical control.
5. Write a short note on minimata disease.
6. What is the base of using neem tree as botanical pesticide?
7. Name various plant/tree species which can serve as bio-pesticides.
8. Write in detail about the bad effects of indiscriminate use of synthetic pesticides.

References

Anonymous. (2006). Pesticide jeopardy - conduct epidemiology study, says court. Down to Earth 14 (22): 7 [April Issue].

Anonymous. (2015). Crop protection - striking a balance between food safety and food security. Agriculture Today 17 (6): 20-24

Chander, S. and Munshi, A.D. (2000). Eco-friendly botanical pesticide. Intensive Agriculture 33 (1-2): 37-39.

Deshmukh, A. and Sabalpara, A.N. (2011). Village level preparation of Neem bio-pesticide for sustainable agriculture. Agrobios Newsletter 10 (5): 27-29.

Gautam, K. and Chauhan, S.V S. (2005). Neem: A treasure house. Indian Farmer Digest 38 (3): 38-39.

Gupta, R., Singh, P.P. and Awasthi, R.P. (2005). Need of IPM in sustainable agriculture. Indian Farmer Digest 38 (12): 29-30.

Gupta, R.D. (2005c). Environmental Degradation of Jammu and Kashmir Himalayas and Their Control. Associated Publishing Company, Karol Bagh, New Delhi (India).

Gupta, R.D. (2006a). Environmental pollution: Hazards and control. Concept Publishing Company, New Delhi (India).

Gupta, R.D. (2010a). Indigenous technical knowledge for farmers. The Daily Excelsior 46 (61): 6 [March 3, Wednesday].

Gupta, R.D. (2010c). Organic agriculture key to food security. The State Times 15 (161): 6 [June 16, Friday].

Gupta. R.D. (2011b). Science and empowerment of women. The Daily Excelsior 47 (46): 6 [February 15, Tuesday]

Gupta, R.D. (2013d). Heavy metals poisoning - A threat to life. The State Times 18 (273): 6 [October 3, Thursday],

Gupta, R.D. and Sharma, A.K. (2005). Indigenous technologies of Siwalik farmers. Indian Farmer's Digest 38 (7): 27-30.

Gupta, R.D. and Singh, H. (2008). Indiscriminate use of pesticides in agriculture: Public issues and their control. Indian Farmers' Digest 41 (1): 8-13.

Gupta, R.D. and Tikoo, R.K. (2005). Neem: A potential organic fertilizer. Command Areas Development Subzar 3 (1): 5.

Gupta, R.D. 2016. Wildlife and Wetland Ecosystems of Jammu and Kashmir Himalayas. NR Books International, Pacca Danga, Jammu Tawi, J&K INdia.

Gupta, R.D. and Sharma, A.K. 2002. Organic farming in J&K. The Daily Excelsior. 38(330):6.

Mahajan, A. Sood, Chitra, Chopra, P., Kalia, B.D. and Padder, B.A. (2004). Pesticide residues in food products. Agrobios Newsletter 2 (10) 20-21.

Mainac, C, Devi, S. and Rajan, A. (2011). Bio-pesticides in sustainable agriculture. Agrobios Newsletter 10 (5): 38-39.

Pandit, J. (2011). Water and sanitation for rural communities. The Daily Excelsior 47 (46): 6 [February 15, Tuesday].

Rajeshwari, E., Krishnaveni, S.A. and Balamurugan, P. (2002). Trichoderma in sustainable agriculture. Intensive Agriculture 30 (1-2): 20-21.

Saxena, R.D. (1996). Report on awareness programme on pesticides and sustainable agriculture. 228, Sidharth Enclave, New Delhi (India).

Sharma, A.B. (2005). Farmer's suicides: Searchlight on the killing fields. Farmers' Forum 5 (12): 14-15.

Sjngh, G.P.I. (2006). Green Revolution and after the public health perspective. Farmers' Forum 6 (6): 16-21.

Singh, Y. (2012). Biological aspects of organic farming for sustainable agriculture. Indian Farmer's Digest 45 (5): 35-39.

Sood, P., Choudhary. A. and Prabhakar, C.S. (2008). Microbial insecticides today's concern for better tomorrow. Agrobios Newsletter 7(5):24-27.

11

Rural Women Role in Sustainable Agriculture

Abstract

Women have always lived in harmony with fields and forests. It is believed that about 12,000 years ago when first agriculture came to fore, the women used to cultivate plants in their dwelling places while men used to hunt the animals. Presently, there are about 6 lakh villages in India and in every village the women folk become the backbone in agriculture. In addition to this, the women are also amenable to bring about sustainable agriculture. How the rural or farm women assist in creating sustainability in agricultural production has been embossed in this chapter of the book? Women and origin of agriculture, their contribution in food grain production, women and farming system approach, livestock rearing, Integrated Nutrient Management (INM). environmental conservation, are the crucial topics which have been prescribed in this chapter of the book. Conclusion and future suggestions/ strategies have also been presented along with study questions and references.

Keywords: Agriculture; agricultural production; food grain proauction; sustainable agriculture: farming system approach; environmental conservation; INM; villages; women; forests; animals; livestock rearng

Agriculture is the main stay of Indian economy as more than 60 per cent of the people of India are dependent upon this evocation and allied disciplines for their livelihood. Even today, at the dawn of 21st century, the Indian agriculture has continued to be the backbone of our National economy, representing country's about one seventh of Gross Domestic Product (GDP). Thus, farming families consisting both of men and women, boys and girls, known as the pillars of our agriculture constitutes backbone of the Nation. The support and contributors made by the rural women in agriculture, consisting 50-60 per cent of farm work force are, however, not less than those performed by the men, and therefore they become spine of the rural economy. Infact, women's role in the economy has been invisible and unappreciated. If we compare the women's work participation rate of some of the selected countries like Indonesia, Bangladesh, China, Pakistan and Sriianka, we find that India ranks second in position (22.7%) atter Indonesia (33.4%). In India, nearly 74 per cent of the entire female working force is engaged in agricultural operations i.e., 28 per cent as cultivators and 48 per cent as agricultural labours. Generally, 60-70 per cent of labour input is provided by women who increase to 80 per cent during agricultural peak season. In fact women participation in economically agricultural productive activities is not new but is a matter of common observation that the women not only engage themselves in many farm operations in their own fields, but also in the fields of others as hired labourers (Gupta, 2013g). It is believed that women's share of labour hours is increased disproportionately to that of men. Women constitute almost half of the agricultural labour force and contribute more towards family income.

The historians believe that it was the women who, for the first time, started growing of crops and domesticating animals, and, thereby, an art and science of agriculture/farming initiated. These days, about half of the available global human resources are women (Singh et al., 2000). The women, contributing to agricultural production, are usually called as "Peasants women" i.e., those women who work in the field but are not the wives of the farmers. On the other hand, those women who are the wives of the farmers are called "Farm Women". On an average, farm women work for longer hours than the men i.e., 69 hours week^{-1} as composed to 57 hours week^{-1} for men. Most of this is classified as house work (Table 11.1) which is unpaid or is subsistence levels economic activity, and brings little remuneration. Then there is the system of "Patriarchy" inheritance whereby daughters, sisters, wives and widows, are entitled to smaller shares of property and decision making (Khetarpaul and Grover, 2001).

Table 11.1: Daily work pattern of rural women (Adapted from Khetarpaul and Grover [2001])

Sr.No.	Activities	Time Spent (hours)
1.	Childcare	1.00
2.	Food activities	2.15
3.	Hauling fuels and water	2.15
4.	Agriculture	5.00 - 6.00
5.	Allied	1.00
6.	Miscellaneous	2.00

Although, India has achieved self-sufficiency in food grain production yet its sustainable supply is a challenge in the light of the burgeoning human population. In developing countries, agricultural production is mainly dependent upon manual labour the lion's share of which comes from the rural women. This situation also holds well in India, where women provide 75 per cent of the agricultural labour. Nearly, 84 per cent of all economically active women in India are engaged in agriculture and allied activities. Roughly, they constitute one-third of the agricultural labour force and 48 per cent of self employed farmers. It is worth mentioning that there are 75 million women as against 15 million men in dairying, while the women engaged in animal husbandry account for 20 million as against 1.5 million men (Meghwal and Singhal, 2000). Not only this, establishment of Kitchen gardening, laying out of lawns, and managing of floriculture, is handled by the women efficiently. The main focus of the chapter is to highlight women's role in "Sustainable Agriculture".

11.1 Women and Origin of Agriculture

History gives evidence that it was the women who started to raise or grow crop plants about 12000 years ago and, thereafter, an art and science of farming or agriculture came into existence. Again, it is believed that they were the women who for the first time began to demonstrate animals about 25,000 years ago (Gupta, 2006). Thereafter, science of rearing of cattle i.e., animal husbandry came into being. It is accepted as true that while men used to go out for hunting in search cf food, the women used to collect seeds from the existing flora and set out cultivating them for obtaining food, fodder, fuel and fibre, which all make up the basic necessities of life. In this way the women have an umbilical attachment with agriculture since time immemorial.

11.2 Contribution of Women in Food Grain Production

Rural areas are most often associated with farm and home activities. The contribution of the farm women in enhancing food grain production has a special significance which could be judged in terms of influencing the farmers with respect to the following:

- Accepting and adopting new technologies in agriculture including horticulture, olericulture, floriculture and animal husbandry to increase their production, and, thereby to enhance income of the farmers
- Modernizing the farms through improved implements, tools and other farm machinery.
- Developing the farms with irrigation facilities depending upon the availability of irrigation sources.

The roles of men in performing such activities are numerically insignificant. In other activities like thinning and gap filling, weeding and hoeing, harvesting and threshing vegetable, picking, processing and storage, the farm women share the work with men folk. Livestock husbandry is an important activity, which can be looked after very nicely and efficiently by the women. It has been found that women can manage the dairy animals and small scale poultry better than men. Women can very easily look after up to 3 cross bred cows in addition to taking care of her family. From these cows, she can earn an annual income of Rs. 15,000-18,000 (Meghwal and Singhal, 2000).

In livestock husbandry or livestock farming, the women are actually involved in operations like feeding, breeding, management, health care and marketing of animals including green fodder production It has been observed that in dairying 75 million women are involved as against 15 million men.

These days the participation of rural women has become an utmost importannt in decision making processes relating to various farm activities. Further, an involvement of rural women in agriculture has bean felt to make them knowledgeable or create awareness about the ill effects produced by the "Green Revolution" and tackling them effectively. For example, the rise of agro-chemical inputs in developing countries under "Green Revolution" and the migration of male heads of households from farmlands to cities, in search of urban jobs leaving the women to take on an increased burden of agricultural operation This has resulted in getting the women exposed to toxic chemicals in increasing numbers leading to endocrine disruption related health problems. Endocrine disruption, the World Watch Institute's State of the world 2002 reports note, "is a far more serious health problem than cancer". Agro-chemicals have also been associated with pregnancy failures and infact developmental problems. It also becomes imperative to make the women familiar with the latest agricultural technology like role of biotechnology, use of bio-pesticides, bio-fertilizers and vermicompost, concept of integrated pest management and integrated nutrient management system, which all are essential for sustainable agriculture.

11.3 Rural Women and Sustainable Agriculture

As rural women prepare FYM/compost, fertilize the fields with these manures and, thereby, protecting the health and fertility of soil, and eventually the crop productivity so the food production is ensured. Apart from this, they provide quality seeds, water and bio-pesticides which don't harm the environment. Thus, the rural women always help in rendering the agriculture sustainable.

11.3.1 Women and Basic Life Support System

"Men and women both are two wheels of society and contribution of both is very essential for building healthy nation. In rural sector 56 per cent of the male and 33 per cent of the female were in the labour force (Badiger et al., 2013). According to the Lawrence et al. (2014), the present statistics shows that inequality of men and women in education, employment, politics and decision making etc. reservation for women in Panchayat Raj institutions has encouraged their participation in the decision making process at the grass root level. Rural women are playing and will play a pivotal role in conserving/preserving basic life support system such as land, soil, water, flora and fauna. In rural areas, they usually get engaged in farm and home activities. The nature and extent of women's involvement in farm activities, however, varies widely not only from region to region but also within the same region, depending upon the differences in the ecological sub-zones, castes and classes, stages in the farm cycle and farming systems. Hardly, there is an activity in agriculture of food production, except heavy physical labour like soil excavation or ploughing the fields where in women is not actively involved. Infact, the rural women are able to perform a number of operations more efficiently than performed by the men.

11.3.2 The Myth of the "Green Revolution"

Most of this mayhem has been carried out in the name of food production. Although the "Green Revolution" increased agricultural production yet it required higher and higher inputs to get the same yield. For example, according to the Economics and Statistical Departments of Punjab, in Ludhiana district during

1970-71, 85.3 kg fertilizers ha^{-1} was applied to get 32.79 q yield of wheat ha^{-1}. Ten years later, during 1981-82, applying 167.2 kg fertilizer gave 33.79 q yield ha^{-1} of the same crop. Thereafter, however, without balanced use of NPK fertilizers, one could not obtain continuously sustained high yields without detriment to soil, so increased use of farm yard manure along with NPK fertilizers was found suitable. Use of nitrogenous fertilizers alone has started giving negative effect and proved most disastrous in low pH soils (Kanwar, 1999). Moreover, the deficiency of micro-nutrients like Zn, Fe, Mn and Cu has become a yield limiting factor in many soils. Above all, indiscriminate use of fertilizers has also created a number of health hazards both to the soils and human beings. The details of such hazards are described elsewhere (Gupta and Singh, 2006) Sickness of soil has been seen in terms of decreasing content of organic carbon, available N, P and K contents particularly in rice-wheat growing soils, exhausting of secondary (Ca, Mg and S) and micro- nutrients (Fe, Mn, Cu, Zn, B and Mo) and loss of macro and micro-organisms. There is contamination of groundwater with NO3 and accumulation of heavy metals like Cd, Cr, Ni, Pb etc.

Presence of flourine (F) above permissible limits has been found in water samples of some parts of 15 Indian states (Gupta and Singh, 2006), causing fluoride poisoning in about 20 million people. Application of nitrogenous fertilizers alone without phosphatic and potassic fertilizers has caused cancer in human beings.

Millions of hectares of once productive farm land have gone, and continues to go, out of production. The elite farmers who could manage their farms scientifically were able to optimize the use of inputs like improved varieties, irrigation facilities and agrochemicals for economic crop production. However, a large number of small farm holders have ended up with miserable failures. There are a number of examples, where farmers have lost their soil productivity and crop yields due to excessive use of water for irrigating the crops or improper use of agrochemicals, particularly crops like cotton. Indiscriminate use of chemicals and pesticides has resulted in reduction in the biodiversity of natural enemies, outbreak of secondary pests, development of resistance to pesticides and contamination of food and the ecosystem as a whole (Gupta and Singh, 2008). Above all, without reliable lean season, water supplies and stable soils, our food security is unquestionably at risk. All this reaffirms that intensive use of external inputs may not be good under all situations.

11.3.3 Agriculture Models Devised by Women

In case of small and marginal farmers, particularly in backward areas or in arid zones, it is necessary to encourage them to build up their farming operations gradually with very little use of external inputs. Here, the rural women can help in devising some agricultural models, which can sustain agricultural production and productivity.

11.3.3.1 Bara Anaj (Big Grains) Mixed Cropping System

The women have less working capacity than man can do. Hence, they devise small and efficiently agricultural models with more sustainability due to having traditional knowledge and skill of various farm operations. For example, "Bara Anaj", the typical mixed cropping system of Uttaranchal (Uttarakhand) hills has been evolved by the rural women. In this farming system, almost all cereals and pulses required in a household are cultivated in the same fields, hence the term "Bara Anaj". This is totally self-sufficient sustainable agricultural practice, preferred by the women in the hills. Thus, keeping in view the traditional knowledge and skill of farm women in different farm activities, it will not be wise to ignore the possibility of increasing avenues, which uphold the country's food status through utilization of this unattended and unrecognized force.

11.3.3.2 Zabo and Apatani Farming System

In the North Eastern Hill region, covering the states of Meghalaya, Nagaland, Manipur, Tripura, Arunachal Pradesh, Sikkim, Assam and Mizoram, there exist viable and sustainable indigenous farming systems, which were developed by the tribal farmers including farm women through their ingenuity and skill. "Zabo" farming system and "Apatani" farming are some of the examples. In Zabo farming system, Chakheson tribal farmers in Phek district (Nagaland) have combined the forestry, agriculture and animal husbandry with soil and water conservation (Thakur, 1999; Sharma, 2007). In Apatani farming system, Apatani tribal farmers have devised rice-based farming combined with finger millet production and fish culture.

Apart from the above, in tribal societies of North Eastern Hill states, where jhuming farm or shifting cultivation is generally practised, women play an important role in agriculture. They clean the jhum land, sow the seeds and harvest the crops (Malik et al., 1999). It is worthwhile to mention that throughout the Himalayan region the major role in agricultural production is played by women. Traditionally, women perform many agricultural activities like transplanting, sowing, weeding, harvesting and threshing.

11.3.4 Close Linkage of the Women in Sustaining Food Cycle and Food Security

Women have played and will continue to play a paramount role, particularly providing a close linkage in sustaining the food cycle They assist in feeding animals, collecting foliage from bushes and trees, gathering crop by-products and cutting the grasses from grasslands. They prepare farmyard manure and compost, and fertilize the fields with these manures. This protects the health and fertility of the soils and, thereby, their productivity. Infact, agriculture based on women's participation is natural, self reproductive and sustainable as the internally recycled sources provide the necessary inputs like good quality seeds, soil nutrients, moisture and bio-pesticide for pest control. This system of agriculture does not harm the environment, rather maintains ecological balance. In reality, it looks for an eco-friendly approach for evolving and popularizing ecology low cost input and environmentally sustainable basis.

Women play a key role in food security by way of devoting greater shape of their income in food and fuel than luxury items. Three important pillars of food security in which the women fully involve are: i) availability of foods, ii) access to food and iii) utilization of food (Gupta, 2009a). In availability of food, women have a crucial role to play in food grain production. They are indeed active food producers. They work day and night in small and marginal farms. The role of women in urban gardens in producing vegetables canot be over emphasized. It is a point to mention that house wives take care of all the family members in respect of availability of food and obtain their food only after getting them eaten and satisfied. Sometimes house wives live themselves on the left of the food.

Women are primarily responsible for buying all food stuffs to which they devote their time and income. They are responsible for utilizing of food in terms of nutritive values. They decide what types of food enterprises have to be purchased and to prepare them, maintain their taste and nutritive values (Gupta, 2009b).

11.3.5 Farm Women and Crop Diversification

The farm women give their full help and cooperation in managing crop diversification methods. Such methods mostly consist of crop rotation, mixed cropping and multiple cropping. The main benefits of diversified agriculture include checking soil erosion, improvement in soil fertility and enhancement of crop yield (Yadav et al., 2005). This also reduces the need for nitrogenous fertilizer in case of legumes rotated fields. In crop failure due to the pests and diseases, where even if one crops fails there remains of other crop in the field that withstands yield and support the farmer.

As the water level of a number of states has gone down so planting area under paddy cultivation should be reduced to save the water. As for example, the water level in Punjab has gone down to 20m or even below in 50 per cent areas of the state. In 1973, there was only 37 per cent area with subsoil water at 10m (Dhaliwal, 2013b). Depletion of water level has not only affected the paddy crop being a water guzzler but has also created scarcity of drinking water for which women have to make a lot of struggle being one of the household responsibility. Hence a special scheme is required to be taken up for diversifying the agriculture. Target area for alternate crops consists of maize, cotton, basmati rice, sugarcane, fruits, vegetables, pulses, agroforestry and green fodder.

Genetic diversity, location specific varieties of various crops and use of local plant species, shrubs and tree species as an inexpensive source of green manure to build-up or maintain soil organic matter and soil fertility are essential for realizing sustainability. The rural women have proved boon in collecting these materials for serving as green manures. In some areas of India, which are still free from Green Revolution, the farm women thus continue to work there as soil builders rather than soil predators. It is from these areas or tracts of natural farming are emerging the ecological struggles to protect the nature.

Horticulture requires intensive and tender care, which can be handled efficiently by women. Establishment of kitchen gardens, nurseries, flowering gardens, laying out of lawns with growing of medicinal plants, vegetables and fruit trees in addition to grasses are now totally managed by the rural as well as urban women.

In Vausda Taluka of Valsad district in Gujarat, BAIF Development Research Foundation, Pune, a voluntary research organization has promoted a nursery of mango and cashew grafts through tribal women. Each group of 4-5 women have able to raise about 5,000 plants and earn a net profit of Rs. 6,000-8,000 every year (Meghwal and Singhal, 2000).

11.4 Women and Forest Products Gathering Activities

Famine has not been a problem in the tribal areas of Bastar (Madhya Pradesh, India) as the tribes have been able to obtain half of their food from innumerable edible forest products. These tribal use about 165 trees, shrubs and climbers. Out of these trees, 31 plants the seeds of which are roasted and eaten by these tribals, 19 plants whose roots and tubers are eaten after baking, boiling or processing, 17 whose juice is taken, 25 whose leaves are taken as vegetables and 65 plants whose fruits are taken raw, ripe or roasted and pickled. They also constitute an important source of dry season fodder. Forest-based gathering and processing enterprises provide seasonal employment and source of income. Women often dominate forest gathering activities, both for household products and income. For example, collection of minor forest produce, particularly oilseeds like neem, mahua, pongamia is an area where women play a significant role. It has been estimated that neem seeds worth rupees one thousand million go waste every year. The main reason for such neglect is lack of awareness about their values and uses as well as inadequate facilities for processing and marketing. If these issues are tackled, oilseed collection is an attractive item for the rural women to generate additional income thereby boosting significantly country's GDP.

Oil (Margosa) extracted from neem seeds is a useful and powerful anti-helminthic and is effective against ringworm, thread worms and scabies. It is the best against lice and dandruff. Another very important use of margosa oil is to help diabetic patients, heart patients and those suffering from arthritis. It is also useful for vaginal itches and uterine inflammations. The residues, after oil is extracted from neem seed, is called as neem cake, which can be used for seed extract (prepared by the women of Butkheda, Jalna), has reduced the number of chemical sprays from 12 to 6 to control the pests of cotton (Sriram, 1999). Neem cake contains more percentage of nitrogen, phosphorous and potassium as compared to farm yard manure, compost and sewage sludge (Gupta and Tikoo, 2005). In addition to this, it has large amount of calcium, magnesium, sulphur and as such neem cake is widely used in India to fertilize cash crops like sugarcane and vegetables.

The women also collect fuel wood, fodder, water and carry them over long distances on hill paths, and graze the cattle over far off grazing lands or pastures. Apart from fire wood and fodder, bamboo, catechu, traditional medicines fruits, honey etc., there are many other forest products gathered by the women for their own consumption and/or for sale.

11.5 Women and Farming System Approach

To make agriculture more sustainable and environmentally sound, it is absolutely essential to adopt, "Farming System Approach" wherein the role of women can not be underestimated.

> "*Farming System Approach refers to a particular arrangement of farm enterprises viz., cropping, livestock rearing, processing of farm products, which are managed in response to the physical. biological and socioeconomic environment, and according to farmer's goals, preferences and resources*".

Farm enterprises such as livestock rearing, poultry farming, dairy farming, are purely, the monopoly of the farm women. These farm activities not only provide food (milk, and its various products, eggs and meat) but also urine/dung to prepare manures of various kinds (farmyard manure, compost and poultry manure) for the fields to enrich the soil fertility. Thus, when all these farm enterprises are integrated in horticulture and olericulture production, the farming system becomes more productive and sustainable.

11.6 Biotechnology and Women's Empowerment

Biotechnology harnesses nature's own biochemical tools to make new products and processes. Who knows it better than house wives who turn milk into curd day in and day out? In scientific parlance, it is called "conventional biotechnology".

Modern biotechnology renders use of living cells and their molecules to modify products. Provided with the knowledge and technical skill of modern biotechnology, a house wife can change the characters of curd such as taste, thickness, nutritive value, colour, texture etc., and can produce the exact kind she desired. This potential has enabled biotechnology to bloom into what it is called "Sunrise Industry". In the coming years we will be able to involve more and more of this technology not only as science, but also as an industry and as supplier of consumer products. During the inauguration of golden Jubilee Biotechnology Park for women at Chennai, Union Minister for Human Resource Development and Science and Technology pointed that it is the woman "who has kept our traditions and values of life intact, which is the envy of the west". Besides, they play a pivotal role in agriculture, cottage industry, health care and education. Unfortunately, this information is not properly documented and, therefore, lacks visibility. This has marginalized the women.

There is livelihood that the emerging field of biotechnology can assist women in securing economic empowerment. Indian women are already making a mark in biology including some areas of biotechnology. This can be evinced by the fact that about 70 per cent of those who have secured admission to higher studies in biological sciences, are the females. In such important area as Molecular Biology, which is the bed rock of modern biotechnology, a basis is working in favour of women. This is because women are methodical, patent and precise which are essential qualities for painstaking research projects in biotechnology.

Establishment of Golden Jubilee Biotechnology Park for women at Chennai is a remarkable step towards promoting lab-to-land programme in biotechnology employing women. This Park will bring together women entrepreneurs and women scientists and technologists. It has also provided for training women, particularly those having rural background. Molecular farming holds the promise of producing crops of greater commercial value. It opens the possibility for farmers to shift cultivation to crops with significantly higher returns. Already engineered corn is being cultivated for antibiotics, and tomato, spinach and banana for vaccines. Edible plant vaccines dispense the requirement for needles and refrigeration, making them particularly suitable for rural women in developing countries. Golden rice rich in vitamin A is another example. Women can take a lead in growing plants as factories of a variety of pharmaceuticals. Biotechnology can also help in improving quality and quantity of milk in cows. As for instance cows injected with engineered version of growth hormone, called Recombinant Bovine Growth Hormone (RBGH), produce 10-25 per cent more milk.

11.7 Farm Women and Other Activities

Advancement in agriculture and allied sectors, confer up on a variety of options to mitigate penury condition of the rural women. What required at moments is its critical evaluation of options depending upon local specific situations and suitability vis-a-vis transfer of the latest agricultural technology as means to the women empowerment. Some such plausible means in which the farm women participate to assist for sustainable agricultural production are as follows:

11.7.1 Livestock Rearing

With recognition of livestock production as an important contributor towards food security and sustainability, the role of women in livestock rearing has also got recognized. It is only for the last 15-20 years that need was felt to study the role of women in livestock production. In view of their contribution, in this respect and other activities of agricultural production, evidence can be seen from FAO publication which speaks "Most farmers in India are women". Ghosh et al. (2000) has reported that the number of male agricultural labours increased by 31.5 per cent but that of female by 31.8 per cent from 1981 to 1991.

In India, rural women have a major place in agrarian scenario, majority of them being actively involved in various operations relating to mixed farming system. Both women and men participate in agricultural and animal husbandry practices as a family unit on the farms. Major share of work on agricultural side is performed by men, while on animal husbandry side, most of the management operations are conducted by women. Such management operations consist of harvesting of fodder and its chaffing, feeding of animals, milking of animals, cleaning of animals and their sheds, and watering of animals (Ghuman and Gill, 2006).

In the Indian Himalayas, a pair of bullocks work 1064 hours, a man 1212 hours and woman 3485 hours in a year on one hectare farm (Jilani et al., 2005). In hilly areas of Uttaranchal, women manage Angora rabbit, goat, sheep, cow, buffalo and backyard poultry. There is general agreement that women handle most of the critical aspects of dairy animal management like feeding, breeding, milking, cleaning, and care of new born and sick animals (Jilani et al., 2005). In hilly areas of Jammu and Kashmir also, a large number of women, while sitting at home have adopted dairy farming, poultry keeping, rabbit and fish farming (Gupta, 2005).

A number of studies to examine the role of farm women in various operations of animal husbandry have revealed that rural women are fully involved in majority of dairy farming operations (Gupta, 2012b), Such operations mainly consist of fodder and its chaffing, feeding of animals and their watering, milking of animals and the sale of milk, cleaning of the cattle, collecting of dung and making cow dung cake, care of sick animals etc. Hence, it can be concluded that dairy farming is in the hands of rural farm women and massive empowerment programme of the rural women in animal husbandry requires to be launched to develop them as progressive dairy farmers/managers.

11.7.2 Agroforestry

Rural women, well known for religious belief, are contemplated to contribute voluntarily in planting various tree species like peepal (*Ficus religiosa),* borh or bhagdhar (*Ficus bengalensis),* neem (*Azadirachta indica),* anola (*Phyllanthus emblica).* Such planting drives can also be helpful in growing medicinal plants like tulsi *(Ocimum sanctum),* mint *(Mentha arvensis),* brahami *(Bacopa monerri),* ashavagandha *(Withania somenifera)* etc. Growing of fruit plants like amb *(Mangifera indica),* jamun *(Syzygium cumini)* and ber *(Zizyphus jujuba)* can also be advocated.

Floriculture, apiculture, sericulture and animal husbandry are the main vocations of the agroforestry systems like agriculture - forestry, horticulture - floriculture - apiculture, agriculture - forestry - animal husbandry and agriculture - forestry - sericulture - animal husbandry - fishery, wherein the farm women can play a pivotal role (Gupta, 2005). Floriculture with immense opportunities for farm women is highly income generating field in the market due to good demand for flowers for temples/mosques/gurudwaras, flower decoration, cut flowers and perfume. Farm women of Ranbir Singh Pura area, are earning a lot by growing marigold and selling @ Rs. 6,000 kanal^{-1} (0.05 ha) as recently stated by Gupta (2019).

11.7.3 Biogas Energy Development Programme

Biogas conversion and utilization of wastes and residues for energy and power has tremendous potential to meet significant part of country's energy requirement through variety of technologies and conversion routes namely:

- Thermal energy recovery from agriculture wastes and residues.
- Energy recovery from agricultural, industrial and urban wastes through bio-methanation.
- Thermo-chemical conversion through gasification and pyrolysis like maintenance of biogas plants The rural house wives can be assigned these jobs after imparting training in the aforesaid technologies and energy routes for energy development.

These sources of energy will not only save cooking time, fuel, cost of fuel, but will also enhance fuel efficiency and reduce soil and environment pollution. Concern of the rural masses, especially of women towards forests by way of illicit cutting of trees for collecting fuel wood, would be reduced, thereby, saving trees and improving the climate and ecological balance.

11.7.4 Integrated Nutrient Management (INM) System

As early as 1974, the need for INM system was realized, when the prices of fertilizers arose very high. INM system envisages use of organic manures, green manures, bio-fertilizers along with chemical fertilizers. At farm level, INM system should be able to maximum the residues from the farm, wastes from the livestock and household before supplementary them with chemical fertilizers. The role of women in INM system can not be overemphasized as entire work of organic manure preparation right from spreading of cattle beneath their feet for absorption of urine and dung, collection of the same early in the morning daily and carrying it to the field in the pit or open heap, is the responsibility of women. To propagate vermicompost, women can be trained vis-a-vis use of bio-fertilizers in compost making. Detailed account of INM in a sustainable rice-wheat cropping system has been recently appeared in literature (Mahajan and Gupta, 2009).

11.7.5 Integrated Pest Management (IPM)

In 1962, Rachel Carson in her book “Silent Spring” aroused worldwide concern about the excessive use of pesticides which eventually led to the concept of IPM as environmentally sound alternative to sole use of chemicals. It encourages the most compatible and ecological sound combination of available pest suppression technique, to keep pest populations below economically damaging level.

Rural farm women, the backbone of labour force, require to be exposed to various IPM techniques in affable manner for effortless implementation. It will enable the women to get rid of, from the insecticides/pesticides assault, who are worst, affected due to their indiscriminate use. This is more so in the developing countries, where the migration of the male members of household from farmlands to cities, in search of urban job, leaving the women to take an increased burden of agricultural operations. Women, thus, get exposed to toxic chemicals in increasing numbers, and this leads to endocrine disruption related health problems in them. Further, a change in social attitude is also mandatory where farm women should be encouraged to participate in pest disease control activities and other preventive measures, who presently are dispirited from such activities because of gender prejudice.

11.7.6 Land Resources

Women are the primary stakeholders in common property land resources, and in rural areas the management of non commercial fuel and fodder is usually their responsibility. It is therefore necessary to give some appropriate rights to the rural women for utilizing produce from the forest to meet their household requirements.

11.7.7 Home Management

The primary duty of the women is to utilize full time household work activities. However, majority of the women are unaware of the scientific way of management with viable resources. By their traditional methods, most of the farm women do not bring out a close ratio between input and output and not get themselves released to take up other economic activities. Now our home science scientists have propounded certain low cost home making processes like low cost nutritive recipes, method of cooking for preservation of nutrients right diet during pregnancy, lactation, infancy and illness, which can be profitably adopted by the farm women and house wives of urbanized areas.

11.7.8 Environmental Conservation

A study of the Indian history has shown that right from the beginning of the civilization, the women have been worshipping the plants and trees. Certain plant species like pipal or barh (*Ficus religiosa),* borh or bayan tree *(Ficus bengalensis)* and tulsi *(Ocimum sanctum)* are considered very pious. Tulsi is grown in and around every house, whereas pipal tree is planted in almost every temple or any other religious place (Gupta, 2009b). Other species of trees which are worshipped in various parts of India are saltree *(Shorea robusta)* in West Bengal, mango tree *(Mangifera indica)* in Uttar Pradesh, neem tree *(Azadirachta indica)* in south India. Thus planting of the above said trees besides performing different religious rituals and traditions also ameliorate and conserve environment.

11.8 Need for Recognizing Women's Contribution

Since the women are important drivers of agricultural economics and ecological sustainability, so they must be given due recognition from this angle. All the policy makers and development experts are required to well understand and look into the connection of women's knowledge, work and skills, ensuring food security and creation of wealth. Like Ladakh region of Jammu & Kashmir (Now Union Territory) and that of Meghalaya, where many women control family property including agriculture land, have full inheritance right, the women of the other states must be given land right (Gupta, 2013b). This will act an incentive to them and eventually to focus the role of women in agriculture visible.

11.9 Conclusion

Women have played and will continue to play a key role in the conservation of basic life supporting systems such as land/soil, water, flora and fauna. They protect the health of the soil through organic cycling and promote crop/food security through the maintenance of varietal diversity and genetic resistance. Rural women may start any one of the enterprises sitting at home like dairy farming, vegetable growing, beekeeping, poultry-keeping, nursery raising, mushroom farming, floriculture, orchard raising, fruit preservation, pickle and jam making etc. Except ploughing, most of the fieid operations like breaking of clods, puddling, transplanting, harvesting, winnowing is done wholly solely by the women. Thus, in sustainable agriculture based on maintaining the integrity and the fertility of the soil, women have played a pivotal role, particularly in work-linked to maintain the food cycle. In feeding animals from trees and crop byproducts, in nurturing cows and animals, in composting and fertilizing fields with organic manure, in managing mixed cropping and, rotation of crops. In addition to this, the critical work of maintaining ecological balances is chiefly done by women. In many areas of India which yet have not been colonized by Green Revolution, women peasants continue to work as soil builders rather than soil predators. And it is from these pockets of natural farming that the ecological struggles to protect nature and is emerging, sustaining, thereby, the crop productivity.

11.10 Future Research Strategies/Priorities

- As agriculture is always a joint activity within a family, hence never ignore women folk. Concerted efforts must be made to empower women to get multiple results in farming.
- Development of self help group viz., Mahila Mandal, Mahila Samities, is a step to strengthen them with common interest.
- Inspite of women's diligence, there has been wide discrimination which calls for immediate attention. This will raise the moral of the rural women to take new initiatives.
- Without any obstacle, the farm women must be allowed to participate in all facets of life, cultural, social, economic and politics as like men.
- They must be made well acquainted with the latest agricultural technology. This can be done by involving them in production-oriented training and extension programmes.

- The farm women including drop out girl students, must be imparted trainings in income generating vocations like mushroom cultivation, bee keeping, poultry and dairy farming, sericulture, horticulture, olericulture, floriculture, and growing of medicinal herbs/plants. After acquiring such training they will be able to start their own income generating units. There should be provision of creches where they can leave their children during training hours.
- It is beyond doubt that strengthening women through various measures in farming would make the goal of national food and nutritional security easier. So, while women of today are emerging from the shadow of indifference and neglect of their personal initiatives, there is an imperative need to empower them by enabling them to graduate to modern farming practices like concept of organic farming, agro-forestry, and watershed management.
- Although women are directly involved in management of livestock, they are hardly aware of the advantages of new farm and livestock technologies. The urgent need is, therefore, to develop animal husbandry extension network at the village and arrange training of women in the technologies for optimizing of output and enhancing from livestock rearing.
- A strategy to achieve the overall development of rural women with special reference to farm and peasant women, consists of:
  a) Formal and informal education of rural girls/women.
  b) Training in farm and home activities
  c) Organizing farm clubs, rural women mandals, rural women television forums etc.
  d) Close linkage of Krishi Vigyan Kendras (KVKs) and Farmer Training Centre's with the women Clubs/Mandals.
  e) Developing co-operative projects/self-help groups' viz., fruit farming, vegetable cultivation, kitchen gardening, poultry rearing.
  f) Implementing poverty eradication programmes designated for rural women.
  g) Involving the rural women in rural institutions and all developmental programmes.
- Women should also be made aware of and trained in growing genetically modified food plants and rearing modified farm animals many of which represent healthier alternatives.

- Proper transport facilities for education and training centres in different rural areas must be provided.
- The plan of training programme must be realistic and practical-oriented. The aim should be to upgrade the knowledge, and improve the skill and attitude of the rural women/girls on their own roles.
- Voluntary groups can be arranged for promoting agroforestry on collaborative basis. Non Government Organizations (NGO's) and State Agricultural Universities can arrange plantation drives to create awareness among the rural masses regarding the importance of agroforestry in maintaining ecological balance and increasing farm production.
- Farm women can be trained by means of vocational courses in skilled jobs such as nursery management, mushroom production, and fruit and vegetable preservation. All these vocations have vast potential of employment as well as income generation.
- Women must be involved in all aspects of land regeneration projects enabling them to divide equal benefits of these projects. Training programmes for farm women should especially include those topics which make them healthier and ameliorate their standards of living efficiency.
- As agriculture is the prime engine of economic growth in developing countries, particularly in India, the roles performed by women must not be neglected.
- All obstacles to the full participation of women must be removed and ensure that women find themselves on equal terms with men in all facets of life - culture, social, economic and political.
- Growth in Indian agricultural needs to be matched with feminization of agriculture for an overall development of rural economy.
- As the role of farm women in the decision-making process helps in toning-up agricultural modernization in the country so this needs to be ascertained.

According to WHO report about 20 million people in our country are affliced by flouride poisoning (Fotedar, 2019). Among the affected people both men women including farm women and children. Among women, flouride (F) not only affects joints, muscles, teeth but also also damage their foetus.

Study Questions

1. Rural farm women constitute the backbone of agriculture - give comments.
2. How rural women are amenable to sustain agricultural production?
3. How farm women assist in livestock rearing?

4. Define farming system approach.
5. Write the role of women in biotechnology.
6. What is the contribution of women in food grain production?
7. Write short notes on the following:
   i. Bara Anaj mixed cropping system
   ii. Zabo farming system
   iii. The Myth of the Green Revolution

References

Badiger, S., Verma, S. and Verma, T. (2013). Entrepreneurship of rural women in India: Challenges. Indian Farmer's Digest 46 (10): 36-37.

Dhaliwal, S. (2013b). Government okays Rs 7,500 crore for diversification plan. The Tribune 133 (145): 6 [December 31, Tuesday].

Ghosh, S., Singh, B. and Ganguly, K. (2000). Role of women in agriculture. Agricultural Extension Review 12 (3): 3-9 & 15.

Ghuman, P.K. and Gill, H.K. (2006). Rural women as dairy manager. Agricultural Extension Review 18 (1): 12-14.

Fotedar, B.L. (2019). Fluoride poisoning Fersomel communication on the address of R.D. Gupta, 258C, Sanik Colony Jammu, 180011.

Gupta, M.P. (2005). Empowerment of hi!! women in agriculture. The Kashmir Times 62 (273): 16 [October 2. Sunday].

Gupta, R.D. (2019). Crop diversification for enhancing income. Success story from J&K. Soil water conservation Today. 14(2): 3-4.

Gupta, R.D. (2006). Environmental Degradation of Jammu and Kashmir Himalayas and Their Control. Associated Publishing Company Karol Bagh, New Delhi (India)

Gupta, R.D. (2009a). Women promote environmental conservation. Ihe DailyJFxceJsior 45 (64): 8 [March 5, Thursday],

Gupta, R.D. (2009b). Women and the food security. The Daily Excelsior 14 '.124): 14 [May'5, Tuesday].

Gupta, R.D. (2010c). Organic agriculture key to food security. The State Times 15 (161): 6 [June 11, Friday].

Gupta, R.D. (2012b). Empowerment of women and co-operatives. The Daily Excelsior 48 (148): 8 [May 31, Thursday].

Gupta, R.D. (2013b). Farm women and future strategies. The Daily Excelsior 49 (361): 6 [December 31, Tuesday].

Gupta, R.D. (2013g). Recognizing women's contribution in agriculture. The Daily Excelsior 49 (160): 6 [June 11, Tuesday],

Gupta, R.D. and Singh, H. (2006). Indiscriminate use of fertilizers poses health hazards. Farmer's Forum 6 (9): 20-24.

Gupta, R.D. and Singh, H. (2008). Indiscriminate use of pesticides in agriculture: Public health issues and their control. Indian Farmer's Digest 41 (1): 8-13.

Gupta, R.D. and Tikoo, R.K. (2005). Neem: A potential organic fertilizer. Command Areas Development Subzar 3 (1): 5.

Jillani, M.H., Singh, C.B. and Tripathi, R.K. (2005). Role of women in livestock farming in Uttaranchal. Indian Farmer's Digest 38 (7): 40-44.

Kanwar, Canwar, J.S. (1999). Soil Science, food security, sustainability and the society. J. Indian Soc. Soil Sci. 47 (4): 634-648.

Kanwar, J.S. 1999. Soil Science, food security, sustainability and the society. J. Indian Soc. Soil Sci. 47(4):634.

Khetarpaul, N. and Grover, I. (2001). Strategic measures for improving nutritional status of women. Intensive Agriculture 39 (3-4): 14-15.

Lawrence Prabhu, C., Esakkimuthu, M. and Kannan, S.T. (2014). Women empowerment in 21st Century. Indian Farmer's Digest 47 (6): 40- 42.

Mahajan, A. and Gupta, R. D. (2009). Integrated nutrient management (INM) in a sustainable rice-wheat cropping system (pp.1-267). Springer Science + Business Media (www. Springer.com).

Malik, J.S., Kumar, R. and Singh, M. (1999). Role of Indian women in different aspects of daily life. Intensive Agriculture 37 (1-2): 22-23.

Meghwal, R.R. and Singhal, S. (2000). Role of women in agricultural development. Intensive Agriculture 37 (11-12): 5-8.

Sharma, U.C. (2007). Indigenous soil and water conservation and management techniques of the tribes of northeast region of India: Case studies (pp. 116-126). In Natural Resource Management for Sustainable Hill Agriculture (Edited by. S. Arora, S.S Kukal and V, Sharma) Soil Conservation Society of India, Jammu Chapter, SKAUST-Jammu, J&K State, India,

Singh, A., Kumar, B. and Kumar, A. (2000). Strategy for training of farm women. Agricultural Extension Review 12 (1): 3-6.

Sriram, C. (1999). Group dynamics in transfer of technology of modern agriculture. In Krishi Vigyan Kendra - A Reality (Eds. P. Das and B.S. Hansra). Division of Agriculture Extension, ICAR, Krishi Anusandhan Bhavan, Pusa (New Delhi), India.

Thakur, P.K. (1999). Indigenous technical knowledge for hill agriculture in Northeast India. In Krishi Vigyan Kendra - A Reality (Eds. P. Das and B.S. Hansra). Division of Agriculture Extension, ICAR, Krishi Anusandhan Bhavan, Pusa (New Delhi), India.

Yadav, V.K., Singh, A.K., Singh, H.V. and Khan, Z.H. (2005). Crop diversification for increasing crop productivity in cold region of Jammu and Kasnmir. Indian Farmer's Digest 38 (9): 12-14.

12

Role of Agroforestry in Sustainable Agriculture

Abstract

There is no doubt "Green Revolution" which ushered during late 1960's brought about a lot of increase in food grains production in India. However, it was confined only to irrigated areas under rice-wheat cropping system. The rain-fed areas of the country which support to about 40 per cent of the total population and constitute nearly 67 per cent of the total cultivated area of about 142 m ha could not reap the benefits of Green Revolution. Moreover, the farmers of rain-fed areas get very low yield of their crops due to erratic rainfall during monsoon. So, an alternative land use system for sustainable agricultural production in rain-fed areas is required to be followed. Such an alternative land use system is known as "Agroforestry", which involves multiple use of land combining the trees with crops, grasses and livestock rearing on the same area of land. The present chapter of the book, therefore, deals with the necessity of agroforestry. Concept of agroforestry including its definition, importance of agroforestry, various agroforestry systems viz., agriculture- silviculture system, agriculture-horticulture system, silviculture-pastoral system, agriculture-fisheries system and macro-micronutrient dynamics in leaf litter decomposition of agroforestry tree species are the main topics which stand elaborated. Conclusion and future research strategies/priorities have also been presented along with study questions and references.

Keywords: Green Revolution; sustainable agricultural production; food grain production; low yield; irrigated areas; rain-fed areas; erratic rainfall; agroforestry; agriculture-silviculture system; agriculture-horticulture system; silviculture-pastoral system; agriculture-fisheries system.

In spite of the fact that the forests are the wealth of a nation and play a vital role in the economy of the country, not much attention has been paid to their promotion, protection and conservation. The forests of India have been reduced to 19.27 per cent (Gupta and Khajuria, 2000) and/or 23 per cent including 20.55 per cent forest cover and 2.48 tree cover (Sharma, 2005). It is much less than as prescribed in the National Forest Policy (1952), according to which one-third i.e. 33 per cent of the total land area of the country must be under forests. This aim is, however, very difficult to achieve due to deforestation which is still taking place on an average rate of more than 25,000. ha^{-1} $year^{-1}$ (Anonymous, 2003). It is mainly due to ever-increasing population pressure both of human beings and livestock. According to 2001 census, the population of India has crossed 1,000 millions which made it the second most populous country in the world after China. Further, in the absence of sufficient pasture lands, the forests are also the main source of grazing and fodder. Out of 450 million cattle in India, about 270 millions graze in the forests (Gupta, 2005c). The world population which was 7 billion up to 31st March 2011, has raised to 7.244 billion (15tn of July 2015). Today India with largest livestock population (512 million) is producing 133 million tonnes of milk (Pasrija, 2015). Accordingly, demands upon natural resources like forests, fossil fuels and others have expanded rapidly

The Indian scriptures and epics are full of references, evincing the role of forests in day to day life of civilization. This can be further substantiated by the words of Lord Budha, which stand mentioned elsewhere (Gupta and Banerjee, 1991 and Gupta, 2011d). According to Him, "*the forest is a peculiar organism of unlimited kindness and benevolence that makes for its sustenance and extends generously the products of its life activity; it affords protection to all beings, offering shade even to an axe man who destroys it*".

In spite of so many advantages, man has left no stone unturned in denuding or deforesting the forests. Deforestation, which means reckless and careless cutting down of forest trees on a large scale, is going on cynically. With the deforestation, degradation and denudation of the forest area in the country, the rural community is facing lot of difficulties in procuring basic necessities of life such as fuel wood, fodder, food, fibre, timber. Deforestation has also become a potent cause of constant weather instability. Drought, flash floods, landslides/landslips and ecological imbalance are its direct consequences. Timber is becoming scarce for industries and construction work. The villagers, especially women folk now have to travel, on an average, about 15 km for gathering a head load of fuel wood for cooking and warming. There is an acute shortage of fodder for the animals. Owing to flash floods there is heavy loss of soil through erosion. From a number of places, erosion has swept away the top fertile soil resulting into low productivity ultimately producing wastelands. Siwaliks of Punjab, Haryana, Himachal Pradesh and Jammu province of Jammu and Kashmir and Karewas of Kashmir are the glaring examples from this angle. Silting up of the water reservoirs due to soil erosion has reduced their life span (Gupta, 2013c and Gupta, 2013e).

Realizing the importance of the forests and saving them from further deforestation and degradation, the Late Prime Minister, Mrs. Indira Gandhi while addressing the State Forest Ministers Conference held during October, 1982 called for a "*ban on the illicit cutting of the forest trees*". She also urged the need for reorienting the forest management with the cooperation of the rural masses through the concept of Social Forestry or Extension Forestry and Farm Forestry or Agroforestry systems. But this remained a politician slogan only. Agroforestry, which is the practice of joint production of wood and food, is the same as agri-silviculture. It has now become an international movement spearheaded by Food and Agriculture Organization (FAO). This is essentially, "a *movement of the people, by the people and for the people*" of the developing and under developed countries where forests have totally been exhausted.

Agroforestry not only helps to increase the tree cover but also makes available to the people fuel wood, fodder, food and timber, for which they traditionally depend upon forests. This, in turn, will assist in reducing their pressure on natural forests, conserving them and eventually ameliorating an environment by way of controlling soil erosion, landslides/landslips, conserving moisture and protecting the crops by serving as wind-breaks/shelter beds and fencings.

Innovations in agroforestry also bring "sustainability" in agriculture. The present chapter briefly pinpoints the performance of various systems of agroforestry viz., Agri-Silviculture System, Agri-Horticulture System and Silvi-Pasture or Silvi-Postoral System. Concept of agroforestry and its necessity have also been dealt with.

12.1 The Necessity of Agroforestry

The farmers have been dependent upon forests from cradle to grave for their sustenance and to meet all their primary needs of grazing, fuel wood, grass, leaf, fodder, flowers, vegetables, fruits and medicinal herbs. The relationship between the rural people and forests has ever remained very sound traditionally and historically and was at its best before the British rule. In the Mughal period and even before that, all forests were private and the villagers were the owners of these forests. As such, they used to protect and manage these forests for meeting their primary needs and lived a happy and prosperous life.

It was during the British rule that the forests were made Government properties and commercialized and exploited ruthlessly. Inheriting the British legacy, the State Governments continued the forest loot in the post-independence periods, also for revenue earning and consequently, depleted/degraded the entire forest resources of the country. This depletion/degradation of the forests has affected the village life badly. Forests have receded far away and villager has to walk a long distance to get a head load of fodder and fuel wood. Among the Western Himalayan States, the Jammu and Kashmir is the worst affected (Gupta, 2005c). Almost all deodar and blue pine forests have been lopped for fuel wood, especially in Doda district of Jammu and Kashmir valley. In Jammu and Kashmir state about 20 to 30 per cent forest cover has been reduced due to illegal felling of trees. Hence the necessity of agroforestry which aims at making farmers self-sufficient on their own farms for meeting their primary needs of fuel wood, fodder and other basic necessities.

12.2 Concept of Agroforestry

The concept of agroforestry and social forestry has been evolved with the objective to bring forests closer to man through people's participation. Most frequently, "*Agroforestry*" includes land management technologies that involve the combination of forest trees with crops including fruit trees and grasses, and raising of livestock or combination of all these enterprises. Contrary to this, "*Social forestry*" lays much more emphasis on the involvement of people in various activities of tree growth and development.

12.2.1 Definition of Agroforestry

Briefly, the agroforestry system may be defined as "*Sustainable land management practice to increase the land utilization by integrating the production of crops and forest trees and/or animal raising simultaneously or sequentially on the same piece of land.*"

12.2.2 Importance of Agroforestry

Various innovations evolved in agroforestry can solve numerous problems which are being presently confronted with by the farming communities. As for instance, shrinking of agricultural holdings due to increasing human population, are the main issues causing great concern to the planners. Hence, adoption of agroforestry based farming system must be encouraged as it not only produces more food, fodder, fuel wood, fruit and fertilizer, but also helps in conserving soil and moisture very effectively. This farming system protects the crops by acting as wind-breaks/shelter belts, provides partial shade, fencing and improves the environment. Moreover, agroforestry maintains soil fertility through recycling of plant nutrients and, thereby sustains the crop productivity, prevents soil erosion and loss of nutrients through leaching and run-off.

The growing of trees in agroforestry delivers multiple benefits for humans, livestock and ecosystem (Sharma, 2015). Tree based farming systems contribute robustly to live-hoods by providing fruit, fodder, fuel, fibre, fertilizer and timber. Thus, agroforestry enhances food security, income generation and provides insurance against crop failure.

Many leguminous tree species like *Dalbergia sissoo*, various species of *Acacia* (*A. nilotica, A. modesta* and *A. catechu*), *Albizia lebbeck, Albizia stipulata* etc., fix nitrogen from the atmosphere and return much more in leaf fall than they take from the soil. Growing of superior types of grasses such as *Cenchrus ciliaris, Cenchrus setigerus*, and legumes like *Stylosanthes scabra, Stylosanthes hamata* etc., boost forage production 2-4 times compared to the natural grasses grown in the grazing lands. Hence, agroforestry farming system improves the productivity of plants and animals since it is based upon sustainable land management and maximum utilization of natural resources to increase the ecological and economic benefits.

12.3 Agroforestry Systems

For the production of food grains, agriculture-silviculture, agriculture-horticulture and silviculture-pasture or silvi-pastoral, are the three most important agroforestry systems which have been widely tested successfully both under “On Farm” and “Research Station” trials, where growing of tree culture has been found to be the dominant component.

12.3.1 Agriculture-Silviculture System

It is the process of combining multipurpose tree species along with food crops. Various tree species are either planted on field boundaries or on the fields. However, planting of the right type of tree species are crucial for the success of the system. The most prominent tree species, which have been found very suitable for growing in the Siwaliks of Jammu, Jammu and Kashmir State, Himachal Pradesh, Punjab, Haryana and parts of Uttaranchal and Uttar Pradesh are: (a) *Leucaena*\ (b) *Aibizia* (c) *Acacia* and (d) *Sesbania*. First two species are preferred as they are quick growing and can attain annual growth rate as high as 2 to 7 m in height and 10-60 m^3 ha^{-1}. Moreover, these species have very good coppicing ability (Gupta, 2009a).

Among six species of Sesbania, Sesbania aculeata (Dhaincha) is tolerant to salinity and alkalinity This species, in addition to serving as green manuring, has also proved beneficial in providing fuel wood to the tune of 15 t ha^{-1} on dry weight basis in about 150 days after its sowing (Gupta, 2005c). Some farmers of Ranbir Singh Pura areas of Jammu are also growing this species for obtaining fibres like sunnhemp, to be used in bed cots (Gupta et al., 1992).

Poplar (*Populus* spp.) has been widely grown in Punjab, Haryana and Tarai area of Uttaranchal, or Uttarakhand and is being used by several industries, e.g. match splints, plywood, packing cases etc Salt tolerant species like *Acacia* and *Eucalyptus* are planted in salt affected soils (Gupta and Arora, 2016). *Acacia* and *Dalergia sissoo* (commonly known as Shisham) have high calorific value and are, therefore, considered very good fuel wood species. These tree species can also be used as timber and grown in India, Pakistan, Nepal, Bangladesh and other Asian countries (Choudhary et al., 2002). Although "Shisham" is mainly grown for timber yet it is an important, multipurpose tree species in Indian Subcontinent. However, the incidence of "Die Back" has threatened the economy of the farmers and has discouraged them planting this species in the agroforestry systems as reported by Choudhary et al. (2002).

Ailanthus excelsa, Grevillea robusta, Melia azederach, Bauhinia variegata and *Grewia optiva* are the other tree species which have been found suitable to grow in the Siwalik Hills. For the temperate hilly terrain, *Robinia pseudocacia* and various species of oaks (*Quercus ieucotrichophora),* are the most suitable The history shows that the farmers of Kashmir valley have been growing forest trees i.e. wiliows (*Salix* spp.), legume fodder (Alfalfa) and fruit plants since long. Three species of poplars viz., *Populus candican, P. euphratica* and *P. nigra* have been planted in the village lands and along the river banks where irrigation is assured in Leh district of Ladakh. However, *P. candican* has performed well in Leh range and *P. euphratica* in Nubra valley. The farmers of Ladakh, infact, are growing these poplar species for their increased demands of fuel wood, timber and fodder for sheep and goats (Gupta et al., 2007). The poles of these poplar species are widely used for constructing bridges and houses. The wood is also used for making quality furniture and barley flour (Satoo) pots.

It is a point to mention that when all the tree species are planted along with food crops, their impact on food production is the reduction in the overall crop yields which lies to the extent of 8-11 per cent. This overall crop yield reduction can, however, be compensated largely with foliage and fuel production due to pruning of the trees annually. Planting on croplands should be in light and only light crown species are selected. Pruning is done for tree canopy. The foliage is rich in proteins and can be used as fodder as well as for green manuring. As for example an investigation conducted in Dry-land Agriculture Research Station under Sher-e-Kashmir University of Agricultural Sciences and Technology, Rakh Dhiansar, Jammu (DARS) indicated 12 kg wood tree^{-1} after 3 years of plantation of *Leucaena leucocephala* at a spacing of 3 m x 3 m in crop fields. Green fodder at the rate of 10 kg tree^{-1} was also obtained during the time of sowing of wheat (Gupta and Raina, 1994). Another study regarding the residual effect of green manure on wheat crop showed that addition of 15 tonnes of *Leucaena leucocephala* leaves in the soil gave the highest yield i.e. 14.56 q ha^{-1} followed by 10 tonnes of *Leucaena leucocephala* leaves applied during *Kharif* in maize.

A study conducted by the scientists of DARS, Rakh Dhiansar showed the highest average yield of green fodder of 285.3 q ha^{-1} in sole fodder crop (Table 12.1) followed by Aonala + green fodder treatment with green fodder yield of 274.8 q ha^{-1}. The lowest green fodder yield 240.2 q ha^{-1} was observed in guava + green fodder. Average yield of leaves and dry fuel wood yields of Leucaena and Albizia were obtained to the extent of 44.6 and 25.1, and 37.8 and 17.2 q ha^{-1}, respectively (Table 12.1).

Table 12.1: Experimental results of two years i.e., 2002 and 2003 (Adapted from Annual Report [2004])

Treatment	Green fodder (q ha^{1})			Green leaves (q ha^{-1})	Dry fuel wood (q ha-^{1})
	2002	2003	Mean		
T_1 – Guava* + green fodder	249.9	230.5	240.2	-	-
T_2 – Leucaena + green fodder	254.6	232.4	243.5	44.6	25.1
T_3 – *Albizia* + green fodder	261.9	236.8	249.3	37.8	17.2
T_4 – *Grewia optiva* + green fodder	285.9	240.7	263.3	-	-
T_5 – Aonla* + green fodder	304.0	245.6	274.8	-	-
T_6 – Green fodder pure	319.9	250.7	285.3	-	-

*Guava and aonla were not in the bearing stage up to 2003 which were planted during 1995 and 1996.

Studies with regard to subabul (Leucaena leucocephala) based agroforestry system initiated at Jhansi (Gill, 1999) with tree/crop interface, indicated that in association with subabul, wheat and chickpea recorded a grain yield of 1.88 t ha^{-1} and 0.43 t ha^{-1}, in addition to grain production, subabul trees were pruned to their 50 per cent height each year just before sowing of monsoon and winter crops to reduce their canopy influence. This gave sufficient amount of fuel wood and forages (Table 12.2).

Table 12.2: Quantitative benefits of various agroforestry systems (on an average basis) (Adapted from Gill [1999])

Sr.No.	System	Subabul biomass production	Field/fruit crop production
1.	Agri-silviculture	19.9 kg $year^{-1}$ $tree^{-1}$ (due to pruning)	1.88 t ha^{-1} grain yield of wheat.
2.	Agri-silvi-horticulture (Ber based)	4.75 t ha^{-1} $year^{-1}$ fresh (due to copping)	a. 1.77 t ha^{-1} grain yield of wheat. b. 0.92 t ha-1 fruit yield.
3.	Agri-silvi-horticulture (Mango based)	1.63 t ha^{-1} $year^{-1}$ dry (due to coppicing)	a. 0.77 t ha^{-1} fruit. b. 1.72 t ha^{-1} grain yield of wheat.

In another study on mango based agroforestry system, subabul proved quite effective. Besides yielding fuel wood and forage (Table 12.2), it protected the young saplings of mango fruit trees during hot summer days in the initial establishment stages. The agri-silvi-horticulture ber based agroforestry system (Table 12.2) gave 1.72 t ha^{-1} grain yield of wheat and 0.92 t ha^{-1} fruit yield (Gill, 1999).

Studies carried out at National Research Centre for Agroforestry, Jhansi (Uttar Pradesh) during 1988-94 with eleven multipurpose tree species and Rabi crops (wheat and chickpea) sown in the interspaces of these trees, indicated average reduction in grain yield of wheat to the order of 13.7 per cent (relative grain yield 86.3%) as compared to when it was sown alone (Table 12.3). The reduction in the yield of chickpea was much higher (41.3%). Reduction in production of the grain yield to a range of 10-12 per cent was, however, compensated with the biomass production due to pruning of trees (Gill, 2002).

Table 12.3: Relative grain yield of *Rabi* crops grown in the interspaces of multipurpose tree species (Adapted from Gill [2002])

Sr.No.	Tree species	Relative grain yield (%)	
		Wheat (Mean - 4 years)	Chick (Mean - 5 years)
1.	Anjan (*Hardwickia binata)*	91.9	76.3
2.	Casuarina (*Casuarina equisetifolia)*	92.7	67.3
3.	Babulorkikar (*Acacia nilotica)*	89.1	62.3
4	Aonla or amla (*Phyllanthus emblica)*	89.1	64.2
5.	Bakain or drek (*Melia azederach)*	90.5	54.7
6.	Ramkanthi *(Acacia cupressiformis)*	87.0	52.8
7.	Sarin or siris (*Aibizia lebbeck)*	89.2	55.6
8.	Jamun (*Syzygium cumini)*	84.6	59.7

9.	Tali or shisham (*Dalbergia sissoo)*	83.5	62.2
10.	Safeda (*Eucalyptus tereticornis)*	75.0	43.8
11.	Subabul (*Leucaena leucocephala)*	77.0	47.0
	Mean	8G.3	58.7

12.3.2 Agriculture-Horticulture System

Raising of fruit trees and exploiting their interspaces, is the main principle of this agroforestry system. Initially fruit trees do not give any returns and hence growing of crops in the interspaces of fruit trees can solve gap problems faced by the orchardists/farmers.

Actually, in this system of agroforestry, approximately one-third of the trees must be of fruit trees. The remaining two-third should be of forest trees for obtaining fuel wood, fodder and timber. The forest trees, richer in crude protein and calcium in their leaves, should be preferred as also for their low tannin contents. *Grewia optiva, Cordia dichotoma, Bauhinia variegata* and *Celtis australis*, are some such species which can be grown from subtropical, sub-mountain low hills of Siwalik to sub-temperate climatic zones. *Grewia optiva* (Dhaman, biul or bhimal), Celtis australis (khirk), *Bauhinia variegata* (Kachnar or katrair), Butea monosperma (Palah or dhak) are, infact, the main indigenous tree species the loppings of which are used as fodder by the farmers of the Siwaliks (Gupta and Sharma, 2005) covering the Jammu region of Jammu and Kashmir, Himachal Pradesh, Haryana and Punjab. Lavania and Singh (2005) reported that *Celtis australis* is a good fodder tree species which can be grown all along agricultural fields with crops under agroforestr/ system throughout the hilly region, having altitude from 500 m to 2500 m. Multipurpose tree species such as *Leucaena leucocephala, Aibizia lebbeck, Acacia nilotica, A. modesta, A. catechu and Sesbania* spp. can also be grown. Tree species such as Robinia and Ulmus have proved very useful for the wastelands of Banihal area of Jammu and Kashmir, where the farmers are getting about Rs. 60,000-80,000 ha^{-1} annually by sale of these planted tree species.

Prominent fruit tree species tested in this system of agroforestry are guava, kinnow, mango, aonla and ber in subtropical and sub-mountain low hills of Siwaliks (Gupta, 2005), while for the temperate agro- climatic zone, the fruits mainly consist of cherry, apricot, pear, apple, walnut, grape and almond. Growing of ber in this system has proved quite remunerative, as beside fruit production, it yields fodder and fuel wood owing to its annual pruning. Umran variety of ber has already been popularized in *Kandi* areas of Jammu and Punjab. In Jammu *Kandi,* it is mostly grown in areas like Raya, Suchani and Ranjri. The villagers, most of whom were ex-servicemen of Dhora village (tehsil Samba now destrict), found horticulture a far more remunerative way than their traditional agriculture and started to grow citrus fruits like kinnow, mosambi and lime for the first time during 1984-85. However, these plantations now require replacement because of age factor. The water stress of fruit plants grown in the area during May and June is met by watering the kinnow plants from the ponds constructed just at the foot of the adjoining hills having a size 26 m x 26 m x 3 m.

A concept of agroforestry has been followed by the peasants of Kandi belt of Jammu under Watershed Management Project sanctioned by Council for Advancement of People's Action and Rural Technology (Jamwal and Gupta, 2007) covering a number of villages like Bari Badhori, Patti, Raya, Suchani etc. This consists of in the form of agriculture-silviculture, agriculture-horticulture, agriculture-silviculture- sericulture and agriculture-horticulture-fish culture systems. Among the various tree species *Grewia optiva, Leucaena Icucocephala* and *Populus* spp. have proved useful, and among fruit trees *Zyzyphus jujuba*, *Mangifera indica*, *Carica papaya* (Papita) and *Phyllanthus emblica* are the promising ones (Sharma et al., 2009). Citrus fruits are also being grown.

The main fruit trees grown in irrigated areas of Jammu like Ranbir Singh Pura, Bishnah and Marh, are mango, litchi, papaya, grapes, loquat, strawberry and citrus. The irrigated areas of Basohli and Billawar blocks of Kathua are congenial for cultivation of mango, litchi and guava whereas, rain-fed areas are best suited for growing of citrus fruits like kinnow, orange, lime and galgal. The temperate zone areas of Jammu region (Bani and Lohai Malhar blocks of Kathua, parts of Udhampur, Doda and Poonch districts) are suitable for growing temperate kind of fruits. Maximum emphasis is given in these areas on cultivation of dry fruits like walnut and almond. Besides, cultivation of apple, pear, apricot, peach and plum is also encouraged. However, in Banihal area, though almond has very good market price, does not thrive well due to frequent hailstorms during its flowering season. Similarly, apples do not grow well. It is because the apple trees are susceptible to scab disease and sanjose scale attack. These plants as such demand frequent sprayings of fungicides and insecticides, and if per chance any unsprayed apple remains, appearance of disease will soon spread in the whole of the area. Vegetables and cash crops like cumin (zeera) and saffron are also being cultivated. Cabbage, cauliflower, knolkhol, cucumbers and brinjals are the major vegetables which are being grown. The results of these crops are encouraging.

Among the field crops, which are found suitable in the interspaces of the fruit trees are legumes such as black gram, green gram and lentil. Although growing of these legumes in the interspaces of the fruit plants has been recommended before the fruit bearing stage i.e. up to 5 years of their plantation yet now, it has become a common practice in Punjab. Besides growing of legumes, the orchardists/farmers of this State are growing wheat and even berseem. In Kashmir, red gram (rajmash) is mostly grown in the apple orchards. The valley of Kashmir is also very suitable for production of off-season vegetables like cauliflower, cabbage, knolkhol, cucumbers, watermelons and brinjal during summer season for plains. Horticulture has also undergone much change after 1947 in cold arid region of Ladakh (Jammu & Kashmir) now Union Territory; and Lahaul Spiti (Himachal Pradesh). Now one can see fruit trees grown even in small attached kitchen gardens. The area is rich in growing apricots and apples (Sharma and Mir, 2000). Strawberry is grown in the vicinity of Leh town. Dha-Haner area of Khaltsi has been found suitable for cultivation of grapes.

The Defence Research and Development Organization (DRDO) has developed agro-food technologies for cold desert areas of Ladakh, Lahaul Spiti and Kinnaur of Himachal Pradesh; and Kumaon and Garhwal hills of Uttaranchal to provide fresh vegetables, fruits, milk and poultry to local people and the armed forces personnel. It has made possible for the farmers to grow green vegetables like coriander, mint, lettuce, celery, parsley and fenugreek by developing polyhouse and trench cultivation techniques using local materials. Cabbage, cauliflower, knolkhol and potato are the main vegetables grown in the area along with field crops like gram, wheat, lesser millets, pulses etc.

12.3.3 Silviculture-Pastoral Agroforestry System

This is also known as silvi-pastoral agroforestry system. It consists of growing trees, shrubs and grasses. Trees are planted at wider spacing ranging from 3mx2mto5mx5m and 9 m x 9 m with an objective to grow grasses and shrubs in between the tree species grown for providing leaf fodder, fuel wood and timber

In arid and semi-arid regions, tree species such as *Acacia albida, Acacia nilotica, Acacia tortilis, Acacia senegal, Albizia lebbeck, Prosopis cineraria, Sesbania aegyptica, Dalbergia sissoo, Melia azederach, Azadirachta indica*, are planted (Ram, 1999) at a spacing of6mx5m,4mx3mor3mx2m depending upon their rotation period. The pit size is normally 50 cm^3 and in each pit 100 gram basal dose of nitrogen, phosphorus and potash is applied in the ratio of 1: 4: 1. In arid region, grasses like *Cenchrus ciliaris, Cenchrus setigerus*, *Panicum antidotale* and legumes like *Stylosanthes scabra* have shown good promise. In semi-arid regions, grasses like *Cenchrus ciliaris*, *Cenchrus glaucus, Dicanthium annulatum, Chrysopogon fulvus* and legumes like *Desmodium intortum*, *Stylosanthes hamata* have been found versatile with high progressive capacity and biomass build-up to provide ideal mix with grasses for forage and controlling other weed species.

Various trees, shrubs and grasses which can be combined, based on agroforestry systems for higher production of fodder, fuel wood, timber for Himalayan rain-fed ecosystem, have been studied and recently appeared in literature (Gupta and Abrol, 2006). Among the several fodder trees (Table 12.4) evaluated for fodder availability duration and production in mid hills of Uttaranchal, *Robinia pseudoacacia, Morus alba, Alnus nepalensis, Melia azedarach* were found to be the fast growing species.

Table 12.4: List of fodder trees available in the mid hills of Uttar Pradesh now Uttaranchal (Adapted from Gupta and Abrol [2006])

Name of the fodder tree species
Celtis australis, Grewia optiva, Morus alba, Robinia pseudoacacia, Quercus glauca, Quercus leucotrichophora, Ailanthus excelsa, *Leucaena leucocephala, Alnus nepalensis, Melia azedarach, Bauhinia spp., Prunus padus.*

In silvi-pastoral system of Jammu hills, tree species such as *Leucaena leucocephala* and *Grewia optiva* have proved useful. These are also found effective in soil erosion control. They act as a shade trees, wind break and source of fence post and charcoal, besides being a source of fodder and forest cover on dry land or rain-fed agriculture. In two years study on the effect of *Leucaena leucocephala* planting at varying spacing with inter-planting of Setaria sphacelata, it was observed that the highest yield of the grass could be obtained to the tune of 469 q ha^{-1} when *Leucaena leucocephala* was planted at a spacing of 1 m x 0.30 m (Khajuria, 1994). In another experiment conducted on the spacing of *Napier, Setaria* and *Leucaena leucocephala* alone and in association, indicated that when *Napier* grass planted at 0.75 m x 0.75 m, *Setaria* at 0.25 m x 0.25 m and *Leucaena* at 0.25 m x 0.25 m, gave the highest green fodder yield while planted alone under Rajouri district, Jammu and Kashmir conditions. However, planting of *Setaria + Leucaena* gave higher green and dry fodder yield than obtained by *Leucaena* when planted alone (Khajuria, 1994). Local grasses grow naturally but their grass slips can also be planted or sown through seeds. Introduction of grass species like Stylosanthes scabra and *Stylosanthes hamata* can improve the soil fertility through atmospheric nitrogen fixation, while providing good quality fodder. These species are bushy in habit and thus protect the soil from erosion. In field trials under dry land farming system, both these species have been found to fix 120-150 kg of nitrogen ha^{-1} over a period of 3 years (Gupta and Banerjee, 1991) average being 40-50 kg ha^{-1} year^{-1}. It has also been observed that *Cenchrus ciliaris* can effectively be grown in Kandi belt of Jammu. In loamy and or sandy loam soils, planting of *Cenchrus setigerus* has given good results. Local shrubs such as *Adhatoda vasica, Carissa spinarum* and *Dodonaea viscosa* help a lot for controlling soil erosion. *Carissa spinarum* and *Dodonaea viscosa* should be encouraged to grow in contour hedges to conserve soil and moisture. Important local grasses are: *Cynodon dactylon*, Panicum antidotale and Chrysopogon fulvus. It is worthwhile to mention that unpalatable bushes and shrubs like *Lantana camara* and *Parthenium* spp. must be removed from the fields manually. These weeds are best controlled by tree planting. They disappear from shady areas.

In the plains and foothills of Jammu like Mandal Phalain, Sunjwan, Saloora, Satani, Chinto, Tutan Di Khui, Sidhra and Basi Jamwal (Hiranagar), silvi-pastoral agroforestry system has very well developed. At Mandal Phalain, the pasture land has been developed through participatory management of the village and Directorate of Social Forestry. At Saloora, a pasture developed in the difficult stony soil with a number of trees is coming up by self-generation. The main trees grown in the area are: *Cassia fistula, C. siameaexotic, Bauhinia variegata, Grevillea robusta and Ficus palmata.* At Samandrani, a good example of a green hedge out of agave was very effective. A number of trees and bushes were also coming up naturally because of the protection provided by the agave fencing. *Grewia elastica* in the midst of orchard grass *(Dactylus giomerata)* was going on very well. At Satani village woodlot, a pasture of elephant grass, *Stylosanthes* and *Cenchrus* grass species has been developed. At Devak Bella, an effort was made to grow forest species which can provide minor forest products along with the pasture. Such forest species are girna (*Carissa spinarum*), ber *(Zyzyphus jujuba)*, aonla (*Phyianthus emblica*) *etc.*

Silvi-pastoral system of agroforestry has also been found suitable in the natural forests of alpine zone, where legumes like red clover (*Trifolium pratense*) and white clover (*Trifolium repens*) have given promising results. Their inclusion in grassland and meadows has not only increased the quantity but also the quality of the forage. The main trees which can be grown to supplement the fodder are: *Celtis australis, Salix* alba (*Kashmir willow*), *S. tetrasperma, Morus serrata, Toona ciliata, Aesculus indica.*

12.3.4 Agriculture-Fisheries Agroforestry System

Three areas for fisheries development in rice based farming systems of Kangra Vailey have been developed (Acharya, 1998). These are block level fish ponds in Khads at diversion sites of major stream (Kuhls) networks, small fish ponds in farmer's field in upper reaches, community fish ponds at diversion sites of supplementary irrigation networks. The creation of such fish ponds will not only help in stabilizing the rice cultivation in the valley but provide immense potential for developing fisheries and reduce irrigation water scarcity during *Kharif* sowing. Existing crop productivity of agro-fisheries models at different locations in kuhl network stand shown in Table 12.5.

Table 12.5: Existing crop productivity of agro-fisheries models at different locations in kuhl network (Adapted from Acharya [1998])

Village cluster	Crop yields (t ha^{-1})			Suitability of incorporating water based system
	Rice	Wheat	Linseed	
Uppermost cluster	1.71*	1.40	0.30	Block level development of high quality fish in ponds at diversion sites Agrifishery farm models (Small fish ponds for all farm families)
Upper cluster	2.31*	1.54	0.32	Agri-fisheries farm models (small fish ponds for all families).
Middle cluster	2.54**	1.67	0.36	Community fsh ponds at suitable sites. Smali fish ponds at farm level depending upon the nature of flow in supplementary irrigation network.
Lower cluster	2.16***	1.82	0.34	Community fish ponds at the diversion sites of supplementary kuhl irrigaton sources. Small fish ponds at farm level depending upon the nature of flow supplementary irrigation network.

* Refers to sprouting seed sown, ** transplanting rice and *** dry seed sown

Agriculture-horticulture-fish culture system of agroforestry has also proved useful in the Kandi belt of Jammu (Jamwal and Gupta, 2007) which was proposed long back by Gupta and Raina (1994).

12.4 Macro and Micro-nutrient Dynamics in Leaf Litter Decomposition of Agroforestry Tree Species

A laboratory study to investigate the macro/micro-nutrient dynamics in decomposing leaf litter of some agroforestry tree species revealed that rate of decomposition of leaf litter was in the order of Acacia>Toona>Eucalyptus>Melia species. The release of the most of the nutrients into available pool was more in *Acacia maliissima* treated soils followed by Toona ciliata and least in *Eucalyptus tereticornis* and *Melia azederach* (Tripathi et al., 2006)

12.5 Conclusion

On the whole, all agroforestry systems are advantageous propositions in the present context. Proper selection of trees, their proper management and right choice of crops raised in combination with various tree species can bring prosperity. Beside food grain production other products like fodder, fuel wood, fruit and timber as well as livestock enterprises can be obtained in conjunction with a number of service benefits. Raising of livestock in the hilly terrain, especially in the Kashmir Valley, a component of mixed farming can be divided into 3 categories viz., sedentary, semi-migratory system (Gupta, 2004). All these three components of livestock rearing and growing of crops like rice, maize, wheat and mustard and cultivation of fruits apple, pear, cherry, walnut, almond stand briefly presented.

12.6 Future Research Strategies/Priorities

- Encouraging broad-leaved species in the conifer forests (Chirpine and blue pine) for fodder and timber is the need of the day. The broad-leaved species must be planted on the basis of agro-climatic conditions
- More research is needed on Subabul (*Leucaena leucocephala)* based agroforestry systems, both in agri-silviculture and agri-silvi-horticulture agroforestry systems in rain-fed areas of Jammu and Kashmir, Himachal Pradesh, Punjab, Haryana and other states as has been done for the semi-arid and humid tropics under Indian conditions (Gill,' 1999).
- The use of loppings of *Leucaena* as leaf green manuring must be advocated among the farmers of dry land agriculture particularly in *Kandi* belt of Jammu. This will prove very effective in conserving soil and moisture as well as plant nutrients in soils of *Kandi* belt which are very poor in soil fertility and water holding capacity.
- Nearly 75 per cent of the people living in hilly areas are wholly solely dependent upon forests for natural fruits like chilgoza, walnut, aonla, mulberry, jamun, ber, girna, karonda etc. These fruit trees are, however, vanishing at an alarming rate. This, therefore, calls for meticulous care and massive extension programmes in planting of these fruit trees and protecting the existing ones.
- Stress must be laid on planting of local fruits like aonla (*Phyllanthus emblica),* dhew (*Atrocarpus lakoocha),* jamun *(Syzygium cumini),* phalsa *(Grewia asiatica),* mango *(Mangifera indica),* girna (*Carissa spinarum),* bael *(Aegle marmelos),* karonda (*Carissa carandes),* mulberry (*Morus alba)* and lasoora (*Cordia dichotoma).* These fruit trees must be planted in deforested/denuded areas of Dansal, Purmandal, Akhnoor, Bhalwal blocks.
- In order to make the fruit growers conversant with modern horticulture technology, demonstration plots should be laid out and maintained under productivity improvement programme. Besides, efforts should be made to transfer new technology to farmers/orchardists by organizing fruit shows exhibitions, Kisan Melas at various places.

- Old orchards in the valley of Kashmir, Himachal Pradesh and Uttarakhand and other states which have now outlined and where the production is showing declining trend should be rejuvenated and new trees require to be planted.
- To educate the rural masses regarding agroforestry, various departments such as Agriculture, Horticulture, Forestry and Animal Husbandry must organize training camps and seminars through joint efforts. Even while celebrating Vanmahotsava fortnight and Environmental weeks, the people can be made aware of this topic.
- The Department of Forestry and Horticulture should make all-out efforts to provide a fruit and forest cover for *Kandi* belt of Jammu falling in the Siwalik range, which is entirely naked and is subjected to all types of soil erosion damages. For this purpose, create at least one nursery in each main village for supplying healthy, well-built tall saplings of various tree species to the villagers. Similarly, nurseries should also be established for supply of horticultural plants.
- A new scheme for boosting horticulture development in hilly and *Kandi* areas of other districts of Jammu region, "on compact area basis" must be introduced like Udhampur covering Ramnagar, Majalta, Gool, Dudu, Chenani, Udhampur, Mahore, Reasi, Arnas, Ghordi and Panchari as well as various *Kandi* areas of Punjab, Haryana, Himachal Pradesh and Uttarakhand.
- Plant neem *(Azadirachta indica)* trees wherever possible as a measure of termite control. Protect the farmers and available land outside the farms effectively for encouraging natural regeneration as it is superior to man-made forests.
- The grass-legume pastures provide a balanced feed to the livestock on a more sustained basis than deteriorated native village grasslands. Hence, this is an area of investigation which requires immediate attention.
- The pasture lands which have been the basic source of nomadic pastoralists, have now degraded to a large extent. These are also infested with heavy weeds. Hence, these should be replanted with suitable species of grasses/ bushes/trees.
- Suitable agroforestry systems must be evolved for temperate and cold arid regions of the country.
- The tree species selected for vaiious agroforestry systems should be fast growing, have efficient nutrient pumps, deep root system, drought and

frost resistant and should be of multipurpose uses and of economic value. Light grown species such as kikar, khair are the most suitable for agroforestry

- As most of the area of Jammu and Kashmir, Himachal Pradesh, Uttarakhand is mountainous and cultivation is done on the mountain slopes or along river banks, so the land area is subjected to heavy erosion, resulting into loss of soil and plant nutrients. It, therefore, becomes imperative to adopt soil and water conservation measures. Soil and water conservation measures are also required for the North East Hill Region.
- These soil and conservation measures consist of land leveling of the sloping farms, bench terracing, bunding, gully plugging, afforestation and other anti-erosion required in the farms or outside,
- The people dwelling in the flood plains of rivers like the Tawi, the Ujh, the Basantar and other streams like Jhajar, Devak etc., should be guided to create embankments on both sides of these rivers/streams so that flood waters carrying silt would make the land more fertile. Similar measure requires to be adopted for other hilly areas. Planting of *Saccharum* species (S. *munja* and S. *spontaneum)* and various species of *Acacia* would act as sieve and not allow the water to carry the silt away
- Since the lands of *Kandi* belt owing to poor management are subjected to the process of degradation, therefore they are not able to sustain agriculture production. Hence, it becomes necessary to think about evolving certain alternate land use systems for such lands. Adoption of agroforestry is one such system which constitutes the need of the day for the *Kandi* dwellers of India. Adopt the principle - agriculture for plains and silviculture for mountains. The entire country's *Kandi* belt should be brought under trees. Adoption of 2/3rd forest trees and 1/3rd fruit trees as per FAO formula would prove an asset.
- Bamboo plantation *(Bambusa arundinacea, Dendrocalamus strictus)* should be encouraged to grow wherever possible. Bamboo, infact, is known as poor man's timber and is used for construction of houses, raw materia! for village industries, tent poles avenues and fodder. It is point to mention that Siwaiiks of north India were hitherto full of bamboos (Gupta, 2005). But because of heavy deforestation, grazing and denudation, most of them vanished and are now confined to parts of Jammu region of Jammu and Kashmir State and Himachal Pradesh only. It is, therefore, suggested that bamboo plants should be planted on large scale.

- There is an acute shortage of water in Kandi belts due to uncertain rainfall which can not fulfill the requirement of agricultural crops. Horticultural crops are therefore, preferred to grow due to their less water requirement. Some of the wild fruits can be grown in these areas as prescribed by Singh (2019).
- Such fruit trees mainly consist of bael, (*Aegle marmelos*) amla (*Phyllanthus emblica*), Ber (*Ziziphus jujuba*), Girna (*Carissa spinarum*), Lasura *(Cordia dichotoma*) etc.

Study Questions

1. What do you mean by deforestation?
2. What is the importance of forests?
3. How we can define agroforestry?
4. Why the necessity of agroforestry felt?
5. Write a paragraph regarding the importance of agroforestry.
6. Name the various systems of agroforestry.
7. Write a short note of Agroforestry-Silviculture agroforestry system.
8. Name the various fodder tree species which are found available in the mid hills of Uttarakhand.
9. Write a brief note of Agriculture-Fishries agroforestry system.
10. What are the future research strategies/priorities?

References

Acharya, C.L. (1998). Rain water conservation and its efficient utilization. In 50 years of Indian Farming (1949-97), ICAR, Krishi Anusandhan Bhavan, Pusa (New Delhi), India.

Anonymous. (2003). Environment in focus. The Daily Excelsior 39 (1): 7 [January 1, Wednesday].

Choudhary, A., Mahajan, A. and Jaggi, R.C. (2002). Dalbergia sissoo Die-back - An emerging threat to sustainability of agroforestry systems in Indian subcontinent. Farmers' Forum 2 (6): 13-14.

Gill, A.S. (1999). Subabul-based agroforestry system. Indian Farming 49 (8): 33-35.

Gill, A.S. (2002). Following agroforestry research system help to achieve successful farming. ICAR News 8 (4): 5-7.

Gupta, D.K., Gupta, R.D. and Singh, H. (1992). Cropping patterns in the Kashmir Himalayas (pp. 78-91). In Conserving Indian Environment (Ed. S.K. Chadha). Pointer Publishers, Jaipur, Rajasthan, (India).

Gupta, R.D. (2004). Agroforestry in Kashmir: An old institution. Command Area Development's Subzarl (3): 10 [November Issue].

Gupta, R.D. (2005c). Environmental degradation of Jammu and Kashmir Himalayas and their control. Associated Publishing Company, New Delhi (India).

Gupta, R.D. (2009a). Agrcforestry for farmers. The Daily Excelsior 45 (180): 6 [June 30, Tuesday].

Gupta, R.D. (2011d). UN calls for conservation of forests. The Daily Excelsior 47 (35): 8 [February 4, Friday].

Gupta, R.D. (2013c). Forest studies of Jammu Siwaiiks. The Daily Excelsior 49 (337): 8 [December 9, Monday].

Gupta, R.D. Arora, S. 2016. Salt affected soils in Jammu and Kashmir (Now Union Teritory) their management for enhancing productivity. Journal of Soil and Water Conservation, 15(3) 199-204.

Gupta, R.D. (2013e). Issues concerning agricultural production. The Daily Excelsior 49 (86): 8 [March 29, Friday].

Gupta, R.D. and Abrol. V. ^2006). Agroforestry - A sound land use system for Himalayan rain-fed ecosystem. Farmers' Forum 6 (3): 19-22.

Gupta, R.D. and Banerjee, S.K. (1991). Problems and management of soil and forest resources of North West Himalayas. Mahajan Book Centre. Last Morh, Gandhi Nacar, Jammu (Jammu and Kashmir State), India.

Gupta, R.D. and Khajuria, M.R. (2000). Environmental holocaust in Himalayan ecosystem and protection. The Kashmir Times 36 (86): 1 [March 26. Sunday).

Gupta, R.D. and Raina, J.L. (1994). Opportunities for agroforestry in the dry-land belt of Jammu region (pp. 250-260). In Dryland Farming in India (Ed. S.K. Chadha,). Pointer Publishers, Jaipur (Rajasthan), India.

Gupta, R.D. and Sharrna. A.K. (2005). Indigenous technologies of Siwalik farmers. Indian Farmers' Digest 38 (7): 27-30.

Gupta, R.D., Mir, A.A. and Mahajan, A. (2007;. Popularizing poplars - Growing poplars in cold arid region of Ladakh. Wastelands News 23 (1): 14-16.

Jamwal, G.S. and Gupta. R.D. (2007). Watershed management project - A case study of Bari-Badhori, Jammu and Kashmir state (pp. 226- 231). In Natural Resource Management for Sustainable Hill Agriculture (Eds. S. Arora, S.S. Kukal and V. Sharrna). Soil Conservation Society of India, Jammu Chapter, SKAUST-Jammu, J&K State, India.

Khajuria, M.R. (1994). Effect of Leucaena planting at varying spacing with interplanting of various grasses. Personnel Communication (while he was acting Associate Director, Regional Agriculture Research Station, Rajouri, under Sher-e-Kashmir University of Agricultural Sciences and Technology, Srinagar).

Lavania, S.K. and Singh, V. (2005). Celtis australis - A versatile tree of Himalayas. Indian Farmers' Digest 38 (7): 19-20.

Pasrija. S. (2015). Rising animal husbandry market. Agriculture Today 18 (7): 41.

Ram, S.N. (1999). Silvi-pastoral system for sustainable development in arid and semi-arid regions of India. Farmers and Parliament 34 (9): 12-13.

Sharrna, A.K., Gupta, R.D. and Rai. P.K. (2009). Crop diversification is the need for conserving soil and water in Kandi belt cf Siwaiiks (pp.69). Soil and Water Conservation. Training Manual (Ed. B.S. Hansra, P. Vijaykumar, R. Nanda and P.K. Rai) organized by School of Agriculture, IGNOU, New Delhi and KVK SKUAST, Jammu, R.S. Pura. Jammu (Jammu and Kashmir State), India.

Sharma, G. (2005). Fast depleting forest cover threat in maintaining ecological balance. The Daily Excelsior 41 (298): 7 [October 26 Wednesday].

Sharma, J.P. and Mir, A.A. (2000). Dynamics of cold arid agriculture. Kalyani Publishers Ludhiana New Delhi Noida (Uttar Pradesh) Hyderabad Chennai Calcutta Cuttack.

Sharma, R. (2015). Agroforestry - The time has come. Agriculture Today 18 (3): 42-43.

Singh, R. (2019). Wild fruits of Mansar Suriansar. The Daily Excelsior. Magazine 55(54): 4 [Sunday, February 24]

Tripathi. D. Sharma, I.P. and Chib, S.S. (2006). Macro and micro-nutrient dynamics in decomposing leaf litter of some agroforestry trees (pp. 59-65). In Balanced Fertilization for Sustainable crop Production (Eds. S.P. Sharma, S.K. Sharma, R.P. Sharma and T.S. Verma). Proceedings National Seminar on INM for Balanced Fertilization held on November 17-18, 2006.

13

Cover Crops and Sustainability of Agriculture

Abstract

Although the Green Revolution, which ushered during (1967-1968). gave a tremendous boost to both production and productivity yet there is obviously stagnation in this respect of agriculture rather a decline. For instance, in the year of the advent of the Green Revolution (1967-68), the productivity of rice was 1.03 t ha^{-1} which rose to 2.09 t ha^{-1} in 2001-02, respectively. Because of severe drought in 2002, the productivity had gone down to 1.9 t ha^{-1} in 2002-03. In case of wheat, the corresponding figures were 1.1 and 2.8 t ha^{-1} . These days, the average yields of these crops have further come down. Of the various factors responsible for causing low yield in these crops, indiscriminate use of fertilizers and pesticides, and neglect of organic fertilizers are the main contributors in this context. It, therefore, becomes an imperative to follow various methods to enhance the production and productivity of rice and wheat. In this chapter of the book, how cover crops assist in sustainability of Agriculture has been embodied. What are cover crops? Characteristics and advantages of these crops are the main themes of this chapter of the book. Future strategies/priorities are also given

Keywords: Green Revolution: sustainability: production: productivity: stagnation: drought; low yield; cover crops: fertilizers; pesticides; organic fertilizers

The Prime Minister of India, Dr. Manmohan Singh has asked the agriculture scientists for speeding up research to boost agricultural production especially in cereals. It is because now the "Green Revolution" has plateaued out to meet the food requirement of the burgeoning human population in the country. The production of wheat crop is stagnating at 70 million tonnes and that of rice had already stagnated or has become low which is about 110 million tonnes over the last five years. Rice yields are hovering between 1.91 ha^{-1} in 2000-01 and 2.1 t ha^{-1} in 2004-05 (Joshi, 2007). Wheat yields are also stagnating about 2.7 t ha^{-1} since 1999-2000. It is further added that our yield levels of majority of the crops are far below than the globs! averages. Infact, in areas where rice and wheat cropping sequence is being followed, their yield has either become stagnant or is in the declining trend (Mahajan and Gupta, 2009). In north India, it is very common in Punjab, Haryana, Himachal Pradesh and Jammu region of Jammu and Kashmir State.

In Punjab, Haryana, Uttar Pradesh, Andhra Pradesh and Maharashtra and in some other states the soil in places has become sick and water that used to flow so noisily has dived deep inside. And the helpless tillers of the land, the faithful sons of the merciless soil, are succumbing to a tempting long noose or to pesticides that proved so ineffective at their real job. Hitherto many farmers relied heavily upon chemical fertilizers but due to rise in their prices and harmful effects as well as purchasing power of the farmers, their use has now become limited. Indiscriminate use of chemical fertilizers has posed a number of health problems both to human beings and animals. It has contaminated the soils, water and air (Gupta, 2005c, Gupta, 2006a; Gupta and Singh, 2006; Mukherjee, 2015). Further these days, there is an increasing demand of green or organically grown foods due to perpetual consumer's requisite and its positive environmental effects.

Repeated crop failure, overexploitation of farming, costly inputs and implements, inadequate irrigation, increasing fragmentation of land holdings, lack of electricity and neglect of cultural techniques are a few of the several other reasons, responsible in bringing about the low yield in rice and wheat cropping sequence. Thus, Indian agriculture which contributes previously a quarter of the Grand Domestic Product (GDP) and is a source of income to more than 60 per cent of the farmers is getting increasingly risky, expensive and water dependent. It is point to mention that the Indian Agriculture which was contributing 18 per cent of GDP (Mahajan and Gupta, 2009) has now reduced to 13 per cent. If we see the record of food grain production (Table 12.1) from 1993-94 to 2004-05, it is observed that production rarely touches 215 million tonnes a year which had tentatively fixed as the target for 2005-06 (Ghose, 2005). Furthermore India's food grain production which comprises of cereals and all pulses including lentils has almost leveled off from 208 to 209 million tonnes during 2005-06 and 2006-

07 (Saha, 2009). During the year 2007-08 and 2008-09 it touches between 230 to 234 million tonnes which fell down to 218 million tonnes during 2009-10 (Table 13.1) and again arose to 241 million tonnes during the year 2010-11. There is no increase in area in the last 15 years for crop production, so the task has become tough but the only option to improve the productivity through science and technology (Sandhu, 2015). India's food grain production was estimated to 257 million tonnes during 2014-15. It was at 265.57 million tonnes during 2013-14 (Sandhu, 2015). Out of this 106.65 million tonnes was of rice. Further, it is added that now-a-days the agriculture is driving the farmers even in the rich states like Punjab, Haryana, Western Uttar Pradesh, and Andhra Pradesh etc. to a riskier remedy, i.e. more and more credit. So it requires number of methods to improve it.

Table 13.1: Food grain production during the year from 1993-94 to 2004-05 (Adapted by Ghose [2005]; Saha [2009] and Sandhu [2015])

Year	Food grain production (million tonnes)
1993-94	184.26
1994-95	191.50
1995-96	180.42
1996-97	199.44
1997-98	192.26
1998-9S	203.61
1999-2000	209.80
2000-01	196.81
2001-02	212.85
2002-03	174.77
2003-04	213.46
2004-05	204.61
2005-06	208.59
2006-07	209.17
2007-08	230.00
2008-09	234.00
2009-10	218.11
2010-11	241.00
2011-12	259.32
2012-13	257.00
2013-14	267.57
2014-15	257.07

In the light of the above, it, therefore, becomes imperative to advocate such methods of cultivation to the farming community which can assist to sustainable food production. These methods may consist of crop diversification from staple food to horticultural crops, vegetables and cash crops as well as animal husbandry, use of organic manures including green manuring, cultural methods, etc. One such method that greatly reduces dependency of the chemical fertilizer as well as helps to increase production is through the growing of "cover crops".

13.1 What are Cover Crops?

Any crop planted to control soil erosion, when regular crops are off the field, is usually referred to as a cover crop. Cover crops for wind erosion control must be drought resistant with enough canopies to cover the soil. The details of cover crops which are responsible for increasing/sustaining crop production have been given elsewhere (Gupta, 2005a and b).

According to United States Department of Agriculture (1969), "any crop while serving as a solid ground cover, whether or not specifically planted for this purpose is a cover crop. Thus, a growing grain crop, grasses in pastures, and crops planted for turning under as green manure is cover crop. However, common customary practice, possibly limits the term more precisely for checking soil erosion, adding organic matter to the soil vis-a-vis other plant nutrients, and ameliorating the crop productivity. It is point to mention that cover crops require more consideration for the cultivated lands though the vegetative cover in any situation is of much importance in soil conservation.

"Annual crop sown to create a favourable soil micro-climate, decrease evaporation and protect soil from erosion is called cover crop (Reijntjes et al., 1992). Cover crops also produce biomass which can be used for soil fertility management" - ILEIA (Information Centre for Low-External-Input & Sustainable Agriculture).

13.2 Growing of Cover Crops without Wasting of Land

"How can one produce cover crops without growing them on land used for food or cash crops", is the common question put by the farmers reluctant to grow cover crops. Following are the ways to avoid this doubt:

- Cover crops can be grown among traditional row crops like maize, pearl millet and sorghum without reducing the production of main crop. This may not be possible in case of farmers intercropping two or three other crops with their main crop.
- Cover crops can be intercropped with the basic grain producing crops towards the middle or end of the growing season, so that their major growth occurs during the dry season.
- Wherever shifting cultivation is still going on, cover crops can be sown on the land the first year it is to be fallowed or abandoned.

- In areas where fruit or coffee trees are common, cover crops can be grown around or under the trees; both will increase the growth and health of the trees.

The ground cover is also developed by growing tree crops to protect the soil from splashing effect (Suresh, 2004). In fact, in this case what happens, when rainfall takes place, the intercepted water in form of large size water drops falls over the ground surface through the plant leaves, which impacting force is much more, as a result there is generated erosion problem. Normally, such covers are not established with the tall crops such as rubber, particularly when the height of fall of water droplets is sufficient to cause the drops to approach terminal velocity. The most common crops used in tropical regions are pueraria, phaseoloides (Suresh, 2004).

The effectiveness of plants and their cover to protect the soil from erosion depends not only on the density of plant population, but also on their growtn stages. The greatest protection is resulted from dense and close growing grasses or legumes or combination of both. However, rows or inter tilled crops also make protection from erosion but in less value than the grasses or legume crops.

13.3 Advantages of Growing Cover Crops

Planting of cover crops is one of the natural and traditional methods to recuperate the soil fertility like other traditional methods such as "Slash and Burn Method", use of organic manures which have always been to enhance the soil fertility. The main advantages of using cover crops are:

- They provide large amount of nitrogen to soil sometimes more than 200 kg ha^{-1}.

  They also add lot of green matter, usually 20 to 30 t ha^{-1} (depending upon the type and nature of the crop chosen) to the soil, thereby, improving nutrients status, moisture retention characteristics and ionic exchange phenomena.

- They improve soil tilth by the addition of organic matter to the soil as stated above and by loosening up the subsurface of the soil with deep-growing roots.

  As the cover crops grow in place, so unlike organic manures or chemical fertilizers, there is no problem of transportation or engaging labours.

- They do not require any kind of capital other than an initial purchase of handful of seed.

Cover crops protect the soil from erosion and mitigate the runoff losses.

- They diminish leaching of plant food, especially of nitrates. It is worthwhile to note that accumulation of nitrate in the underground water is a big problem in many states of India like Uttar Pradesh, Haryana, Gujarat, Andhra Pradesh etc., as it is responsible for causing blue baby disease in young babies.
- Cover crops apart from protecting the soil from erosion and mitigating the run-off losses also protect newly constructed terraces and other mechanical erosion control devices.
- They supply protein rich fodder for animals.
- In many cases cover crops provide human food i.e., various edible beans, peas and pods when legumes are grown.
- Besides preventing water erosion, the cover crops also control wind erosion.
- The farmers can earn cash by selling seeds and other products like firewood or fuel wood.
- They increase the yield of maize, cotton and other regular farm crops.
- When intercropped with basic food grain crops can help in checking the weeds, thereby eliminating costly weeding operations.
- Growing of legumes as cover crops will also cut the shortage of fodder of the hilly remote areas like Warwan in Kishtwar district, Mahore in Reasi district and Banihal of Ramban district in Jammu Region of Jammu and Kashmir State.

The advantages or benefits of cover crops are more valuable than any reasonable objections raised to them. Inspite of the fact, for practical information, the principal features which might be disadvantages using cover crops, are as follows:

- The cost of seed.
- Cost of labour in the preparation of the seed-bed and seeding.
- Uncertainty of getting a crop stand.
- A possible refuges for insect pests, such as cutworms that later attack corn or other crops. Despite of the facts as stated above, the cover crops will ultimately increase crop production as recently reported by Kumar and Namgyal (2020).

13.4 Soil Cover Rating

In order to calculate or evaluate various soil covers for their suitability to reduce the soil detachment, a term referred to as "Soil Cover Rating", has been introduced which denotes the effectiveness of a cover in preventing the detachment of soil particles due to rain drop impact, during storm. The following formula as reported by Suresh (2004) can be used to evaluate this, numerically:

$$SCR = [(W_b - W_e) / Wb]*100$$

Where, SCR indicates soil cover rating (%), W_b and W_e are the weight of soil splashed from the bare land and that of under-cover.

In accordance with this formula, the worst condition in so far as erosion is concerned, would be bare soil which will show the value of "soil cover rating" as zero. By knowing the SCR of different covers, the effectiveness of them to check the soil detachment can be studied. The "soil cover rating" of different cover crops at various growth stages can also be evaluated, which would be a valuable information for choosing the cover crops to be maintained into the cropping system.

13.5 Importance of Legumes in Cover Crops

Most of the legumes like black gram, green gram, soybean, french bean, berseem etc., act as cover crops. It is because they cover and protect the soil, and add nitrogen for improving its fertility. In fact, the *Rhizobia,* which are generally referred to as the root nodule bacteria, live in the nodules of these legumes and fix atmospheric nitrogen in them. This nitrogen eventually enters into the soil either through ex-osmosis or after death and decay of the nodules and/or *Rhizobia.* In this way the soil becomes rich in nitrogen. Thus, many farmers have found cover crops as an ideal solution for the infertile soils. Some of the farmers, however, have a prejudice that they will not be able to spend little amount required to grow the cover crops and/or can not spare the labour meant for this purpose.

Mixtures of legumes and grasses are particularly desirable for erosion control and soil improvement. When used for checking soil erosion and improving soil tilth, these crops should be seeded and cultivated in a manner recommended locally, when they are grown as forage. Red clover seeded in wheat, red clover and other legumes, and grass seeded in small grain food production prevents erosion.

Leguminous green manuring crops and cover crops can function directly when used as fodder for animals which provide manure for crops. Some green manure and cover crops, for example, *Crotalaria* spp and *Tephrosia candidat* are known to posses' pesticide effects (Reijntjes et al., 1992). And as such can be used for plant protection.

Cover crops (trees, shrubs, legumes, grasses, weeds, ferns etc.) supply green manure which can constitute an in extensive source of organic matter. Organic matter on decomposition produces humus which acts as store house of plant nutrients and increases ionic exchange phenomenon and improves physical and chemical conditions of soils. Some leguminous cover crops such as *Centrosema pureria* and *Crotalaria* spp. are capable of developing deep root system in humid tropics having acidic soil reaction.

In the "Jhum" or "Shifting Cultivation", the peasants use cover crops to suppress the weeds by way of shortening or abandoning the fallow period.

13.6 Characteristics of Cover Crops

What characteristics should be looked for in a cover crop to produce fertilizer and forage, with no cash costs and a minimum of labour?

- It should produce large amounts of green matter (about 25 t ha^{-1}).
- It should grow vigorously in poor soils without any assistance of fertilizers.
- It should be planted with the minimum of soil preparation, and either with a dibble stick or broadcasting the seed.
- The plant should have minimum of natural enemies i.e., it should grow without using any pesticide There should not involve any major labour requirement.
- The legume should be shade resistant (for intercropping or growing under trees) or drought resistant (for growing into and through the dry season) or both.
- It should have the capacity to fix large amount of atmospheric nitrogen.

13.7 Constraints/Limitations of Cover Crops

- Cover cropping is not a cure to all. It represents only one of the recommended practices of a well rounded soil conservation programmes.
- Crops grown in wide rows with cultivation like cotton, corn, potatoes and tobacco, are highly conducive to soil loss. All of these crops must be cultivated with systematic crop rotations. and the rotations of these crops require being supplement with cover crops if soil loss has to be reduced to a minimum and crop yields sustained which is difficult to follow especially in the hilly terrain.

- Wheat and other small grain crops, when grown on steep slopes, especially under summer fallow system, are also require to be given a special consideration if soil loss and reduced yield have to be shun.
- Many permanent cover crops such as kudzu, grasses and various shrubs are commonly not grown with cultivated crops.

13.8 Conclusion

Excessive use of chemical or inorganic fertilizers has damaged the soils by way of producing deficiency in Ca, Mg and S (secondary) and Fe, Zn, Cu, Mn, B & Mo (micro-nutrients) and acidity in many soils, which has resulted in stagnation in productivity especially in rice and wheat crops. To sustain the yield of these crops new methods/strategies are required to be applied. In this chapter of the book, growing of cover crops without wastelands, their advantages and characteristics have been highlighted which enrich the soil fertility, conserve the soil and moisture especially of sloppy lands, and eventually the productivity of the crops grown.

13.9 Future Research Strategies/Priorities

- Soil depleting crops like maize, potato, cotton are highly conducive to soil loss. All of these crops, therefore, should be grown in systematic crop rotations, and the rotation supplemented with cover crop if soil loss has to be reduced a minimum and crop yields has to be maintained or sustained.
- Wheat and other small grain crops, when cultivated on steep slopes, require to be given special consideration if excessive soil loss and reduced yields are to be avoided.
- Mixture of legumes and grasses are particularly desirable for erosion control and soil improvement.
- Wherever it is possible, legumes such as sweet clover, alfalfa, red clover, field peas are grown to increase the soil fertility vis-a-vis to check the soil erosion and runoff losses.
- To have more growth of the cover crops for taking their full advantage, one has to go into the balanced use of nutrients. The recommended application of organic fertilizers in association with inorganic nutrients such as urea, DAP and MOP. Incase of leguminous crops recommended dose of phosphatic fertilizer is a must. It will enhance biological nitrogen fixation phenomenon.

Study Questions

1. What are the cover crops?
2. Write down the main advantages of the cover crops?
3. What is the importance of legumes in cover crops?
4. Name a few cover crops which fix atmospheric nitrogen in soils?
5. What are the limitations of cover crops? Write briefly?
6. How much quantity of nitrogen and organic matter is added after growing cover crops in soils?
7. How cover crops help in controlling soil erosion and conserving water?
8. Write a brief note on soil cover rating?

References

Ghose, A. (2005). Second green revolution should aim at raising crop yield per unit area. The Kashmir Times 62 (293): 13 (October 21 Friday].

Gupta, R.D. (2005a). Cover crops enhance food production-l. The Kashmir Times 62 (293): 13 [October 22, Saturday].

Gupta, R.D. (2005b). Cover crops enhance food production-l I. The Kashmir Times 62 (300): 13 [October 29, Saturday].

Gupta, R.D. (2005c). Environmental degradation of Jammu and Kashmir Himalayas and their control. Associated Publishing Company, New Delhi (India).

Gupta, R.D. (2006a). Environment Pollution: Hazards and Control. Concept Publishing Company, New Delhi, India.

Gupta, R.D. and Singh, H. (2006). Indiscriminate use of fertilizer poses health hazards. Farmers' Forum 6 (9): 20-24.

Joshi, P.K. (2007). Indian agriculture: problems and remedies. Agriculture Today 10 (3): 36-37.

Kumar P. and Namgyal, D. (2020). Regenerative agriculture. The State Times, 25(6):6

Mahajan, A. and Gupta, R. D. (2009). Integrated nutrient management (INM) in a sustainable rice-wheat cropping system (pp.1-267). Springer Science + Business Media (www. Springer.com).

Mukherjee, D. (2015). Water Contamination Integrated Strategy must. The Daily excelsior. 51(97): 6.

Reijntjes, C., Haverkort, B. and Bayer, A.W. (1992). Farming for the future - An introduction to low external input and sustainable agriculture. The Macmillan Press Limited London and Bastingstoke, pp. 211.

Saha, B.K. (2009). Stagnation in agricultural output: An introspection. Agriculture Today 12 (2): 46.

Sandhu, J.S. (2015). Nutrient rich varieties of rice in the making. Agriculture Today 18 (5): 16.

Suresh, R. (2004). Soil and water conservation engineering. Standard Publishers Distributors, 1705-B, Nai Sarak, New Delhi -11006.

USDA. (1969). A Manual on Conservation of Soil and Water. Oxford and IBH Publishing Company, Calcutta Bombay New Delhi, pp. 117.

14

Indigenous Technological Knowledge Concept and Importance in Sustainable Agriculture

Abstract

Indigenous technology is the knowledge of the people or farming population, residing in a certain specific area. This knowledge concerning with local farming system, lies in its functional integration of different resources and farming techniques. By integrating various land use functions viz., producing food, fodder, fuel wood, fibre and fertilizer, conserving vis-a-vis use of different components, the stability and productivity of the farm can be enhanced. Accordingly, the indigenous technologies followed by the farmers of Siwaliks including *Kandi* belt of Jammu and Kashmir, Himacfial Pradesh, Uttarakhand, Haryana and Punjab as well as south India are imparted in this chapter of the book.

Keywords: Indigenous technology; farm; farming population; local farming system; farming techniques; productivity; resources, food, fodder; fuel wood; fibre; fertilizer; *Kandi* belt

Indigenous knowledge is predominantly embedded in the practices and experiences. It is generally exchanged through personnel communications and demonstrations from master to apprentice, from parents to children, frem neighbour to neighbour etc. It is stored in people's memories and activities, and is expressed in the form of stories, songs, folklore, proverbs, cultural values, beliefs and rituals, agricultural practices, equipments, materials, plant species and animal breeds. Indigenous knowledge is shared and communicated orally by specific examples and through culture.

14.1 Concept of Indigenous Knowledge

The concept of indigenous knowledge or indigenous knowledge system (IKS) delineates a cognitive structure in which theories and perceptions of nature and culture are conceptualized. Thus, it includes definitions, classifications and concepts of the physical, natural, social land economic environments. The dynamics of IKS takes places on two different levels, the cognitive and empirical. On the empirical levels, IKS is visible in institutions, artifacts and technologies.

Indigenous farm technological knowledge is of practical nature concerned with operationalized local thinking in such fields as agriculture, olericulture, horticulture, forestry, fisheries as well as animal husbandry and human health.

14.2 What is Indigenous Knowledge (IK) or Indigenous Technological Knowledge (ITK) and Indigenous Knowledge System (IKS)?

It refers to unique, traditional local knowledge existing within and developed around at the specific conditions of women and men, which is indigenous to a particular geographical area. The development of IK system, covering all the aspects of life including management of the natural resources, has been a matter of survival to the people who generated these systems. Such knowledgeable systems are cumulative, representing generations to generations of experience, and trial and error experiments.

It is point to mention that IK, IKS and ITK, is not confined only to tribal groups, the original inhabitants of an area or to rural people but can be represented by any community - rural and urban, settled and nomadic or migrants. Other names of the indigenous knowledge are "local knowledge", "indigenous technical knowledge" and "traditional knowledge".

Indigenous knowledge in the 'form of communication and organization are vital to local-level decision making process and to the preservation, development and its spread. However, World Bank states that IK systems are at risk of becoming extinct.

14.3 Advantages of Indigenous Knowledge

Indigenous knowledge has two powerful advantages over outside knowledge. It has little or no cost and is easily available. Learning for IK can improve understanding of local conditions and provide a productive context of activities designed to help the communities. Apart from this, the use of IK assures that the end users of agricultural development projects are involved in developing technologies appropriate to their needs in sustaining agricultural production.

IK is still underutilized resource in the development activities. It needs to be intensively and extensively studied and incorporated into formal research and extension practices in order to make agricuuural and rural development strategies more sustainable. Specific efforts are needed to understand, document and disseminate IK for preservation, transfer or adoption elsewhere.

IKS and technologies are found to be socially desirable, economically affordable and sustainable and involve minimum risk to rural farmers and procedures, and above all they are widely believed to conserve various natural resources.

There are institutions in which modern science is not appropriate and use of simpler technologies and procedures are required to be adopted. The IK provides basis for problem solving strategies for local communities, especially for the poor.

14.4 Importance of Extension in ITK for Achieving Sustainable Agricultural Development

Extension organization, especially concerning with agricuilure, is essential in identifying, documenting and incorporating indigenous knowledge to achieve sustainable agricultural development. Systems relating to IK provide a framework for strengthening agricultural extension programmes, which finally lead to reorganization of interventions made by extension personnel. The participatory technologies which are developed through IK integration will help to:

- Provide diversified technological option which enable farmers to choose using their own decisionmaking systems;
- Originate from the farmer's own knowledge; and
- Use of diversified sources in active participation of research minded farmers (Sharma et al., 2009).

14.5 Strategy of Extension Education

Extension strategy should now be focused on improving current benchmark of IK practices. The strategy should also concentrate on facilitating skills to build capacity of producers of innovations to formulate questions, which they use to engage specialists or to utilize information systems. Indigenous knowledge is more just technologies and practices. It includes information, practices and technologies, beliefs, education, communication and experiments on farms.

14.6 Documentation of IK and Its Practices in Indian Agriculture

Unless IK is properly documented, analyzed and disseminated, there is a risk that within one generation the knowledge could be lost forever. Data based resource centre's would help to exchange IK from one community to another and promote integration of IK into development process. Identification of IK can at times prove difficult. It may be embedded in a mix of technologies or in cultural values, rendering them unrecognizable at first glance to the external observer. Technical and social analyses may, therefore, be required to identify IK.

The use of modern information and communication technology for the exchange of IK is very exceptional. As the countries established connectivity to reach rural people, modern information centre technology can become a powerful to enable for dissemination of IK. In the near future, more traditional and suitable channels of dissemination of IK could be used. Depending upon the availability of infrastructure, accessibility and connectivity of communities in a country, the channels like radio, television, telecasters/tele-learning centre's etc., could be used for broadcasting in local languages.

In India, attempts like Warna Wired Village (Maharashtra) and Info Villages (Pondicherry) are still in the initial stages to take note of relevance of transfer of IK. The information and communication technologies would definitely promote activities related to collecting processing and dissemination of IK globally. In fact, the agricultural practices in India date back to over 5,000 years. A number of indigenous practices have been in vogue in Indian agriculture since time immemorial. Use of organic manures and other organic inputs is very much native to this country. As mentioned in Arthashastra, farmers in Vedic Period possessed a very good knowledge of replenishing soil fertility through the use of organic manures (cow dung, green manure)

Apart from this, traditional methods of integrated pest management have been found to be more in tune with the nature and quite effective in minimizing damage to crops. Traditional or the indigenous methods bring out the sound knowledge of host-plant and insect pest interaction possessed by our ancestors. These time-honored practices were not only found eco-friendly but were also cost-effective. Similarly use of natural enemies devoured the crop pests. The use of spiders in trapping houseflies and cattle flies was another testimony to the wisdom of traditional farmers.

In our country, indigenous practices of safe storage of grains had been prevailed since long among the farming communities in many states of the country. It was because the indigenous practices have proved more useful in controlling food grain loss and preserving the quality of grain stored in the household.

Indigenous Technological Knowledge, imparting insight into the management of soil fertility, nutrients and crop management for sustainable agriculture, was also available. Sustainability was due to the fact that this knowledge has been testified time and again and carried on from generation to generation. Infact, the farmers of an ancient India used to adhere to the natural laws while developing farming systems and practices. This has helped to sustain soil fertility, for long of their farms. Moreover, this knowledge is generally eco-friendly and does not need much "off farm" inputs. However, in the post independence period, the most important challenge for the country was to produce more food grains for the growing population.

This challenge was overcome through introducing of high yielding varieties of various crops with infusion of irrigation water, fertilizers and chemical pesticides, which got a major fillip. Use of traditional practices (organic manuring, botanic pesticides), however, remained uncared. Though the package of high yielding production technology, so called the "Green Revolution", has helped the country to become food surplus yet it resulted in soil health problems, pesticide toxicity and environmental pollution, which finally hampered overall sustainability of agricultural production. So the scientists and the policy-makers have now again relooked at the traditional agricultural practices which rely more on biological inputs rather than heavy usage of chemical fertilizers and synthetic pesticides. Such agricultural practices are detailed in the following specific lines:

14.6.1 Traditional Methods of Integrated Pest Management

As already pointed out that the traditional methods are quite effective in minimizing damage caused by some pests to crops, so these are being adopted in some of the areas. According to Mr. P.S, Balathandayuthanpani - a progressive farmer of Pappampati village near Ondipudur in Coimbatore district (Tamil Nadu) - the community action and group endeavours are the strength of these traditional methods. Some festivals such as the "light festival" are celebrated during the seasons when the population of insects increases. In the Tamil month "Karthigai" when the weather is conducive for the proliferation of insect pests, people set-up oil lamps and fireworks. Sometimes bonfires are lit using crop residues. The adult insects attracted by the lights, get burnt by naked flame.

The entire village would be aglow with lamps, and the setting-up of massive light traps would eliminate the entire range of pests. The farmers themselves know when the pupation in several insects would be over and adults, thereafter, would emerge in swarms. They synchronize the "light festival" to it and protect the crops accordingly.

Dragon flies and damsel flies were encouraged in crop fields, as the natural enemies devoured the crop pests,-especially mites, aphids and scales checking their population. The use of spiders in trapping houseflies and cattle flies was another testimony to the wisdom of traditional farmers. "Spider nests in tamarind twigs would be brought to the houses and kept over sugar solution. The flies lured by it would get entangled in the web, and fall a prey", said Mr. Balathandayuthanpani.

14.6.2 Eco-Friendly Botanical Pesticides

Eco-friendly botanical pesticides like pyrethrum, nicotine, rotenone and ryania were in vogue before the advent of synthetic pesticides. But, thereafter, these botanicals were relegated to insignificant position in pest control (Chander and Munshi, 1995).

14.6.2.1 What Are Botanical Pesticides?

Though insecticides including repellents obtained from various species of plants have been around for years yet their increased awareness about insect resistance and resurgence problems has given new dimension in crop production. “For those who wish to switch over to ecological farming as an alternative to chemical plant protection is now available than ever before. An organic farmer has just to pick up the right methods and package them nearly to suit specific needs of his crop”, says an eco-farming expert.

14.6.2.2 Advantages of Botanical Pesticides

Botanical pesticides are the derivatives of a large number of plants and being less stable in environment have, therefore, become the main focus of more attention these days. They are safer to mammals as well as other non-target organisms. They are easily available in a number of places and are often very cheaper than their synthetic counterparts. Moreover, their crude extracts can be prepared very easily. Even the farmers can prepare various extracts from the plants to be used in the fields for controlling insect pests.

As a matter of fact, the most of the botanical insecticides have relatively short field-life and being highly biodegradable, they leave no toxic residues. Since most of the plant-based pesticides have repellent and anti-feedant properties, so they have been used as a prophylactic. The only problem the farmers will face is that they may have to spray the crop too often which might push up the cost of plant protection. According to experts as most of the botanical pesticides are not market purchased inputs, the cost may not be that prohibitive. Moreover, if a farmer uses a solar powered sprayer, there will be substantial savings in energy as well.

14.6.2.3 Various Kinds of Botanical Pesticides and Why Their Necessity Came into Fore?

Insect pests, diseases and weeds are important biotic constraints inflicting with 20-25 per cent loss in agricultural production. Although synthetic pesticides played a significant role in restricting many pest problems yet their indiscriminate use took its toll by creating several problems like contamination of food affecting thereby, health’s of foetus, children, women and men, resurgence of pests and detrimental affects on non-target organisms. Owing to ill effects of synthetic pesticides there aroused interest in alternate methods of plant protection. Bioactive products of plant origin being less persistent in environment and safe to mammals and other non-target organisms have, therefore, rendered the focus of attention these days. Botanical pesticides are easily and cheaply available in a number of places.

There are a number of bio-pesticides. Neem, drek, vana, tulsi, lantana etc., are some of the typical examples of trees/bushes, which behave as bio-pesticides. Pyrethrum, rotenone, ryania and nicotine are some of the other examples in case of botanical pesticides. Pyrethrum is extracted from derived flowers of chrysanthemum. Rotenone is derived from the rhizomes of tropical legumes Derris and Lonchocarpus. Nicotine is obtained from tobacco (*Nicotiana or Nicotiana tabacum*).

14.6.2.3.1 Neem

Out of the various kinds of botanicals, neem *(Azadiractha indica)* is the most promising source of bio-pesticides of plant origin and occupies pivotal position today. Neemows its toxic effects due to the presence of a number of bitter compounds called meliacins like azadirachtin, nimbin, salanin etc. The bioactivity of neem products can control more than 300 insect species, many nematodes and a number of plant pathogens. Neem has 127 different ingredients to which insect pests affecting crops cannot become resistant. Spraying found to repel chewing and sucking insects. Its products besides affecting crops are also found effective against different pests of vegetables. Extracts of neem can provide more than 90 per cent control of hadda beetles in brinjal.

Collect ripe neem fruits and clean them thoroughly. Remove the skin and prepare one litre of the spray. Grind seeds flesh. Sun dry the seeds for 4 hours. Dry them in the shade afterwards. Continue this process for a week. Store the dried seeds in baskets or sacks in dry, airy place. Do not use plastic bags.

Take two handfuls of neem seeds to coarsely. Avoid dust or mould contamination. Mix the powder with water and let it stand overnight. Filter the solution next day. The spray is ready. The effect of the spray is least on the first day. Use a fresh spray every day. Spray the solution using a sprayer or water can.

Neem products exert behavioural, physiological and biological effects on insects. As they are environment friendly and leave no residual toxic effects, so neem products are considered the best for integrated pest management programmes. The neem products, infact, are harmless to non-target organisms such as pollinators and beneficial microorganisms as well as spiders and mammals. Unlike other synthetic insecticides, neem derivatives are systemic and do not induce pest resurgence. Insects developing resistance to single active component of synthetic pesticides over a period of time may be unable to develop resistancc to neem compounds because they possess numerous bioactive principles. Generally, neem derivatives are greatly unstable and persist for a short while in fields when compared to conventional insecticides.

Now-a-days, many neem-based formulations are available in the market. They are margocide, neembicidine, bioneem. Neemazal F, which can be used to control a number of pests of different crops. Even freshly prepared neem seed kernel extract can be recommend for checking various harmful pests. Neem has demonstrated to show considerable potential as fertilizer. For this purpose, neem cake and neem leaves have been found more promising. In some of the tribal areas of Sambalpur district (Orissa), the farmers collect ripe neem fruits. The oil is extracted from the fruits for their use. The leftover cake is added in the paddy fields (500-1000 kg ha^{-1}). Farmers are of firm belief that this practice in addition to providing plant nutrients also controls crop pests.

14.6.2.3.2 Use of Neem Leaves as Safe Storage of Food Grains

Safe storage of grains is very important as insect pests some times cause a loss of about 30 per cent in total grain production both in fields and storage. Use of dried leaves of neem or their powder after wrapping in the papers, placing them at the upper, middle and bottom layers of granaries and close the granaries air tightly (Gupta and Sharma, 2005; Gupta, 2010a). This practice was hitherto very common in many farming villages of Jammu and Akhnoor tehsils of Jammu districts vis-a-vis in those of Samba district. But with the advent of Green Revolution where under the use of insecticides became necessary, use of indigenous technique came to halt.

14.6.2.3.3 Others

The leaf extracts of lantana (*Lantana camara),* tulsi (*Ocimum basilicum* synonym *sanctum)* and vetiver (*Vetiveria zizanoides)* are found useful in controlling leaf minors infesting potato, beans, brinjals, tomato, chilies, cucurbits and onion. Botanical product RD-repelin, a combination of extracts from the seed of neem, pongamia, mahua and sesame can be effectively used to control the incidence of aphids in pea and whitefly in okra (Chander and Munshi, 2000).

The roots of *soanjna* or drumstick (*Moringa oleifera)* contain an active principal called pterygospermin which can be extracted as an unstable low melting substance using organic solvents. This compound with a characteristic odour is sparingly soluble in water and inhibits the growth of several bacteria and fungi. Hence, it can be employed to preserve fruits and vegetables and to treat the seeds (Gupta and Sharma, 2005).

Leaf and fruit extracts of bael (*Aegle marmelos)* have proved beneficial to protect crops from various kinds of pests. Farmers of *Kandi* belt of Jammu say, "Feeding of bael leaves and bael kath (fruits) to cows and buffaloes, enhance their milk production and increase milk fat also." The leaves of neem, drek (*Melia azedarach)* and wana (*Vixtex negundo)* are used to check insect infestation during grain storage.

14.6.3 Traditional Use of Organic Manures

In India, the use of dung as manure and its usefulness seems to have been practised since the Rig Vedic Age i.e., 2500-1500 BC. The practice of application of bones as phosphatic fertilizers also dates back which was about 300 BC. Contrary to this, magnitude of excreta of goats and sheep was recognized in post-Vedic Age i.e., 300 BC to 500 AD which has been mentioned in Kautilya Arthashastra. Other manures like oilcakes appear to have come into use in the country from 1008 to 1400 AD.

Agriculture during the Vedic periods had a lot of knowledge about fertility of soil which was mainly divided into two classes: Urvara (fertile) and Un-urvara or Usara (sterile). The ancient Indian cultivators were a wealthy and respectable section of the people, and possessed also a good knowledge of climate, plant physiology, rotation and protection of crops. Indeed, one will be surprised if one cares only a look into the detailed accounts have been elucidated in Arthashastra and Agnipurana, regarding the selection of treatment of seeds and use of animal excreta, fish and bones, beef and fish washings and various kinds of mixtures and decoctions as manures/organic fertilizers. Smearing of seedlings with mixtures of sesamum, ghee, honey, cow dung and milk prior to transplanting is the main example of traditional agricultural practice to control diseases caused by cold and hot sun. Vriksh Ayurvedic (Indian Plant Science) is of immense relevance in agriculture, horticulture that provides knowledge about a agronomical practices, soil classification and control measures of cultivation and grain storage practices (Gupta, 2011c).

It has been already stated that Indian farmers had the knowledge to enhance soil fertility through the use of organic manure since Vedic Period. As a matter of fact, the organic manures had been classified into two main groups' viz., (i) Bulky organic manures such as FYM, compost, green manure etc.; and (ii) Concentrated organic manures such as groundnut cake, castor cake, linseed and neem seed cakes. FYM involves preparation in pits or trenches, which contains mainly animal dung, soil, slurry and dry litter soaked in cattle's urine (Mahajan and Gupta, 2009).

Green manuring is the practice of applying of large quantities of green matter such as leaves, twigs of plants and green plants as manures by two ways - in situ and green leaf manuring. Green manuring crops are such as dhaincha sunnhemp, mung, indigo, glyricida and karang (leaf manuring crops). Growing of dhaincha *(Sesbania aculeata)* in sandy and loamy sand soils and burying it into the soil before flowering stage during transplanting of rice has been found useful not only to increase but also enhanced the yield of wheat (Gupta et al. 2005). The main effect of green manuring is to increase the availability of soil nitrogen and other nutrients (Gupta and Sharma, 2004).

Organic manures play a vital role in fertilizing the crops naturally as they bind soil particles into aggregates, serve as a source of energy to growing microbes, reduce the soil by wind and water erosion, increase the water holding capacity of soil, enhance soil fertility and, thus, sustain the productivity (Mahajan et al,, 2004).

Considering the fact that India is quite rich in organic manures with around 87.5 million tonnes availability, which is just equivalent to 18.45 million tonnes of plant nutrients annually, the stress is, therefore, more and more on indigenous technology for nutrient management developed by the farmers with some modifications. Thus, Indian farmers were traditionally in favour of application of FYM, compost and/or green manure for maintaining physical and biological conditions of the soils, and ameliorating their macro/micro-nutrients status. Application of FYM is kept at very high esteem by the peasants of all Indian states in general and those dwelling in the hilly terrains in particular. It is applied to maize in maize-wheat cropping sequence and to wheat in rice-wheat cropping pattern in higher reaches of Jammu and Kashmir, Himachal Pradesh and Uttarakhand. In hills of Chhotanagpur (Jharkhand) forest leaf is traditionally used as component of compost or FYM. This is also the common practice of the farmers living in various states of North Eastern Hill region.

Application of FYM and compost not only provide essential plant nutrients for the growth of plants, but are also known to reduce parasitic nematodes of crops (Gupta, 2005). For example, their addition reduces the infestation of *Globodera restochiensis* in potato roots. It is ascribed to some type of biochemical resistance which is imparted in the plant by nutrients from these manures. Reduction in population of citrus nematode, *Tylenchulus semipenetrans* with steer manure amendment was also observed. Organic nitrogen in the form of sewage sludge was more effective than ammonium nitrate against *Belonolaimus longicaudatus* in turf grass.

Green manuring with leaves of *Melilotus alba* and sorghum reduce root knot incidence. Addition of chopped leaves of *Pongamia glabra* and *Crotalaria juncea* in soil also reduced root knot disease. The green leaves of neem are also effective against root knot.

14.6.4 Indigenous Fodder Tree Species

A number of fodder trees grow in Siwalik and other Himalayan regions, the leaves of which are used as fodder for their cattle. Biul, bhimal or dhaman (*Grewla optiva),* khirk (*Celtis australis),* mulberry (*Morus alba),* kachnar or katrer (*Bauhinia variegata),* palah or dhak *(Butea monosperma)* and different species of *Acacia (A. nilotica, A. catechu, A. modesta),* are the main indigenous trees whose loppings are used as fodder.

Out of these tree species, biul or dhaman has been rated to be the best as green fodder for cattle due to the presence of higher content of N (2.70%) and Ca (4.07%) as compared to other fodder trees (Gupta et al., 1993-94) and negligible amount of tannin, and almost free from oxalates. The leaves of the dhaman are used as fodder during winter i.e., from November to March as they have warming effect. During summer, the leaves of this tree are not liked by the cattle as most of them feel drowsiness if these are fed during summer. The leaves of palah are mostly fed to the buffaloes during winter. Feeding of ber (*Zizyphus jujuba)* leaves to cows and buffaloes are known to enhance milk production and make butter greasy (Gupta and Sharma, 2005). When cows and buffaloes are fed with leaves of bael (*Aegle marmelos)* and its fruit bael kath, their milk becomes more thick i.e., milk possesses more fat.

14.6.5 Traditional Methods of Rain Water Harvesting

Water is a scare commodity in Kandi belt of Jammu, Punjab, Haryana, Himachal Pradesh and Uttarakhand. Before 1947, many people used to store water in tanks locally called talabs. According to official data, the *Kandi* belt of Jammu district and that of Udhampur possessed 335 or 336 ponds with an average size ranging from about 2 to 5 kanals (0.10 to 0.25 ha) as reported by Gupta (2010b). These ponds till as late as about 60 years ago i.e., up to 1960 were the major sources of drinking water for the area. Not only this, "Many of the peasants, particularly the mango growers used to irrigate their mango trees grown in the fields or in kitchen gardens through pitchers after filling them with water from the ponds during morning hours", as the elders of *Kandi* belt of Jammu say. The people of this belt also used to utilize the water for domestic animals and for washing the clothes as recently reported by Jamwal (2019).

As the first and the foremost requirement of any crop is the water so the farmers must harvest the rainwater after constructing the ponds wherever it is possible (Gupta, 2013h). The water so harvested and stored in the ponds can be utilized not only for irrigating the crops but also for other purposes bathing and drinking. Farmers may join together to form groups known as Self Help Group (10-15 farmers in each group) and jointly construct the ponds, reservoirs and share the rain water harvested in them for irrigating their crops. By doing this, the farmers will not only be able to irrigate their crops at right and proper time but can also get loan from the banks to construct the ponds. Harvesting and conserving rain water for irrigation is not a new phenomenon especially in India. The digging and constructing ponds in Indus Valley has unearthed a number of startling facts about the water management of Gujarat (Gupta, 2013f). Thus, construction of ponds for harvesting rain water is required to be resorted in Gujarat state (Charan Dass, 2015).

14.6.6 Use of Organics in Ginger Cultivation

Jammu region of Jammu and Kashmir State is considered the best for cultivation of ginger after Himachal Pradesh. The predominant areas for growing ginger are Reasi, Arnas, Pouni, Katra, Tanda, Ghord and Tikkri. It is also grown in parts of Basohli and Bhadu Billawar. Maize ginger intercropping system with an application of FYM is a popular practice in the aforesaid areas. The quantity of FYM used varies from 20-50 t ha^{-1} depending upon its availability. The un-decomposed materials in the form of dry leaves, farm wastes and crop residues used under the feet of the cattle as beds in the cattle shed after growing ginger. All these organic materials act as mulches. In Himachal Pradesh, the ginger is mainly cultivated in parts of Sirmour, Solan and Mandi district area. About 50-60 tonnes of well decomposed FYM are applied with substantial amount of mulch materials (Gupta and Sharma, 2005).

14.6.7 Use of Ash

Many of the vegetable growers of Jammu and Samba districts scatter ash in the fields. Ash primarily meets the deficiency of potash and enhances availability of phosphorous. Some of the farmers prior to adding ash to the fields mix it with household wastes as household waste possesses good amount of plant nutrients both macro (N, P, K, Ca, Mg, S) and micro (Fe, Cu, Mn, Zn, Mo, B and Cl). Sprinkling or sprawling of ash in vegetables like pumpkin, gourd, muskmelon and watermelons is resorted to repel insect pests of these vegetables.

14.6.8 Use of Cow's Urine

Many of the traditional farmers still use the urine of cows to combat crop pests and diseases. When cow's urine is sprayed at 10 per cent concentration to diseased plants, they soon recover and become healthy. Cow's urine is also served as foliar application of nitrogen to withering crop plants.

14.6.9 Technology for Preventing Soil Erosion

As the Siwaliks are liable to suffer from severe soil erosion and run-off loss, so the major stress of the farmers is to conserve soil and moisture. The indigenous technology used by the farmers is to follow field leveling, contour bunding and terracing of the land (Gupta, 2010b). They also use the vegetative barriers on the terraced bunds with suitable growing bushes, shrubs and fodder trees.

In the degraded, eroded and denuded lands, shrubs such as *Vitex negundo* (wana), *Carissa spinarum* (girna), *Adhatoda vasica* (basuti or brainkar), and *Dodonaea viscosa* (santha) have been found useful in checking soil erosion and run-off loss. Grewia optiva, *Bauhinia variegata, Butea monosperma* and different species of *Acacia (A. nilotica, A. catechu, A. modesta)* and *Albizia* spp. are some of the preferred fodder tree species (Gupta, 2012b). Growing of black gram on the raised bunds of paddy and maize fields is also an old age practice which not only helps in checking soil erosion but also enhances soil fertility by fixing atmospheric nitrogen.

Organic farming, in fact, has been found an important kind of eco-friendly system of food grain production vis-a-vis agro-ecosystem management, especially for rain-fed agro-ecosystem (Gupta, 2013b: Mahajan et al., 2007b). Organic techniques such as use of organic manures (farm yard manure, compost, vermicompost, green manures) and botanical pesticides i.e., neem based, (neem seed kernel extract, neem leaf extract, neem oil spray, neem cake suspension) and others have been found very beneficial for rain-fed ecosystems of Arunachal Pradesh, Himachal Pradesh and Sikkim. For the last 10 years, many farmers of these states have adopted organic farming and improved the yields not only of crops but also of vegetables and fruits (Gupta, 2013h)

14.7 Conclusion

Indigenous technical knowledge usually propounded by the ancestors of the peasants based upon a particular experience or practice, has been highlighted in this chapter of the book. Traditional methods of integrated pest management including botanical pesticides, traditional use of organic manures, indigenous fodder tree species, traditional methods of rain water harvesting, use of ash, cow's urine, technology for checking soil erosion and organic cultivation of ginger, are the crucial topics discussed in this chapter of the book. Concept of indigenous knowledge, its advantages and importance, is also detailed. Future research strategies/priorities and study questions have been given in the end.

14.8 Future Research Strategies/Priorities

- Considering the importance of indigenous technical knowledge, multidisciplinary centres by State Agricultural Universities and Indian Council of Agricultural Research Institutes require to be started.
- Keeping in view the cost of chemical fertilizers, soil health, animal health, quality of produce and resources of the small and marginal farmers, more research on green manuring, FYM/compost is the need of the day.
- There is a dire need to monitor the use of bio-pesticides to check insect pests of various crops.
- More research is required to be done with regard to various fodder tree species in respect of soil and water conservation. Similar kind of research must be resorted in case of certain bushes like santha, wana, brainker etc.
- More use of ash and cow's urine should be made in various crops, vegetables and fruit trees.

Study Questions

1. What is meant by indigenous technical knowledge?
2. Why indigenous technology is considered more appropriate?
3. Name few methods of indigenous technology. Write details of any one of them.
4. Name the crops on which use of ash has been found useful.
5. Write in details about the concept of indigenous technology.

References

Charan Dass, S. (2015). Conserve the sacred water. The Daily Excelsior. 51(55):8

Chander, S. and Munshi, A.D. (1995). Eco-friendly botanical pesticides. Intensive Agriculture 33 (1-2): 37-39.

Gupta, M.C. (2005). Cultural practices for management of plant parasitic nematodes of crops. Indian Farmers' Digest 38 (11): 22-25.

Gupta, R.D. (2010a). Indigenous technical knowledge for farmers. The Daily Excelsior 46 (61): 6 [March 3, Wednesday].

Gupta, R.D. (2010b). Soil conservation: Need of the hour. The Daily Excelsior 46 (115): 8 [April 26, Monday].

Gupta, R.D. (2011c). Traditional organic farming. The Daily Excelsior 47 (109): 6 [April 11, Monday].

Gupta, R.D. (2012b). Indigenous plant nutrients technology in Siwaliks. The Daily Excelsior 48 (211): 8 [August 2, Thursday].

Gupta, R.D. (2013f). Rain water harvesting recalls Indus valley civilization. The StateTimes 16 (312): 6 [November 12, Tuesday].

Gupta, R.D. (2013h). Strategies for rain-fed agroecosystem. The Daily Excelsior 49 (342): 8 [December 14, Saturday].

Gupta, R.D. and Sharma, A.K. (2004). Revive green manuring practice for higher food production. The Kashmir Times 61 (290): 7 [October 16, Saturday].

Gupta, R.D. and Sharma, A.K. (2005). Indigenous technologies of Siwalik farmers. Indian Farmers' Digest 38 (7): 27-30.

Gupta, R.D., Khajuria, M.R., Gupta, J.P. and Bali. S.V. (2001). Indigenous technology for nutrient management in Jammu and Kashmir (pp. 147-164). In Indigenous Nutrient Management practices - Wisdom Alive in India (Eds. C.L. Acharya, P.K. Ghosh and P. Subba Rao). Indian Institute of Soil Science, Bhopal (Madhya Pradesh), India.

Gupta, R.D., Khajuria, M.R., Gupta, J.P., Khanna, Y.P. and Leharia, S.K. (1993-94). Common fodder trees of Jammu and nutrient content of important species amongst them. Dharti 3 (1): 40-47.

Gupta, R.D., Kher, D. and Jalali, V.K. (2005). Organic farming: Concept and its prospective in Jammu and Kashmir. J. Res., SKUAST-J 4 (1): 25-37.

Jamwal, G.S.(2019). Shivaliks and the Cimate change. Personnel communication on the address of Dr. R.D. Gupta Ex-Associate Dean, SKUAST-J C/O, 258-C Sainik Colony, Jammu - 18001.

Mahajan, A. and Gupta, R. D. (2009). Integrated nutrient management (INM) in a sustainable rice-wheat cropping system (pp. 1-267). Springer Science + Business Media (www. Springer.com).

Mahajan, A., Bhagat, R.M. and Sharma, S. (2004). Conservation tillage - An ecological approach for better soil management to improve crop productivity. Indian Farmers' Digest 37 (3): 16-20.

Mahajan, A., Sharma, R., Gupta, R.D. and Trikha, A. (2007b). Organic farming and sustainable agriculture. Agriculture Today 10 (3): 48-50.

Sharma, A.K., Gupta, R.D. and Rai, P.K. (2009). Crop diversification is the need of the hour for conserving soil and water in Kandi belt of Siwaliks (pp. 69-71). In Soil and Water Conservation Training Manual (Eds. B.S. Hansra, P. Vijaykumar, R. Nanda and P.K. Rai) organized by School of Agriculture, IGNOU, New Delhi and KVK, SKUAST of Jammu, R.S. Pura Jammu (Jammu and Kashmir State), India.

15

Management of Wastelands

Abstract

Management of our land resources is one of the most important areas where over the last 4-5 decades excellent information and research work in various aspects has been generated. Such aspects may consist of promoting agro-forestry, control of water and wind erosion through adoption of cultural, mechanical and biological measures. All these aspects have been dealt with in this chapter for reclaiming of wastelands. Apart from this, concept of wastelands, factors amenable for converting productive lands into wastelands, have also been, stated. Role of the council for advancement of people's action and rural technology (CAPART) in respect of watershed management stands given. In the end conclusion, future research strategies, study questions and references are quoted

Keywords: Wastelands; watershed management; land resources; productive lands; agro-forestry; reclaiming; water erosion; wind erosion; cultural measures; mechanical measures; biological measures; CAPART

In whole of the world, about 6 to 7 m ha of agricultural land is becoming unproductive every year due to one or the other reasons (Gupta, 2018a). This is, however, more so in case of developing countries (Gupta, 2003). Productive lands are utmost important to meet the country's growing needs for food, fodder, fuel wood, medicinal herbs and timber. They also assist in conserving water resources and sheltering biodiversity. However, India that supports 16 per cent of the world's human and 18 per cent cattle population with just 2.45 per cent of the total land mass has vast stretches of wastelands. While there are varying estimates of the total area of wastelands, not less than 37.5 m ha is available for viable treatment. According to "Wasteland Atlas of India", out of country's 32,87,263 km^2, about 6,38,518 km^2 is wasteland which constitutes 20.17 per cent of the total area (Anonymous, 2000). however, keeping in view the whole of the picture of degraded lands of India, the estimates vary from 75 to 175 m ha or even up to 200 m ha. While Rajasthan has the greatest acreage of wasteland - twice as much as any other State, Madhya Pradesh, Andhra Pradesh, Uttar Pradesh, Maharashtra and Gujarat have large tracts of wasteland. Also, a high proportion of the geographical area of the hill states is having a wasteland. It is pertinent to state that Kerala has the minimum wasteland and the Jammu & Kashmir the maximum like the Rajasthan.

The common picture of wasteland is that of a vast uncultivated, dreary, inhabited expanse - the areas devastated by mining, the land turned into true desert-like condition, deforested mountain slopes, reduced to bare rock, even the areas subjected to nuclear test explosions or chemical warfare. But the largest category of wasteland is soil eroded cultivated land. The per capita availability of land has declined from 1.37 ha in 1901 to 0.39 in 1951, which has further reduced to 0.15 ha during 2000. Today each Indian possesses less than half of the arable land than was inherited from his elders about 18 years ago. If this trend is not reversed soon, we would face food insecurity and as also problems in sustaining ourselves in future. The wastelands, therefore, call for their management or reclamation urgently as these lands can be made productive again following suitable technologies. Land regeneration programmes will not only increase land productivity but also boost the rural economy as they are highly labour intensive and can alleviate rural poverty through employment.

15.1 Concept of Wastelands

Soil degradation mainly responsible for converting land resources into wastelands as it exists today - Soil scientists will have to play a crucial role in generating and promoting successful technologies for rehabilitation of wastelands and for preventing of already rendered wastelands.

15.1.1 Definition of Wastelands

Briefly, any land that is unproductive is called wasteland. Sometimes the wasteland is also known as degraded land. However, land degradation is considered a natural phenomenon as erosion cycle, involving degradation and aggradation which results into peneplanation all over the world. The following definitions of wasteland would clearly explain the whole gamut of the problem of land degradation:

- The Webster's encyclopedic dictionary defines wasteland as "*land that is uncultivated, barren and the area of which is being developed by floods, storms and other natural calamities. Apart from it, wastelands also connote lands, which have been devoid of natural greenery over a number of years due to ruthless cutting, lopping and biotic pressure above all levels.*"
- Although there is no definite accepted definition of wasteland yet at one time a definition based on economic potential and actual returns from land had been found accepted one. This definition explains that "*any land which is giving less than 20 per cent of its economic potential is wasteland.*" Though there are some advantages of this definition yet for the purpose of estimating the extent of wasteland it is not very useful.
- According to D.R. Bhumbla, a well-known soil scientist, "*Wastelands are those lands which are virtually unstable ecologically, whose top soil is completely lost, devoid of any vegetative cover. Such lands have developed toxicity in the root zones for the growth of most plants, both for growth of annual crops and trees.*" Further it has been added that wastelands cover all those lands which are affected by erosion (water and wind), floods, soil salinization and alkalization,

acidification, water-logging and so on. In the hilly terrains, wastelands are snow-clad mountains, bare and denuded fragile rocks which over the years are geologically unstable.

- National Remote Sensing Agency defined, "*Wasteland as that land, which is presently lying unused to it optimum potential due to certain constraints.*"
- National Wasteland Development Board (NWDB) as quoted by a Planning Commission Committee in 1987 has considered wasteland as, "*land which can be brought under vegetative cover with reasonable efforts, and which is currently underutilized, and lands which are deteriorating for lack of appropriate water and soil management or on account of natural causes.*"

15.1.2 Limitations of Wasteland Definitions

The above definitions have some limitations such as "Optimum Potentials" and "Underutilized" words are most subjective and could lead to any interpretation and inference. Even the irrigated lands having no proper management practices may yield below the optimum potential and several lands may be underutilized due to economic or social reasons. Some lands may be "presently lying unutilized" due to legal complications and/or absence of landlords. Such lands can neither be taken as wastelands requiring reclamation measures, nor as degraded land. Another lacuna is that lands at false colour composites have been used for preparing maps in scale of 1:50,000 and "the different good lands like arable and forestlands are shown as blanks in the wastelands map." Hence, to overcome the above limitations and to get the whole of the picture of degraded lands, the Ministry of Agriculture has taken into account the following definition which would include all kinds of land degradation.

15.1.3 What is Degraded Land?

"Degraded land is that land surface, which was worn down qualitatively and quantitatively by any or all epigene agents like wind, water or glacier and is a locked-up production that could be retrieved back to normal production through economically viable reclamation measures".

15.2 Classification of Wastelands

Wastelands can be broadly classified based on soil erosion i.e. water erosion and/or wind erosion, and based on other causative agents, responsible for soil degradation (Table 15.1).

Table 15.1: Wastelands and their causative agents (Prepared by the authors themselves)

Sr.No.	Causative agent	Type of wasteland or degraded land
1.	Water erosion	Sheet erosion, rill erosion, gullied land, ravines land
2.	Wind erosion	Sand dunes, sand bare and sandy tracts of coastal areas
3.	Chemical degradation	Saline soils, sodic or black alkali soils, acidic soils
4.	Man made	Industrial wasteland, degradated forestlands, degraded pastures, mine spoils, shitted cultivated lands
5.	Other agents	Shallow lands, waterlogged area, land slide areas

15.2.1 The largest category of wasteland is soil-eroded arable land (Gupta, 2018a). Most of this category of land is rainfed or dry-farming area. The next largest category is degraded forest land. In South India, when the forest soils become exposed to the sun they quickly turn into laterite, a kind of acidic soil (Gupta, 2018b).

15.2.2 Wastelands have further been divided into: (a) culturable; and (b) un-culturable wastelands. The former class of wastelands comprises mainly of gullied land and undulating upland, water-logged land and salt-affected land, shifting cultivated area, sandy area, overgrazed pastures and degraded forests, vis-a-vis mining and industrial wastelands. Contrary to this, barren rocky areas, steep slopes, snow-covered areas and glacial areas constitute the latter kind of wastelands.

15.2.3 Wastelands have also been Grouped Under the Following Four Broad Categories

15.2.3.1 Cultivated Land Affected by Soil Erosion

Soil erosion in cultivated land is a very gradual process. The extent of soil loss depends upon the climatic conditions, the hilly terrain of the land and nature of the soil vis-a-vis the ongoing agricultural practices in the area (Gupta, 2009c). Agricultural practices such as land cultivation on steep slopes hastens the process of soil erosion. On the other hand, traditional practice of growing trees and crops together on the same field helps to protect land because the trees provide a cover to the soil (Mahajan et al., 2004). When the land is cleared of trees and exposed to sun, some forest soils especially those containing iron and aluminum dry out and become hard. Such hardened soils are called laterites which are found in South India and Eastern India, and are virtually incapable of supporting vegetation due to acidic problem.

The deforestation in Siwaliks of India has made the position worst in creating the wastelands. The entire Siwaliks belt from Assam to Jammu including Uttarakhand, Uttar Pradesh, Haryana, Chandigarh, Himachal Pradesh and Punjab is almost barren and unproductive and degraded (Gupta, 2009d). There is a recurrence of landslides all along the Siwaliks or outer Himalayas as well as inner Himalayas. The naked soils promote landslides or landslips.

15.2.3.2 Degraded Forests and Overgrazed Pastures

Forests play an important role in stabilizing the economy of a region or a country and therefore, these natural resources must be utilized in the best optimum possible manner to meet the objective of upliftment of the poor. However, overexploitation of this natural wealth is a continuous process leading to decrease in the forest area of our country year after year. Hitherto the forests of our country were about 23 per cent of the land area, covering nearly 75 m ha which has reduced to about 20 per cent (Gupta, 2005c). But according to the National Remote Sensing Agency, Hyderabad there was a considerable loss of forest cover reducing it to only 14 per cent over the years. Indiscriminate harvesting of timber, construction of roads deep inside the forests, commercial exploitation of existing forests at an unprecedented scale and uncontrolled grazing have contributed to the reduction of forest cover rendering vast tracts of forest lands to deforested, degraded and denuded ones.

Due to deforestation quite akin to valley of Kashmir, dunes and dales of higher reaches of Jammu region like Bhadarwah, Bhalesa, Dudu Basantgarh, Dansal, Jindrah as well as those of *Kandi* belt of Jammu region, Jammu and Kashmir State (Now Union territory) have lost their charm and beauty due to degradation of their environment including forests and natural pastures (Gupta, 2010a). Sustained lack of green cover causes desertification i.e., waste land (Sriraj, 2019).

15.2.3.3 Degraded Land Due to Accumulation of Salts

Such lands constitute part of wastelands and include saline, sodic or black alkali and saline and alkaline soils. In these lands, soluble salts accumulate in the form of Na, Ca, Mg as cations and CO3, HCO3, Cl, SO_4 as anions. Such soils are called *User Lands* in Uttar Pradesh; and *Kalar Aali Mitti* in Punjab and Jammu region of Jammu & Kashmir State. The main sources of salts are rock weathering, ground water and water from canals, streams and rivulets, especially in command area. About 1.07 lakh ha of the Bhakra command area in Haryana and 2 lakh ha of land in Punjab is affected with saline and alkaline soils. Salinity is also being developed in command area of Jammu region, Jammu and Kashmir State.

Salt affected soils due to their high pH and more salt content accompanied with exchangeable Na^+ provide unfavourable environment for plant growth. Although such soils are potentially productive yet without suitable technology for their reclamation, it becomes cumbersome to harvest production at desired level. A number of organic and inorganic amendments tried by various scientists to improve such soils for increasing food production, have been reviewed (Najjar and Gupta, 1996). The results of the study regarding the effect of gypsum, FYM, dhaincha, gypsum + dhaincha and gypsum + FYM on physico-chemical properties of soil and yield of rice in salt affected area, indicated improvement in soil properties with concomitant increase in yield of rice (Table 15.2).

Table 15.2: Effect of amendments on the composition of saturation extracts (Adapted from Najar and Gupta [1996])

Treatments	PH	EC (ds m^{-1})	Ionic composition (m.e.L ')					Yield (q ha^{-1})		Increase over control (%)	
			CO_3^{2}	HCO_3^-	NO_3^-	Cl^-	SO_4^{2-}	Grain	Straw	Grain	Straw
Dhaincha	7.9	0.19	2.0	5.0	0.56	8.0	1.76	13.2	27.8	13.6	17.4
Dhaincha + gypsum	7.5	0.14	1.0	2.9	0.47	5.0	1.78	14.7	28.7	25.8	21.2
FYM	8.0	0.19	2.2	6.6	0.51	6.0	1.54	13.1	27.2	12.2	15.0
FYM + gypsum	7.7	1.8	6.0	0.49	3.0	1.77	13.3	28.1	14.0	18.4	
Gypsum	7.8	0.15	2.0	6.5	0.53	5.2	1.84	12.6	24.9	8.3	4.9
Control	8.3	0.34	2.6	8.7	0.60	9.5	1.86	11.7	23.7	-	-
LSD (0.05)		0.11	0.63	1.45	0.038	1.47	0.17	0.63	0.6	-	-

Sandy soils with high pH value can be found along the banks of major rivers. Such soils are characterized by pH range from 9 to 10. They are also suffered from severe drought during summer and heavy frost during winter with temperature going down to zero degree Celsius vis-a-vis unprecedented floods and sand drifts. On these soils, planting of poplar *(Populus* spp.) at 5 m interval in rows, 4 m apart is best suited. However, *Syzygium cumini, Terminalia arjuna* and *Dalbergia sissoo* can also be grown.

According to Remote Sensing (RS) and Geographic Information System (GIS) as well as ground truthing, more than 25,000 ha land has been found to be affected by salt affected soils in Jammu region of Jammu and Kashmir state (Now Union territory) (Gupta and Arora, 2016).

15.2.3.4 Degraded Lands Due to Acidic Problem

Acidic soils are formed due to acidic parent material, leaching of bases caused by heavy precipitation and indiscriminate use of nitrogenous fertilizers (urea, ammonium sulphate). Crop production in these soils suffer from poor availability of plant nutrients like P, Ca, Mg, K and toxicity of Fe and Al. Such soils can be reclaimed by application of liming. Recently, Sarkar and Sharma (2006) have reported the management of acidic soils as well as current issues and policy guidelines in this regard. Based upon response to liming arhar, soybean and cotton have shown considerable increase in their yield. The yield was further improved when balanced use of fertilizers (N, P and K) was made in conjunction with lime (Sarkar and Sharma, 2006),

15.3 Extent of Land Degradation and Its Categories

Based on various definitions, there are varying estimates of the extent of land degradation. According to Indira Gandhi Institute of Development Research (2001), there are 14 different categories of wastelands in India which require different strategies for their reclamation. For instance, hill slopes in Siwaliks from where top soil has been badly eroded and requires biological measures or engineering structures (Gupta, 2002). Others which may be for salt affected and water-logged areas need to be treated using latest technologies developed by the State Agricultural Universities and Soil Salinity Research Institute, Karnal.

The National Wasteland Board in the Ministry of Rural Areas and Employment has estimated 175 m ha of land as degraded, which comprised approximate 53 per cent of the country's total geographical area. The National Afforestation and Eco-development Board under the Ministry of Environment and Forests, reckoned 130 m ha as degraded land. The National Remote Sensing Agency, however, estimated that merely 75 m ha of land constituted wasteland. According to the estimates given by National Bureau of Soil Science and Land Use Planning

(Sehgal and Abrol, 1994) as reported by Sharma et al. (2004), 57 per cent of land (187.7 m ha) in the country is potentially exposed to various degradation forces. Of which major fraction is under water erosion (148.9 m ha) and the rest of about 12 per cent is suffering from wind erosion (13.5 m ha); chemical deterioration (13.8 m ha) and physical deterioration (11.0 m ha).

The overall land degradation figures furnished by various agencies (as stated above) have been at variance due to differences in approaches, methodologies and criteria for assessment, but the fact remains true that large area is under degradation due to various factors. The data sets on land degradation due to various factors available with different agencies (NBSS & LUP, Nagpur, CSWCR and TI, Dehradun, CAZRI, Jodhpur, CSSRI, Karnal and FSI, Dehradun) were harmonized in GSI environment for the whole country (Minhas, 2012). As per new estimates, 120.8 M ha, constituting 36.5 per cent of total geographical are degraded due to soil erosion, salinity/alkalinity, soil acidity, water-logging and some other complex problems (Magi et al,, 2010).

15.4 Factors Responsible for Turning Productive Lands into Wasteland

A wasteland is the product of soil erosion, hostile climate, misuse of land, forest fires, floods, droughts, over- exploitation of land resources, faulty methods of cultivation, loss of soil fertility, denudation of rocks and deforestation.

15.4.1 Denudation of Rocks and Deforestation

Denudation of rocks is the major factor for converting productive lands into wastelands. It occurs mainly due to deforestation. Continuous deforestation has landed us into severe ecological and socio-economic crises. The haphazard development is the principal cause for deforestation. The trees are felled to construct roads, bridges, buildings, dams, hydroelectric projects, tunnels, railways tracks etc. This trend accompanied with encroachment of the forest areas, has reduced the forest cover and has turned land into wastelands.

As continued deforestation has landed us into severe ecological and socio-economic crisis so the efforts to recover wastelands should be undertaken at war-footing. It is worth mentioning that actual forest cover of Jammu & Kashmir has left over less than 10 per cent of the total area including Ladakh now Union Territory.

15.4.2 Population Explosion

The world population, which was 7 billion up to March 31, 2011, which stands become 7.244 billion up to July 15, 2015 and 7.70 billion during August, 2019. It is worthwhile to mention that world population is not going to decline with the coming years, the food demands are also increasing, the UN Food and Agriculture Organization (FAO) estimates that farmers will have to produce 70 per cent more food by 2050 to meet the needs of the world's expected 9 billion strong populations (Anonymous, 2014b). Similarly, the livestock sector has been growing faster than crop production sector. India's livestock sector is one of the largest in the world. It has 57.3 per cent of World's buffaloes, 14.7 per cent cattle, 20.4 per cent ruminants. For this increased population of both human beings and livestock population, more productive lands are required to meet the country's growing needs for food, fuel, fodder and fibre. However, owing to more biotic pressure on these lands, they have become degraded.

Population explosion of humans is also responsible for deforestation for residential quarters and agricultural lands. It is point to mention that human population of the country has risen from about 320 millions during 1947 to more than 1000 millions during 2001. The population in Jammu and Kashmir like other Himalayan States has also risen. It was 59,37,389 in 1980 and arose to 100,69,917 in 2001. According to data, the state's population has grown from 1.01 to 1.25 crore during 2011 (Fazili, 2011). The process of converting forest lands into agricultural and residential lands, have greatly changed the forest status of the Himalayas in general and Siwaliks in particular.

Deforestation is rendering the soils totally naked especially those of Siwaliks of Jammu. Rain water runs swiftly over the surface of these soils, resulting into its no percolation in the subsoil. As forest status of Jammu Siwaliks has deforested heavily, so its soils have become very low in organic matter content and other plant nutrients (Gupta, 2013c). Moreover, there is lot of soil erosion from the existing lands of Siwaliks. As such these lands have become degraded. The burgeoning human population is out stripping the earth's capacity to sustain it, the World Wide Fund for Nature has warned (Gupta, 2011a).

15.4.3 Soil Erosion

In India based upon the sedimentatin data from rivers and reservoirs, it has been found that about 5334 million tonnes of soil gets eroded annually in the country which works out to 16.34 t ha^{-1} year (Sikka, 2013).

The quick run-off results into soil erosion and the eroded debris deposit in the river beds and basins. Once the basins get filled with sediments the water retention capacity of the rivers reduces and consecutive rains for few days may result into flash floods.

Water erosion and wind erosion (in arid regions of India - cold arid zone of Ladakh and its adjoining areas of Warwan in Doda district now Kishtwr District in Jammu & Kashmir and Lahaul Spiti in Himachal Pradesh are considered the main reasons for wasteland development. Water erosion includes sheet erosion, rill erosion, gully erosion, ravine land, while wind erosion comprises mainly shifting of sand dunes due to high wind velocity in the arid and cold arid zone of Ladakh and Lahaul Spiti of the country.

Among the north Western Himalayan states, the forests of Jammu & Kashmir are the worst affected by deforestation. This has not only created bareness in the hills, especially Siwaliks but has rendered barren fields below in the plains (Gupta, 2011 d) because the erosion of up hills creates havoc by causing flash flood which changes the green fertile lands into unproductive sandy area, eventually affecting the food production.

15.4.4 Flash Floods

The flash floods further promote erosion in denuded areas and also increase landslides/landslips. Landslides and flash floods may turn green fields into barren or degraded lands.

Outer hill ranges including Siwaliks are the areas mostly affected by the hill torrents subjected to flash floods which are responsible for the formation of wastelands. Soil erosion in these areas goes unabated (Gupta, 2009c). Once the natural equilibrium becomes disturbed between resources and their demand, the surface loses its canopy and direct impact of rain drop initiates the erosion. This, in turn, gets the land converted into gullies and wastelands as recently communicated (Anonymous, 2019).

15.4.5 Movement of Nomads

In whole of the Himalayas, the centuries old practices of biannual movement of Bakerwal, Gujjar, Gaddis known as nomads and shepherds with their livestock appears to be a major reason of the heavy or overgrazing of the pastures. Overgrazing of the pasture lands is due to increased livestock population. Even the nomads intrude their livestock in enclosures (Gupta, 2009).

15.4.6 Over-Exploitation of Agricultural Lands

The agricultural lands in the country being under severe pressure are undergoing a constant process of degradation and the fast turning into wastelands.

With the increasing human and livestock pressure on land, production of vegetation for food and other uses has extended to the areas under great ecological stress and unfavourable environment. Growing demand for food, fodder, fuel wood and timber have extensively depleted plant cover and exposed surface soils to the processes of degradation. This has resulted in partial to complete loss of productivity. The arable land of Jammu & Kashmir is becoming wasteland at such an alarming rate that concerted efforts are required to recover wastelands. Where the land is degraded, eroded by water or wind, poor people are forced to cultivate even more marginal land which may be too infertile, too steep or swampy to cultivate. This is happening, for example, in the *Kandi* belt of Jammu and south-western border areas near Pakistan as well as in the Chambal Valley where the spread of the ravines has resulted in fragmentation of land holdings and cultivation is becoming uneconomical.

15.5 Management of Wastelands

It is suffice to say that various biotic activities like felling of trees, their lopping, browsing etc. are seemingly the handiwork of man in so far as most part of devastation is concerned. We have far exceeded and at this juncture it is rather cumbersome to apply the reverse gear to bring back prestige grandeur and glory. Though the adage, "it is better to be late then never" and going by the connotation it is amply imperative to put together all out efforts to revive forest cover and greenery to save the leftover wastelands from being further degraded.

Wasteland treatment is a mega venture and many agencies are to be involved to-design workable strategies in a variety of ways by the thread ware discussions and deliberations The frontline agencies mainly comprise the forestry, soil conservation, flood control, agriculture and allied disciplines. And in its formulation some basic data are pertinent which include climatic factors, soil profile vis-a-vis local factors, livestock and human population.

Turning wastelands into lush green fields is an uphill task but undaunted by the difficulties, the NWDB placed under Department of Wasteland Development setup during 1992, had been performing a pioneering role in reclaiming wastelands. The Board was reconstituted to provide fresh impetus for checking degradation and putting wastelands to sustainable use. Up to July, 2000 as many as 247 integrated projects in 25 states with outlay of Rs. 7.78 billion were at various stages of implementation covering an area of lakhs of hectares. The government is also playing considerable emphasis on the involvement of Non-Government Organization (NGO) in this work. The Board has been asked to adopt a mission approach for enlisting people's participation for planning and implementing the wastelands programme.

15.5.1 Promoting of Agroforestry

The Board is laying emphasis on agriculture and agro-forestry vis-a-vis soil and water conservation, in which both ecological and economic concerns have to be met to increase crop yields, generate income and create employment. In rural areas, about 70 per cent of the cooking energy is still fulfilled by field wood collected from the forests. Adoption of various agro-forestry systems i.e., agri-silviculture, agri-horticulture, agri- pastrol, agri-silvi-horticulture, agri-silvi-animal husbandry will prove a boon for the rural masses as these systems of agro-forestry not only provide fuel wood but will also supply fodder, food, fibre and fertilizer to the rural masses. Their pressure towards forests cutting will be thus reduced in the long run.

Gupta (2006b) has described a few methods for greening the cold desert wasteland of Himachal Pradesh. These consist of growing fruitful crops like hops, seed potato and unseasonal vegetables. A new deciduous shrub known as "sea buck thorn" having unique quality to withstand adverse climatic condition including low temperature up to 60°C, has been found to grow in Kinnaur, Pangi and Lahaul Spiti. The fruits of "sea buck thorn" contain a sufficient quantity of Vitamin C, high quality oil, fuel wood and fodder.

Recebtly Sood (2019) has reported that in Jammu and Kashmir agroforestry is mainly practised in the traditional form by planting trees in home gardens, on bunds of cropland, boundaries of agricultural holdings.

15.5.2 Proposed Land Use Pattern

A scientific land use pattern suggests that the land can be used on the system of land capability classification. In our State and other Hill States, wastelands, whether are under croplands, forestlands or grasslands, village common lands, roadsides, by and large, fall under Class V, VI, VII and VIII land capability classes. It is owing to the occurrence of serious limitations of these land capability classes. These limitations consist of slope, stoniness, soil depth and erosion, and are presently on the process of denudation and degradation due to overgrazing and deforestation apart from natural process of erosion.

Hence, major part of these lands call for permanent vegetation cover such as plantation of fuel and fodder species, grasses and forest species or combination of all the three. Such a land use pattern will improve the infiltration rate in soil due to enhancing large quantities of organic matter and plant nutrients to the soil and reduce peak rate of run-off losses. What is more important, is that this system would besides conserving soil and moisture, meet the requirement of forage for livestock and plant based occupation of the people living around.

15.5.3 Checking of Water Erosion

Water erosion in the Outer or Siwalik Hills and other areas has to be checked by training the water channels and dry nallah beds (khads) by reducing the speed of water and the slope by check dams. A model of this is at Bari Badhori, as has been explained in a case study of Jammu Siwaliks.

To check the gully erosion, provision of mechanical measures like building of bunds, spur and/or vegetative cover are urgently required by way of planting fuel wood and fodder producing species or mixture of grasses and fodder trees and bushes along the gullied areas.

As already stated, hill slopes in Siwaliks have their own sad stories from where most of the top soil has been badly eroded, so it creates a hindrance for planting on such barren and denuded areas. Such areas can, therefore, be treated by introducing various shrubs. The following kinds of shrubs, which have proved beneficial for rehabilitation of eroded areas, should be preferred to grow:

- *Girna* - Carissa spinarum
- *Santha* - Dodonaea viscosa
- Basuti or Branker - *Adhatoda vasica*
- *Bana* - Vitex negundo
- *Phagwari* - Flacourtia indica
- *Dhatura* - Dhatura innoxia

After the soil becomes stabilized, the most suited local quick growing tree species require to be planted (Table 15.3)

Table 15.3: Quick growing tree/shrub species for the rehabilitation of wastelands

Quick growing species	Fodder tree species	Fuel wood species	Bushes/Shrubs
Leucaena leucocephala *Albizzia* spp. *Sesbania* spp. *Eucalyptus* spp. (on suitable sites only)	*Grewia optiva* *Albizzia* spp. *Aegle marmelos* *Bauhinia variegate*	*Acacia nilotica* *Albizzia lebbeck* *Albizzia procera* *Albizzia stipulata*	*Carissa spinarum* *Dodonaea viscosa* *Adhatoda vasica* *Vitex negundo*
Populas spp. (on sub-temperate and temperate climatic zones besides subtropical irrigated)	*Morus alba*	*Acacia catechu*	*Flacourtia indica*
Black locust and honey locust (suitable for sub-temperate zones)	*Cordia dichotoma* *Celtis australis* *Leucaena leucocephalla* (All the above said species are suitable for Siwalik Himalay as including *Kandi* belts of Jammu, Punjab and Himachal Pradesh)	*Dalbergia sissoo* (suitable on Siwalik hills) *Aesculus indica* *Curcus* spp. (suitable for lesser to great Himalayas)	*Datura innoxia* (suitable for denuded areas of Siwalik Himalayas)

The major problem in the ravines is that these occupy vast tracks and further these areas are prone to soil erosion. On such areas, the main thrust should be on gully plugging. Once the gullies are plugged, root cuttings are strongly recommended. Thereafter, the most local suitable species should be planted. Bamboos, grasses need preference especially at places where soil binding is the urgent need. Planting of different species of *Acacia* viz. *Acacia nilotica, A. catechu, A. modesta, Dalbergia sissoo, Syzygium cumini* have shown promising results on flat-top areas. Apart from *Vitex* negundo and *Acacia* species, *Ipomea carnea, Agave americana, Prosopis juliflora* and *Dendrocalamus strictus* have also proved the most suitable species for reclaiming the gully beds and ravines.

Closure of the areas from increased human and animal influences of over-exploitation should be the first and foremost step to take care of. Besides, an integrated approach of watershed management would be an appropriate step for controlling such areas. Further, ever-grazed lands can be regenerated by providing alternative sources of fuel wood and fodder. An intensive programme of afforestation, fodder and pasture development has to be planned.

15.5.4 Tackling of Water-Logged Areas

The quantum of water in the water-logged areas has to be reduced by checking and if need arises it may be decreased by stopping this water in the outer hills or anywhere en-route before it reaches the waterlogged areas. Similarly in the areas, where the water-logged areas are due to faulty irrigation, can be checked by improving the drainage.

15.6.5 Landslides and Landslips

Recent studies have indicated that landslides in the hilly areas of Jammu region led to disruption on the Jammu-Srinagar and the Batote-Doda-Kishtwar highway for a number of hours due to heavy rains and snow during March 18, 2014 (Anonymous, 2014c). Hundreds of vehicles were stranded on the Jammu-Srinagar region on March 19, 2014 and Batote-Doda-Kishtwar highway. The traffic was, however, restored later in the day. Again Batote-Pull Doda National Highway was closed on March 31, 2014 when a long retaining wall suddenly collapsed near Pull Doda due to landslides triggered by continued rains during the past few days (Rangrez and Singh, 2014). Although traffic was diverted to newly constructed road link near Pull Doda to resume transport service on Doda-Batote link road but the risk accidents and tumbling down the rest of the road portion continue to hunt passengers traveling on Doda-Batote National Highway.

In the hilly areas of Jammu and Kashmir, large scale landslides and landslips on the steep slopes under croplands, forestlands and grasslands have rendered very large areas as wastelands. Factors like the road construction, mismanagement of vegetative cover, overgrazing of pastures, unstable geological formations and seismic disturbance by blasting operation and excessive water discharge are mainly responsible for this process of waste. Such lands though need provision of mechanical structure but vegetative measures have far reaching effect. Package of practices recommended for controlling landslides/ landslips are as follows:

- Enclosure of the area to check uncontrolled and excessive grazing and lopping and cutting of trees
- Planting of the area with quick growing species of fuel wood, fodder and grasses
- Mulching of the area apart from watling on contour at suitable intervals
- Such areas need not to be disturbed for exploration in the form of grassing. However, grass cutting after 2-3 years of establishment can be permitted.

In addition to the methods stated above regarding the control of landslides, a number of other techniques/methods have been presented elsewhere (Gupta, 2005c). They consist of checking of water courses, reduction of slope, proper drainage, canal lining, engineering structures etc.

15.5.6 Afforestation

Greening or so to say afforestation of all wastelands is the paramount need of the hour and covering of all such lands with suitable vegetative species preventing soil erosion. It will also help to conserve water, restore soil fertility, increase agricultural productivity of adjoining lands, improve soil microbiological activities and remove soil salinity/sodicity, and ameliorate an environment.

Since the landslides at a number of places in the Himalayas are due to the diminishing of forest cover and declining productivity of forestlands so afforestation is an imperative for such areas (Gupta et al., 2013) The entire territory of Jammu & Kashmir Himalayas in the high risk of landslides is due to the deforestation and overgrazing of the pastures. Landslides are putting down mountain slopes of deforested areas and arable lands. Sometimes whole of the villages are devastated. For example, the denuded and deforested slopes along the river Chenab are slipping into it, right from the source of origin to International border with Pakistan. The entire Chir pine from Doda to Reasi along with the sides is slipping into the river Chenab.

In order to attain meanigful progress in controlling desertification, instead of treating the symptoms of land degradation, India must focus on its main causes (Sriraj, 2019). The immediate cause that typically triggers desertification is the loss of primary vegetation which must be reviwed.

The methodology of afforestation is, however, site specific as some of the mostly degraded wastelands are totally salt affected - saline, sodic and other are acidic for which the suitable measures have already been detailed. Some are waterlogged, coastal areas, river banks having lot of sand, silt or clay, sand dunes and others possess brackish water. And as such different categories of wastelands demand various ways of treatments. Afforestation and creation of good tree/plant cover further warrants to meet the needs of the farmers/villagers of fuel wood, timber and fodder.

Some of the plant/tree species which play a significant role in providing leaf fodder, timber besides controlling soil erosion, are given in Table 15.3.

15.5.7 Planting of Grasses

Planting of grasses plays an important role as these serve twin purpose. First, soil conservation and second to mitigate local stall feeding for the livestocks. The most favourable grasses which act as soil binder are: bans (*Bambusa arundinacea),* lemon grass (*Cymopogon fulvus),* swankh *(Panicum colonum),* water grass *(Phragmites karka*), bhabor grass (*Eulaliopsis binata),* etc. In addition to these grasses, *Vetivera zizanioides* commonly known as vetiver grass has variously been described by the researchers with regard to its role in preventing soil erosion. Being serving as good soil binder, drought proving, nurse plant, first aid plant, grass plant and silent sentinel, it is known as "botanical dam". Planting of this grass in Dansal and Jindrah eroded areas of Jammu, Jammu & Kashmir State, has shown very good results in checking soil erosion.

15.5.8 Water Harvesting Structures

Wherever it is possible, water harvesting structures i.e., watertanks (ponds) be made and old structures like water tanks (ponds), baualis and wells be renovated. This will not only assist in preventing the flow of water which is going waste for irrigating the crops/orchards and creating large fish ponds.

15.5.9 Socio-Economic Consideration - People's Participation

One of the very important aspects of improving production base from wastelands is the people's participation in social and democratic management of the community lands. Public cooperation in the protection of closures and plantation on community lands is particularly important because nobody has a personal slake for such lands. Experience has shown that afforestation and protection works carried out on these lands have failed to produce any worthwhile results because of continued degradation of the local population. Poor economic condition and social behaviour/attitude towards such lands are the main constraints in their development. This would, therefore, require a change in approach by the Soil Conservation Department Staff which should have sympathetic understanding of ignorance and genuine difficulties of the locals and have faith on them. Once they are convinced that restriction on their present activities are ultimately to their own advantages.

15.6 Role of the CAPART in Developing Wastelands

More than 500 volunteer organizations have been recognized by the Ministry of Forests and Environment. All these organizations including NGOs are closely associated with wasteland development and other rural development programmes run by the government through a number of departments. The CAPART (Council for Advancement of People's Action and Rural Technology) is one such government department that has been supporting many NGOs primarily meant for rural development through wasteland management and other activities connected with the programme.

15.6.1 A Case Study of Jammu-Siwaliks through CAPART in Developing Wastelands

The watershed management project at Bari-Badhori, first of its kind was sanctioned by the CAPART for Shivaliks or Siwaliks of Jammu, Jammu and Kashmir State. It was undertaken with the objective to reverse the wasteland of the area lying in the *Kandi* belt of low hill Siwaliks, which had occurred due to massive deforestation and denudation accompanied with heavy soil erosion.

15.6.1.1 Extent and Location of Wastelands in Jammu and Kashmir

Although still no exact information is available on the extent of the wastelands yet on the basis of the land use statistics, Gupta (1994) reported that nearly 1.17 m ha could be grouped under the category of wastelands in the State. But according to Jamwal (1993), in Jammu and Kashmir, the wasteland was worked out to be 3.72 m ha - non-forested areas. In Jammu region, it was approximately to the extent of 1.5 m ha. Gupta (2005c) reported that out of total geographical area of 22.22 m ha of Jammu and Kashmir, roughly 32 per cent is in highly degraded conditions. According to Sharma (2004), the total degraded area of Jammu and Kashmir is around 7.02 m ha of the total geographical area, constituting roughly about 31 per cent.

Soil erosion - the physical removal of soil by the movement of water or the wind, is the most widespread cause of the land degradation. Of the total degraded area of about 7 m ha in Jammu and Kashmir as stated above, 5.46 m ha, 1.36 m ha and 0.20 m ha, are afflicted with water erosion, wind erosion and physical deterioration, respectively. This wasteland of Jammu and Kashmir can be utilized for multiple ourpose by adopting scientific and modern system of irrigation like drip irrigation system which consumes less than 70 per cent water and power in comparison with old conventional methods. All our production can be increased more than 30 to 38 per cent as implemented in India for horticulture, social forestry and cash crops.

This project can yield a good amount of production on wasteland which is lying vacant in Jammu and Kashmir since decades. It can generate employment to the rural youth. Agroforestry and social forestry programmes can also be taken up on these lands for increasing food, fodder, fuel wood, fertilizer and fibre production.

The CAPART sanctioned, "Watershed Development Project for Bari-Badhori during 1996-97 under the aegis of Major General Goverdhan Singh Jamwal, AVSM (Retd.), President J&K Paryavaran Sanstha, Jammu, This project sprawling over an area of about 2,500 ha, covering 6 villages viz., Bari, Badhori, Mahin Charkan, Suchani, Raya and Patti, was a portion of *Kandi* belt of Jammu which was started in December 1998. The *Kandi* belt of Jammu falling in outer hills, experiencing subtropical climate, having coarse to very coarse textured soils often mixed with gravels pebbles, and low in water holding capacity, cation exchange capacity and fertility (Gupta et al., 2005a), was mostly denuded and eroded (Gupta et al., 2005b). The aim of the project was to bring back the greenery in the selected area to produce an example before the *Kandi* belt dwellers for learning and follow-up the same in other areas also.

The concept of agro-forestry and soil/water conservation was taken into account in this project also. Accordingly, the beneficiaries were apprised about planting of various forest tree species along with food crops. Owing to very poor rains during rainy seasons from 1996 to 2002, the survival rate of the planted trees was not conducive However, due to occurrence of good rains during 2003 rainy season lot of plantation, consisting of forest and fruit tree species has been done. Among the medicinal plants, preference was given to grow amla,' harar and bahera. To grow plants of various species, nurseries have been setup.

To reduce soil erosion and conserve the water, engineering structures such as trenches, gully plugging, check dams stand established. Bunding, land leveling and renovation of ponds, are the other works done under the project Among all these works, check dams have proved a boon as their construction has not only improved an environment through storing of water but has also helped in using the water for irrigating the crops. Income generating units like rearing of jersey cows, backyard poultry, apiculture, fish farming, handicraft goods from sisal fibre etc. have given good results The salient findings of this project have been mentioned in Table 15.4.

Table 15.4: Salient findings of Bari Badhori Watershed Development Project (Adapted from Paryavaran Sanstha [1998-2004])

Type of work	Achievements
A. Soil and Water Conservation	
Land leveling	It was operated in about 6 ha gently sloping area covering 3 villages viz., Badhori, Patti and Rahya.
Bunding	It was tried in 3 villages - Bari, Rahaya and Suchani covering 8 beneficiaries.
Gully plugging	Six villages viz., Bari, Badhori, Patti, Rahya, Suchani and Meen Charkan were covered under this control measure to check soil erosion in about 867 m^3.
Check dams	This is the biggest achievement where under 17 check dams were built. And now they arc storing thousands of litres of water at time of every rain. Among them, 2 are concrete and 15 earthen.
B. Renovation Work	
Ponds	Renovation of ponds was another fascinating area of the project against the target of 32 ponds, 61 were renovated. The water so collected made fantastic revelation that how much could be done with little efforts and amount. Badhori pond had now large capacity to store water.
Dug wells or Baulis	A few baulis were also renovated. Among them revival of the perennial water on 100 years old bauli for the villagers is the worth mentioning. It is now catering the drinking water needs of the entire Gujjar Basti of village Bari, comprising 200 families (1000 members).
C. Protection of Cultivable Land	
Crate works	As many as 555 crate works stand completed in 6 villages viz., Bari Badhori, Meen Charkan, Patti, Rahya and Suchani, each having 100, 150, 28, 28, 83 and 166, respectively against the target of 163. Most of these works are in the shape of spurs to protect the land. The beneficiaries are happy to this achievement.
D. Water Resource Development for Plantation	
Trenches	This was the most important activity from the angle of water resource development and plantation. More than 15000 such trenches were dug against the target of 10,952.
E. Plantation Work Raising of small nurseries	Three nurseries were established at Badhori, Sukhianee and Suchani villages. They produced innumerable number of saplings. These saplings were less prone to mortality than saplings from Government nurseries. Indigenous species like phalsa (*Grewia asiatica*), dhiu (*Atrocarpus lakoocha*), jamun (*Syzygium cumini*) and mango (*Mangifera indica*) were grown.

Herbal gardens	Inspired by the emphasis on Ayurveda system of medicine, a number of herbal gardens were setup. In these herbal gardens, harar (*Terminalia chebula),* bahera (T. *belerica),* aonla (*Phyllanthus emblica),* neem (*Azadirachta indica)* and ashwagandha (*Withania somnifera)* have been planted and are going up well (Gupta et al. 2016).
F. Innovative Income Generating Schemes	
Mushroom cultivation, beekeeping, rearing jersey cows and silkworm, leaf cups making	After acquiring training in these avocations, many of the farmers including women have now started their own income generating units.

The Agriculture Minister, Jammu and Kashmir, Mr. Abdul Aziz Zargar visited the area along with General Secretary and President of J&K Cooperative Union on May 1, 2003. They were highly impressed to see the work of project. It is worthwhile to mention that although the work of this project has almost been completed, yet a number of dignitaries pay a visit from time to time to see the wonderful work done under this wasteland development project.

15.7 Conclusion

It is emerged from the chapter titled management of wasteland that about 6 to 7 m ha of land is becoming unproductive in the world. Out of 32,87263 km^z, 638518 is wasteland in India. Concept of wasteland, classification of wastelands, factors affecting productive lands to convert them into wastelands and management of wastelands are detailed in this chapter of the book. The main factors causing productive lands into wastelands consist of denudation of rocks and deforestation, population explosion, soil erosion, flash floods, movement of nomads and over-exploitation of agricultural lands. In the management of wastelands topic promoting of agroforestry, proposed land use pattern, checking of water erosion, tackling of water-logged areas; landslides and landslips are described. Future research strategies/priorities and study questions are also given.

15.8 Future Research Strategies/Priorities

- As the process of conversion of forest lands into agricultural lands and further into residential lands has greatly changed the position of the green cover - making the soils naked through erosion and quick run-off loss, so deforestation must be stopped immediately.
- Although the need for converting wastelands into forests and green lands is being stressed at the highest level in the country yet very little has so far been done to enact or even to formulate the basic policies for wastelands recovery. It is, therefore, suggested to frame strong policies in this regard.

- Similarly, nothing has been done to give the country a National greening policy or National land use pattern. Millions of ha continue to be wastelands and it is impossible to get such lands to promote afforestation. No one can explain why such lands are not made available to those, who can convert them into forests without much costing to the Government. Hence the Government should go beyond formulating the wasteland developments.
- The Planning Commission and other agencies require playing a dynamic role in making use of wastelands. A Master Plan for the purpose may, therefore, be drawn up to bring these lands under cultivation or green cover through various methods so that they become productive.
- Likewise, a National Water Grid System should be brought into being for providing irrigation to all parts of the country, where there is existence of wastelands. Water flowing from the major rivers of the country and going to the seas need to be diverted to various parts of the country for optimal use.
- Strict legislation requires to be framed with regard to putting ban on the felling of forest trees. We must stick to the target of developing particular area; say 5 m ha of wastelands annually.
- Wastelands under the Revenue Department must be allotted to the poor under anti-poverty programmes. Similarly, a large chunk grabbed by the people with clout at various levels of administration should be leased out of these chunks to the poor for fodder cultivation and reforestation.
- Forest fires which take place wethout reason or in simpler parlance cause a colossal loss of our green gold trees, resulting in, thereby creation of lot of waste lands (Anonymous, 2019).
- Time has now come when the nodal agencies should be given a free hand to reclaim the degraded lands of various kinds. The actual implementation of various schemes may be decentralized through the State Governments. Co-operation even may be sought from voluntary organizations, for the purpose of making wastelands into productive ones.
- As a number of NGOs, social and voluntary organizations are collaborating with the government agencies to develop the wastelands, so it is suggested that there must be close coordination of the Government and NGOs to help in reclaiming wastelands.
- Keeping in view the urgency of the development of degraded lands, it is desired that the social and voluntary organizations be entrusted with the task of motivating and organizing communities for protection, promoting

afforestation and development of wastelands, especially in the vicinity of habitation.

- According to the noted environmentalists such like Chandi Parsad Bhatt, Shri Sunder Lai Bahuguna, Major General (Retd.) Goverdhan Singh Jamwal, NGOs can play a very constructive role in improving the state of degraded lands. Hence the Government departments must give them full cooperation as these departments themselves cannot do much in this direction unless they receive cooperation from social and voluntary organizations.
- What Government can do at the best is the financial assistance or the expertise. The Government should propose to the NGOs for identifying the projects and funding must be done without any delay. In this line, the Finance Ministry, Government of India, should be requested to give green signal to allocate the funds immediately for the rehabilitation of the wastelands of the country
- To boost rural economy of the hilly areas, especially of Jammu and Kashmir, there is an urgent call for massive willow *(Salix* spp.) plantation on wastelands. Thrust must be given to those species which have economic importance and are able to adaptable to the climatic conditions of the Kashmir Valley.
- Since chunk of cultivable lands in Ladakh, and Lahaul Spiti and Kinnaur, are left as wastelands for want of irrigation facilities; efforts should, therefore, be made to plant various species of poplars *(Populus* spp. - *Populus nigra, P. ciliata, P. candicans* etc.) depending upon their suitability to grow in the area.
- Recently, a new deciduous shrub known as "sea buck thorn", has been discovered to grow at Kinnaur, Lahaul Spiti and Pangi areas of Himachal Pradesh as well as Ladakh (Union Territory). So it must be grown in the wastelands of these areas as this shrub has unique quality to withstand adverse climatic conditions including low temperature up to 60 degree below freezing point.
- It is suggested that wherever wasteland is present, gardens may be established out of the wastelands as have been built in Chandigarh and Pune. Not only this, but more important of all is the unique experiment in reversal of the environmental degradation that the "Osho Teerth" represents. It was almost a "garbage to garden" story. And perhaps it is why the Puneites would like to call it "Nallah Park" for reasons of nostalgia or contrast or both.

- The Osho Teerth is spread over an area of about 5 ha of land behind the Osho Commune International. About 16 years ago, it was a wasteland through which flew a highly polluted stream, sewage ridden and frothing. Moreover, the land was heavily overgrazed and soil had been eroded and depleted to the point of uselessness. But with the efforts of Shunyo Foundation, the area has now been restored into its natural regeneration capacity to the soil and re-established a balance between a clean, healthy environment and needs of a modern city.

Study Questions

1. What do you mean by wastelands?
2. What is the difference between degraded land and wasteland?
3. How we can define wasteland?
4. Write in detail the concept of wastelands.
5. What are the various factors responsible for rendering wastelands?
6. What do you mean by CAPART?
7. Importance of planting grasses in reclamation of wastelands.
8. Write paragraphs on the following:
   i. Afforestation
   ii. Water harvesting
   iii. Checking of water erosion

References

Anonymous. 2019. State actionplan on forest time. The Daily Excelsior. 55 (181):6 [July 2, Tuesday]

Anonymous. (2000). Wastelands or wasted lands. The Daily Excelsior 36 (203): 6 [July 22, Saturday].

Anonymous. (2014b). Livestock - The bank on hooves. Agriculture Today 17 (5): 20-27.

Anonymous. (2014c). Landslides disrupt traffic on highway. The Tribune 134 (77): 13 [March 19, Wednesday].

Anonymous (2019). Traffic blocked in Kinnour Shimla interior parts. The Tribune 4 (345):6.

Fazili, E. (2011). Literacy rate rises, sex ratio falls. The Tribune 131 (9G): 3 [April 8, Friday].

Gupta, D.P. (1994). Management of degraded and denuded wastelands. Dehat Sudhar, Jammu and Kashmir 3 (8): 14-17.

Gupta, G.D., Gupta, R.D., Khar, D. and Sumberia, N.M. (2005b). Studies erodibility indices in relation to soil properties in Kandi belt of Jammu district. J. Res., SKUAST-J 4 (2): 209-213.

Gupta, G.D., Gupta, R.D., Kher, D., Gupta, J.P. and Sumberia, N.M. (2005a). Studies on the fertility status in relation to soil properties in Kandi belt of Jammu district. J. Res., SKUAST-J 4 (1): 102-110.

Gupta, RD. (2009). Glimpses of Northwest Himalayas. Published and Printed by Radhakrishanan Anand and Co, Pacca Dhanga, Jammu Tawi, J&K, India

Gupta, R.D. and Arora, S (2016), Sold affected soils in J&K-their Management for enhancing productivity. Soil water conservation, 15(3):199.

Gupta, R.D. (2002). Save the soil save the nation. Environment and People 8 (12): 20-25.

Gupta, R.D. (2003). Wastelands call for reclamation. The Kashmir Times 60 (320): 11 [November 17, Monday].

Gupta, R.D. (2005c). Environmental Degradation of Jammu and Kashmir Himalayas and Their Control. Associated Publishing Company,' New Delhi (India).

Gupta, R.D. (2006b). Regreening the cold desert Himalayan wastelands. The Glimpses of Future 21 (144): 4 [May 26, Friday].

Gupta, R.D. (2009c). Introduction to soil erosion and its classification, water erosion mechanism, types and factors affecting water erosion (pp. 23-28). In Soil and Water Conservation Training Manual (Eds. B.S. Hansra, P. Vijaykumar, R. Nanda and P.K. Rai) organized by IGNOU, New Delhi and KVK, SKUAST, Jammu.

Gupta, R.D. (2009d) Make Siwaliks Green. The Daily Excelsior 45 (354): 6 [December 22, Tuesday].

Gupta, R.D. (2010a). Checking the depleting environment of J&K. The Daily Excelsior 46 (88): 6 [March 30, Tuesday].

Gupta, R.D. (2011a). Population pressure on agriculture - A grim scenario. The State Times 16 (223): 6 [August 11. Thursday].

Gupta, R.D. (2011d). UN calls for conservation of forests. The Daily Excelsior 47 (35): 8 [February 4, Friday].

Gupta, R.D. (2013c). Forest studies of Jammu Siwaliks. The Daily Excelsior 49 (337): 8 [December 9, Monday].

Gupta, R.D., Arora, S. and Sharma, A. (2013). Landslide hazards in Himalayan region (pp. 368-385). In Mountain Ecosystem and Man (Eds. S. Arora, B.S. Negi, S. Bhan, J.S. Bali and V.K. Bharti). Soil and Conservation Society of India, New Delhi.

Gupta R.D. and Arora Sanjay(2016). Salt affected soils in Jammu and Kashmir: Their management for enhancing productivity. Journal of Soil and water conservation. 15(3): 192-204.

Gupta, R.D., Gupta S.K. and Bhardwaj S.D. (2016). Agrotechniques and uses of Medicinal. A Division of Astral International Pvt. Ltd. New Delhi - 110002.

Gupta, R.D. (2018b) Managing wastelands for sustaining food production. The State Times. 23(248):6 [September 18, Saturday].

Indira Gandhi Institute of Development Research, Mumbai. (2001). Sustainable Land and Forest Regeneration Environmental Governance- 3, pp. 1-9 (Personal Communication).

Jamswal, J. 1993. Watershed in Jammu region. The Kashmir Times (April 9, Fridary)

Magi, A.K., Obi Reddy, G.P. and Sarkar, D. (2010). Degraded and wastelands of India; Status and Spatial Distribution, ICAR, New Delhi, pp. 158.

Mahajan, A., Bhagat, R.M. and Sharma, S. (2004). Conservation tillage - An ecological approach for better soil management to improve crop productivity. Indian Farmers' Digest 37 (3): 16-20.

Minhas, P.S. (2012). Edaptic disabilities of degraded soils. J Indian Soc. Soil Sci. 60 (Supplement): 588-590.

Najjar, G.R. and Gupta, R.D. (1996). Effect of organic and inorganic amendments on soil properties and yield of rice. Agropedoloay 6 (2): 83-86.

Paryavaran Sanstha, J.K. (1998-2004). Impact analysis of Bari Badhori Watershed Conservation and Development Project (Funded by CAPART), Tehsil Samba, District Jammu, J&K. Paryavaran Kendra, Mubarak Mandi, Jammu. June 2004.

Rangrez, S. and Singh, S. (2014). Slides block Batote-Doda road. The Kashmir Times 71 (90): 12 [April 2, Wednesday].

Sarkar, A.K. and Sharma, S.P. (2006). Management of acidic soils - Current issues and policy guidelines (pp. 18-22). In Balanced Fertilization for Sustainable crop Production (Eds. S.P. Sharma, S.K. Sharma, R.P. Sharma and T.S. Verma). Proceedings National Seminar on INM for Balanced Fertilization held on November 17-18, 2006.

Sehgal, J.L. and Abrol, I.P. (1994). Soil degradation in India; Status and impact. Oxford and IBH Publishing Co. Pvt. Ltd., New Delhi (India), pp. 80.

Sharma, K.L., Srinivas, K., Mandal, U.K., Ramakrishna, Y.S., Vittal, K.P.R., Venkateswarlu, R.V. and Grace, J.K. (2004). Strategies and approaches to improve and asses soil quality. Agro-ecosystem Directorate (Rain-fed) National Agricultural Institute for Dry land Agriculture Santoshnagar, Hyderabad (Andhra Pradesh), India.

Sharma, P.D. (2004). Managing natural resources in Indian Himalayas. J. Indian Soc. Soil Sci. 52 (4): 314-331.

Sikka, A.K. 2013. Soil erasion research in India: An overview. Journal Indian Society Soil Science. 61(4):569-583 (Supplement)

Sood, K.K. (2019). Agroforestry for boosting farmers income. The State Time. 24(210): 6 [July 31, Wednesday]

Sriraj, K. 2019. Desertification: A Real time threat. The State Time. 24(38):36 [February 8, Friday]

16

Some Alternatives to Sustain Productivity of "Jhum" Cultivation

Abstract

"Jhum" or shifting cultivation is prevalent mostly in tropical/subtropical countries of the world. In India, the peasants of eastern and north eastern regions practise shifting cultivation on hill slopes. In north eastern region, about 85 per cent of the total cultivation is by shifting cultivation. Owing to burgeoning human population, requirement of cultivation of land vis-a-vis cycle of its cultivation has decreased from 25-30 years to 2 to 3 years. This has caused a great impact on environment and ecology which has been elucidated in this chapter of the book. In addition to this, some alternatives for sustaining Jhum cultivation such as agroforestry, terrace cultivation, contour cultivation, and replacement have also been stated. Myths and realities of Jhum cultivation, its constraints, main reasons causing low agricultural productivity, future research strategies/priorities, are the topics which have been elucidated vis-a-vis what constitutes "Jhum" cultivation?, its sequence and classification.

Keywords: Jhum or shifting cultivation; tropical/subtropical countries; India; peasants; eastern regions; north eastern regions: hill slopes; human population; environment; ecology; agroforestry; terrace cultivation; contour cultivation

North Eastern Himalayan Region comprises mostly of seven states viz. Arunachal Pradesh, Assam, Manipur, Meghalaya, Mizoram, Nagaland and Tripura (Gill, 2005). This Himalayan Region stands entirely different from the rest of the country not only in language, dress, food habits and festivals but also differs in agricultural practices. Among agricultural practices shifting cultivation locally known as "Jhum" or "Bun" is the most important. According to Ministry of Agriculture, Arunachal Pradesh stands first in the "Jhum" cultivation followed by Assam, Manipur, Meghalaya (Agnihotri, 2003). In Meghalaya, the problem has assumed an epidemic proportion in Gargo hills where tracts of forests are burnt. The farmers of Khasi and Jainta Hills usually practise the "Bun" cultivation.

"Jhum" or "Bun" cultivation is not only confined to North Eastern Himalayan Region but also exists in other states of India viz. Andhra Pradesh, Bihar, Madhya Pradesh and Orissa. About 4,92000 tribal families of North Eastern Himalayas are involved in "Jhum" or "Bun" cultivation and the total area cultivated by this practice is about 26,94000 ha (Borthakur, 1987). It has been recently stated that about 300 to 500 million people are practising shifting cultivation worldwide (Anonymous, 2006). In India, millions of tribal farmers (making 12% of the total population here) are a part of the global community of "Shifting Cultivators".

"Jhum" cultivation is the most primitive system of farming, which is believed to have been originated in the Neolithic period around 700 BC and is dangerous to the resources degradation like forests, lands, soils etc. Moreover, this system of farming does not provide at subsistence level to the farming community, hence, it becomes an imperative to have some alternatives to control "Jhum" cultivation and to sustain the agricultural productivity (Borthakur, 2001b). Such alternatives have been mentioned in this chapter of the book. Apart from this, what constitutes shifting cultivation and their impact on ecology has also been accounted vis-a-vis suggestions for the future improvement.

16.1 What Constitutes Shifting Cultivation?

When man learnt about agriculture nearly 10,000 or 12,000 years ago, he used to clear a patch of forestland for cultivating crops with whatever experience he then had. Thereafter, he moved on to clear the next patch of forestland. This is what comes to be known as "Shifting Cultivation". "Shifting Cultivation" is, indeed, one which does not require ploughs, bullocks and tractors. Just an axe and a toiling farmer are required for "Shifting Cultivation".

This primitive system of farming is regarded as the first step in the transition from food gathering or hunting to food production. This farming system is believed to have been originated in the Neolithic Period about 7,000 BC or 6,500 to 7,500 BC (Reddy, 2001) and is mainly confined to North Eastern Hill Region or North Eastern Himalayan Region of the country.

Apart from "Jhum Cultivation" or "Bun Cultivation", its other names called by different communities living in various corners of this planet to describe a particular art of agriculture are: "Khallu", "Podu", "Bagoda", "Kumri", "Walt", "Swidden Farming", "Slash and Burn agriculture".

16.2 Sequence of "Jhum" or Shifting Cultivation

The shifting cultivation is generally in the following sequence or order:

- After selecting a forest patch, it is cleared and vegetation is felled usually in the month of December or January.
- Small cut trunks portion and roots are normally not removed. Branches, twigs, herbs and shrubs (slashed vegetation) are burnt during February and March.
- Sowing of seeds by dibbling generally consisting of cereals, vegetables and oilseeds, is done in April- May.
- Cultivation is continued for a few years. Hitherto "Jhum" cycle varied from 20-30 years or 20-40 years but now it has reduced to 2-3 years.
- Abandoning the cultivated site and shifting to another forest site is common phenomenon.
- Returning to the former site and once again practise on it - shifting cultivation.

With reduction in "Jhum" cycle from 20-30 years to 2-3 years, the land after shifting cultivation loses its nutrients and top soil resulting in low productivity. With reduction in crop yield, the tribal families start moving to other virgin areas. Now a stage come that it has been already affected 2.7 m ha of land, and each year 0.45 ha of land falls under shifting cultivation, in North Eastern Hills (Sharma et al., 2006).

16.3 Classification of Shifting Cultivation

Shifting cultivation in the seven sister states as cited above in the introduction part of this chapter, can be broadly classified into "Jhum" - Slash and Burn and "Bun" - Ploughing Slashed Biomass Types. In "Jhum", the local people slash and burn a patch of forestland which mostly consists of thick bamboos growth and other wild vegetation. Contrary to this, in "Bun" the peasants do not burn forests. They rather plough back whatever vegetation they cut with the objective to use the biomass as a source of organic manure.

16.4 Cycles of Jhum Cultivation and Their Impact on Environment and Ecology

The current practice of shifting cultivation in the eastern and north eastern regions of India is an extravagant and unscientific form of land use. The bad influences of this type of cultivation are many, which are amenable to degrade the environment and ecology of these regions. Such effects stand elaborated in the following paragraphs:

16.4.1 Unproductive System of Farming

"Jhum" Cultivation system of farming hitherto worked well when the population was low and "Jhum" cycle was long i.e., either 15 to 20 years or 30 to 40 years. It helped to restore fertility of the soil sufficiently. However, with the increasing population pressure on the land, the "Jhum" cycle has now been reduced to 2 to 5 years or 2 to 3 years. As such, this system of farming has become one of the most unproductive. More over, it has hit the biodiversity badly (Gupta, 2009).

16.4.2 Reduction in the Food Grain Production

Owing to un-productivity of "Jhum Cultivation", the growth rate in the food production in the north eastern part of India is not keeping with the rapid rise in the population. While the food grain production was increasing at an annual compound growth rate of 2.18 per cent, the population in the region was growing faster at the rate of 2.43 per cent.

16.4.3 Massive Ecological and Environmental Degradation

Shortening of "Jhum" or "Bun" cycle of 2 to 5 years due to burgeoning human population and paucity of cultivable land, have led to massive ecological and environmental degradation. Soil erosion is soaring the hills due to depletion of forests. Records show that on an average every year 5 to 45 tonnes of top soil is lost ha^{-1}. These figures are the minimum which actually may be much more and can be accelerated by the high rainfall

Continuous "Jhum" cultivation has led to a large scale deforestation and denudation of hill slopes, causing siltation of water reservoirs, thereby degrading the environment largely. Besides creating siltation and recurring flash floods in the plains, "Jhum" cultivation is also responsible for loss of soil fertility since the top soil which is more fertile had already been washed away from a number of "Jhum" cultivated places.

16.4.4 Land Degradation

From land degradation point of view, "Jhum" and "Bun" cultivation can be regarded as one of the most destructive forms of land use as well as wasteful practice. It is because this system of farming not only involves large scale deforestation of land leading to accelerated soil erosion and low productivity but also causes ecological imbalance affecting hydrological cycle, nutrient recycling, microclimate and biotic environment i.e., animals and plants.

With reduction in "Jhum" cycle from 20-30 years to 2-3 years, the land under shifting cultivation has lost its nutrients and top soil has depleted the Humus or Humic Substances. Humus is the well decomposed organic matter which is very essential in increasing the cation exchange capacity and water holding capacity of the soil. It is also a source of all plant nutrients i.e. macro (N, P, K, Ca, Mg and S) and micro (Fe, Cu, Mn, Zn, Mo, Cl & B). Humus in fact, acts as chelating agents of the micro-nutrient cation and renders their availability slowly and uniformly. The depletion of humus has, thus, reduced the yield of the crops. With reduction in crop yield, the families of "Jhum" and "Bun" cultivation start moving to other virgin areas. The total degraded area of North Eastern Hill States is presented in Table 16.1 as follows:

Table 16.1: Total degraded area of North Eastern Hill States (in 1,000 ha) (Adapted from Sharma [2004])

Sr.No.	State	Water erosion	Complex problem	Physical deterioration	Total degraded area
1.	Arunachal Pradesh	4327	-	176	4503 (54)
2.	Mizoram	1187	694	-	1881 (89)
3.	Manipur	133	708	111	952 (42)
4.	Nagaland	390	605	-	995 (60)
5.	Tripura	425	-	203	628 (60)
6.	Meghalaya	1168	34	146	1208 (53)

Adapted from Sharma (2004)

The major causes of land degradation as stated above are water erosion, complex problem and physical deterioration. All north eastern hill states have badly suffered due to soil erosion of more than 20 t ha^{-1} $year^{-1}$ and in shifting cultivation regions, it is more than 40 t ha^{-1} $year^{-1}$. The productivity of crops, which is at present low, could be increased by following proper management practices i.e., soil water conservation measures, starting of different components of agriculture, horticulture, animal husbandry etc. (Sharma, 2004).

According to Singh and Kukal (2013), the north-eastern states of India have severe water erosion because of the prevalent practices of shifting (Jhum) cultivation. In the past, the practice of shifting cultivation worked well because of long fallow cycle (20-30 years). But due to increasing population pressure, the cycle has been narrowed to 3 to 6 years and thus aggravated erosion and degradation problems. Forest cutting, burning, cleaning and dibbling of seeds causes nearly 4 t ha^{-1} $year^{-1}$ of soil material to slide/roll down to foot hills. Soil erosion from hill slopes of 60 to 75 per cent during the first, second and third years has been reported to the extent of 146.6, 170.2 and 30.2 t ha^{-1} $year^{-1}$, respectively.

16.4.5 Air Pollution and Loss of Flora and Fauna

The burning of biomass during "Jhum" cultivation creates air pollution, loss of soil fauna and micro-fauna (nematodes and protozoa) vis-a-vis micro-flora (bacteria, actinomycetes and fungi). There is also loss of primary forests and generation of secondary forests. As per scientific records, the largest wastelands are in the north eastern hill region, where "Jhum" cultivation is done. "Jhum" cultivation has also rendered a lot of flora and fauna as endangered.

16.5 Initiatives and Strategies for Checking "Jhum" Cultivation

Clandestinely, "Jhum" cultivation is being practised on the Revenue Forests, Reserve Forests and Protected forests. Although "Jhum" cultivation is a non-viable resource-utilization system of farming yet Tribals of North Eastern Hills are still doing this primitive farming practice to sustain themselves and their families. It is attributed to non availability of timely employment avenues. So from time to time various attempts have been made by the Government to settle the tribals involved in shifting cultivation.

16.5.1 Shifting Cultivation and Loss of Forest Cover

Quantum jump of population pressure of tribal societies practising shifting cultivation and declaration of reserve forests by government, have shortened the fallow periods, which now on an average is 3-4 years in comparison to 25-30 years half a century ago. Forest survey of India (1999) estimated the cumulative area affected by shifting cultivation during 1987-97 was found to be 1.73 Mha and maximum affected by this practice of agriculture was in Nagaland and Mizoram (Patiram and Choudhary, 2013).

16.5.2 Constitution of Task Force

The union Government constituted two task forces to study the "Jhum" cultivator's menace. For example, the first one such task force was established in 1983 under the Ministry of Agriculture and the second in 2001 under the Ministry of Environment and Forests. They recommended that a sustainable approach to manage the shifting cultivation should be adopted, using holistic and integrated approaches or means. However, inspite of Government Endeavour's most of the "Jhum" rehabilitation schemes have failed due to the following reasons:

- It is mainly because many of the families who do "Jhum" cultivation were not involved when the schemes were formulated.
- Most of the projects were not allegedly tested before formulating/ implementing. Infact, the whole of the approach lacked of research back up and intensive monitoring and evolution. As long as adamant farmers and half hearted implementation of Government policies continue to exist in the North East Hills, "Jhum" and "Bun" will stay.

A few more schemes were implemented under integrated tribal development programme in the districts of Koraput, Keonjhar and Phulbani in Orissa. These schemes also did not yield the desired results. It was perhaps because of the ignorance by the authorities on the socio-economic and agro-ecological conditions of the shifting cultivation. It is also due to minimum involvement of the Forest Department officials who are much more informed about the above said factors, in implementation of the scheme. Failure of the scheme led to the National Commission of Agriculture to reformulate the schemes only after considering the impact of the forest management.

16.5.3 Setting Up of Agroforestry

Agroforestry i.e., the joint production of food and wood has to be exploited to its fullest potential. Agriculture is the subservient to trees and forests. Agroforestry can go a long ***way*** to meet our needs of food, fruit, fuel wood, fodder, fertilizers, fibres, timber besides creating suitable microclimate for individual farms favourable to larger production of grains ha^{-1}. China has shown us the way. Between 1975 and 1990, agroforestry has helped China to increase grain productivity by 50 per cent besides other crops.

Adoption of agroforestry is also the need of the hour for the area of the "Jhum" Cultivation. An agroforestry Project known as Nagaland Environment Protection for Economic Development (NEPED) funded by Canadian International Development Agency through India-Canada Environment Facility was initiated in 1995 to make Nagaland self-sufficient in firewood, timber, fibre, food etc. The main objectives of this project were:

- Identification of local trees for cultivation by the tribes and their demonstration method leading to more sustainable resource management.
- Meeting and addressing people's need and evolving better management systems.
- Promoting marketing facilities.

Apart from the above, under this project experienced Government officials after convincing the villagers 6 ha of land was also set a side, known as test plot, in "Jhum" areas for over two and half years. Owner of this plot is required to plant 1200 seedlings along with sowing of usual crop under the supervision of village council and project team. These tested plots became open school cum research station as well as demonstration plots to each new technology. Thus, NEPED project did not aim at eliminating "Jhum" cultivation but making it more stable and profitable. This novel project may give us a more scientific way to tackle tribal forest conflict.

With the concerted efforts of the scientists of Indian Council of Agricultural Research (ICAR) it has been found that on the top of the hills, the pine trees were planted and the lower portion was covered with grasses and in the villages crops were grown.

In abandoned "Jhum" lands, perennial plants and leguminous trees like *Cajanus cajan, Albizzia lebbeck, Albizzia procera* and *Tephrosia candida* have been planted and are found suitable for the welfare of the farmers.

16.5.4 Terrace Cultivation

Alternative measures may consist of terrace cultivation which unlike North Eastern Himalayan Region is very common among the farming communities of North Western Himalayan States - Uttaranchal, Himachal Pradesh and Jammu and Kashmir in the sloppy lands. Terracing cultivation will not only help in conserving the moisture in soils but also checks soil erosion.

The terrace risers which constitute 33 to 40 per cent of the total, have never utilized for cropping in the past. It has been found out that terrace risers can be very effectively utilized for growing perennial fodder grass and legumes. This will assist not only to help in conservation of soil and water but also in providing enough fodder to maintain animal husbandry as a subsidiary source of income (Tables 16.2 and 16.3)

Table 16.2: Number of animals which can be recorded under various farming systems (Adapted from Sharma and Pareek [1999])

On the basis of	Number of animals which can be reared		
	W-1	W-2	W-3
Dry matter	0.78	2.73	3.26
Digestible crude protein	0.66	1.18	1.05
Total digestible nutrient	0.50	1.78	2 27
Digestible energy	0.52	1.76	2.36
Metabolizable energy	0.52	1.72	2.35

W-1: Jhum cultivation; W-2: Agro-Horticulture and W-3: Agriculture

Table 16.3: Livestock production potentiality (Adapted from Sharma a nd Pareek [1999])

Land treatments	Area for livestock produce	Type of feed	Livestock unit
Shifting	No space available	Crop residue by-product	0.50
Agri- horticulture	Terrace riser/horticulture interspaces	Annual perennial fodder crop residue	1.18
Agri-bench terraces	Terrace riser	-do—	1.05
Agri-puetrorican terraces	Bunds and emerging terrace riser	-do-	1.07

While planting crops in the terraces, the concept of topo-sequences has been developed so that crops requiring well drained conditions, such as ragi, seasmum, and maize can be raised with assured production in the upper terraces while crops like rice can be grown in the lower terraces with favourable moisture regimes.

16.5.5 Contour Cultivation

Contour bunds in "Jhum" land will check the soil loss greatly. Further, it will facilitate continuous cropping which in turn will help in converting slope into bench terraces with 8 to 10 years.

For growing crops within contour bunds topo-sequence should be followed taking advantage of moisture regime and fertility of soil. Crops like cassava, maize intercropped with soybean/pigeon pea, finger millet, sesame etc., which requires less water can be grown in the upper part of the sloppy land. On the other hand, lower portion of the land should be used for growing rice.

16.5.6 Replacement Approach

The studies on replacement were aimed at developing alternative farming systems to replace "Jhum" cultivation to permanent one. Results have shown that mixed land use is the most ideal land use system for hilly slopes. In this system, the trees in the higher ridges, horticultural crops in the middle portion and field crops in the lower traces can assist not only in increasing productivity, preserving ecology but also help in maintaining an ideal hydrological behaviour of the area.

Pineapple as well as mandarin oranges can be successfully grown. However, due to citrus decline, there is reduction in the production of mandarin oranges. Maize, ragi, sesame and rice are the main crops grown.

Among the various trees found suitable for growing are *Albizzia lebbeck, Cassia fistula, Cassia siama, Dalbergia sissoo, Robinia pseudoacacia* etc. Various fruits, vegetables and flowers found suitable for growing in the "Jhum" cultivated areas of North Eastern Hills and other similar areas are shown in Table . 16.4.

Table 16.4: Suitable horticultural/olericulture and floricultural plants for shifting cultivation (Adapted from Sharma and Pareek [1999])

Suitable plants	Region	
	North East Region	Parts of Orissa, AP, MP and Bihar
Fruit	Apple, pear, peach, orange, lemon, banana, papaya, pineapple, breadfruit, jackfruits	Jackfruit, grape, mango, citrus, ber, onola, khirni, guava, papaya, fig
Vegetables	Potato, brinjal, elephant foot yam, colocacia, cole crops, bitter gourd, parwal, chilies, tomato, root crops	Potato, tomato, cabbage, radish, cauliflower,chilies, colocacia, bitter gourd
Flower	Rose, gladiolus, tuberose, orchids, carnation	Rose, gladiolus, tuberose, jasmine, chrysanthemum, marigold,
Others	Tea	Petunia coffee

After adopting the above said alternative cultivation practices will definitely serve as a boon for the sustainability of the "Jhum" cultivated areas.

16.6 Myths of "Jhum" Cultivation

- Shifting cultivation is considered not only destructive but it is also wasteful and unsustainable. And as such cause forest and soil degradation.
- Shifting cultivation possesses low productivity.
- In shifting cultivation, the farmers use primitive technology and are less acquainted with agricultural and environmental knowledge.

16.7 Realities of "Jhum" Cultivation

- Shifting cultivation is not the major source of forest and soil degradation. Commercial logging is a far serious offender.
- Shifting cultivation makes good use of the resources and is adaptable to changing conditions. The productivity is high because a mixed crop is traditionally planted. The crops complement each other's growth and also the nutrients in the soil.
- Shifting cultivators have a great understanding of their environment, land use pattern and other resources. Although the technology is not modern yet it is an appropriate for the ecological space they work in.

16.8 Constraints of "Jhum" or Shifting Cultivation

- Tribals of Bastar known as "Marias" and those of Koraput called as "Saoras" were forced to cultivate terraced fields. But their cropping pattern changed due to dominant cultivation of rice. This completely disrupted the pattern of food consumption of the Saoras, who used to grow 10 to 11 varieties of crops especially millet, their prime cereals.
- In Agartala (Tripura), 60 per cent of the forests have been "reserved". Some of the families - Tripuris were herded down to settle in the plains, with promise of leased plots in lieu of the forestland they have left behind.
- These tribals later on built their own bamboo huts and waited in vain for the Government to make good its promise (Anonymous, 2006). They waited for a long but all in vain. And as such some of these tribals are now inclined to go back. They say, "If we do not go back, we will starve".
- Some of the Environmentalists/Ecologists are opined that the Forest/ Agriculture Departmental officers/officials should not impose a total ban on "Jhum" or shifting cultivation. If it is forcibly banned, there would

certainly be a recipe for disaster. Because aided by modern technical support, "Jhum" does not have the potential to become viable, sustainable production tool for these tribal farmers, provided the local communities are involved and use of their age-old wisdom is done to find a way. However, the "models built in experimental plots, far removed from real life situation will just not work.

16.9 Shortage of Food in North Eastern Part of India

North Eastern part of India is facing for food crisis. The food shortage in this region was about 2 million tonnes during 2001 and will further rise 2.82 million tonnes by the year 2010-11, mounting to 3.95 million tonnes by the year 2020-25. By the year 2025-26 the food deficient in the region is expected to reach a level of 4.55 million tonnes. This has been predicted by the Scientists of ICAR, North Eastern Hill Region

16.9.1 Main Reasons for Low Agricultural Productivity

Shifting cultivation, faulty land ownership pattern, predominantly hilly terrain ill managed and indiscriminate use of natural resources, poor infrastructure and marketing facilities, lack of finance and low absorption of technologies are some of the main reasons for low agricultural productivity in the region.

16.9.2 Main Concerns in the Region

The two main concerns in the region are low productivity and low cropping intensity. The average yield of rice, maize and wheat are far below the National average. To meet the projected food demands of the region, an average yield ha^{-1} of food grains would have to be increased by 55 per cent and nearly 100 per cent by 2010 and 190 per cent by 2025 taking into account 1993-94 as the base year.

Large scale of firewood, collection of plants, orchids, medicinal herbs and ferns for commercial purposes, use of rare species of plants as decorative, use of chemicals for cultivation and fishing, and unplanned development, have posed severe threats to the Environment and Ecology of north eastern states.

Many people after cutting down natural forests, they have started to plant trees of commercial value in an unplanned manner. In most of the cases, a single species monoculture was adopted leading to disease and insect pest attack in the trees.

16.10 Conclusion

North Eastern Himalayan Region like North West Himalayas is also endowed with unique physiography, soil, water, biotic and tourism resources. However, with the increased anthropogenic pressure accompanied with reduction of "Jhum Cultivation" cycle from 20-40 years to 2-3 years has led to their degradation, posing, thereby, many environmental problems. Water erosion is the major cause of soil erosion. The productivity of the crops is low due to large number of constraints. So the area requires high level of management and alternatives to sustain the crop productivity. Setting up of agroforestry is one of the alternatives which have been detailed. In addition to this, terracing cultivation and contour cultivation, have been mentioned which require to be followed immediately. Future research strategies/priorities are also presented. Apart from setting up of agroforestry, crop diversification is the need of the day to enhance income of the farm (Gupta, 2019). This will definitely assist in managing soil and water resources (Surajbhan, 2019).

16.11 Future Research Strategies/Priorities

- A sustainable approach to manage the shifting cultivation should be adopted using holistic and integrated means.
- Notwithstanding Government Endeavour's, most of "Jhum" rehabilitation schemes have failed. It was because "Jhum" cultivable families were not involved when such schemes were formulated. It is, therefore, humbly suggested that while formulating any programme for the peasants who do shifting cultivation must be involved to make it a more success.
- One can not accept any substantial increase in food grain production on sustainable basis in any region of the country unless the balanced and adequate dose of nutrients is made available for the crops grown by the farmers. The farmers should also be made aware of the importance of the organic manures and use them along with the fertilizers i.e., INM system. The details of INM have recently been appeared in the literature (Mahajan and Gupta, 2009).
- Appropriate Government policies, new relevant technology and good soil management practices are called for to achieve self sufficiency in food grain production in the area.
- Many of the soils of North Eastern Hill States are acidic in character and have been found deficient in phosphorous, calcium and to some extent in magnesium. Liming should be done in such soils as per the recommendations made by the Scientists of ICAR working at Shilong to increase soil productivity
- It appears that the economy of the region will continue to remain dependent up on agriculture for many years to come as there are less industries and the tourism potential had also not been exploited much as yet.

- All the three kinds of soil and water conservation measures viz. agronomic techniques, engineering practices and biological methods require to be followed.
- Biological methods like integrated programme of afforestation i.e. planting of forest trees, fruit trees and grasses have to be planned according to slope and land capability.
- While planting the forest trees local species including medicinal plants should be preferred to grow as has been done incase of ***Kandi*** belt of Jammu (Gupta, 2010d, Jamwal and Gupta, 2007).
- Adoption of diversified agriculture is the need of the day and must be advocated to follow as is now being followed in Kandi belts of Siwaliks (Gupta, 2009b, Arora and Gupta, 2019).

16.12 Employment Opportunities

- Providing employment opportunities and income generation schemes on regular basis through proper utilization of the land resources, by equitable distribution of wasteland among the tribal, is the need of the hour.
- Co-operative efforts for carrying out forest based activities viz., basket making rope, cane furniture processing of minor forest produce, honey collection should be made commercially viable by providing proper marketing facilities. This will enable the tribals to become richer monetarily and leave an idea to render "Jhum" cultivation.
- Village Forest Committees should be constituted for protecting and developing of the degraded forests. These communities by providing suitable incentives to the tribals can divert their attention away from the "Jhum" cultivation. Generating employment opportunities during the lean season or forestry operations will also prevent tribal's from shifting to other areas.
- Employing the tribals for collecting the kendu leaves and soil seeds as well as to involve them in various rural employment schemes is also the need of the hour.
- Another important need of the hour is to educate the tribal women and children vis-a-vis men about the ill effects of the "Jhum" cultivation and technical know how to check this system of farming. For this purpose, training programmes should be arranged frequently to change their attitude. *"Learning by doing"* and *"Seeing by believing"*, are some of the agenda items which can be perceived easily through laying out certain kinds of demonstrations as an alternative to "Jhum" farming.

- Wherever possible edible bamboo farming, organic farming and cultivation of aromatic and medicinal plants are required to be undertaken.

Study Questions

1. What is meant by Jhum or shifting cultivation?
2. Write the local names of shifting cultivation.
3. Write in brief about the consequences which have taken place due to reduction in Jhum cycle from 20-30 years to 2-3 years.
4. What are the various alternatives to sustain productivity of Jhum cultivation? Write the detailed account of anyone of them?
5. What is meant by agroforestry?
6. Differentiate between biological and mechanical measures taken to check soil erosion?

References

Arora, Sanjay and Gupra, R.D. (2019). Soil and water resources conservation and management for sustainable for rural livelihood and foot hills of north west Himalayas India. Souvenir A joint International Conference of ISCO and SCSI, New Delhi, India

Agnihotri, P. (2003). Farmers resist government efforts to check the menace - Jhum plays havoc in Meghalaya hills. The Glimpses of Future 8 (277): 4.

Anonymous (2006). Lost in the woods - The science and art of slash and burn. Gobar Times (May 31, 2006) Down To Earth 59 (Supplement): 62-68.

Borthakur, D.N. (1987). Scientific land use and watershed management. ICAR, Research Complex for North Eastern Hill Region, Shilong.

Borthakur, D.N. (2001b). Alternatives for controlling shifting cultivation. Ex-Vice Chancellor, Assam Agricultural University (Personnel Communication).

Gill, C. (2005). North east: Ravaging Paradise. The State Times 10 (281): 6 [October 8, Saturday].

Gupta, R.D. (2009b). Diversified agriculture in Kandi belts. The Daily Excelsior 45 (320): 6 [November 18, Wednesday].

Gupta, R.D. (2010d). Preventing soil erosion in Kandi belt. The Daily Excelsior 46 (8): 6 [January 8, Friday].

Gupta, R.D. (2009) Glimpses of North West Himalayas. Radha Krishan Ananad Co. Pacca Danga Jammu (1800) J&K INdia

Gupta, R.D. (2019). Crop diversification for enhancing income. Success Story from J&K. Soil and Water Conservation Today. 14(2):3-4.

Jamwal, G.S. and Gupta, R.D. (2007). Watershed management project - A case study of Bari-Badhori, J&K state (pp. 226-231). In Natural Resource Management for Sustainable Hill Agriculture (Eds. S. Arora, S.S. Kukal and V. Sharma). Soil Conservation Society of India, Jammu Chapter SKUAST Jammu, J&K State, India.

Mahajan, A. and Gupta, R. D. (2009). Integrated nutrient management (INM) in a sustainable rice-wheat cropping system (pp.1-267). Springer Science + Business Media (www.Springer.com).

Patiram and Choudhary, B.U. (2013). Soil water resources and its conservation in Himalayan region of India (pp. 37). In Mountain Ecosystem and Man (Eds. S. Arora, B.S. Negi, S. Bhan, J.S. Bali and V.K. Bharti). Soil Conservation Society of India, New Delhi.

Reddy, S.R. (2001). Principles of Agronomy. Kalyani Publishers, Ludhiana, New Delhi, Nodia (Uttar Pradesh), Hyderabad, Chennai, Calcutta, Cuttack.

Sharma, P.D. (2004). Managing natural resources in the Indian Himalayas. J. Indian Soc. Soil Sci. 52 (4): 314-331.

Sharma, S. and Pareek, O.P. (1999). Horticulture paves the way to better days for them. Intensive Agriculture 30 (11-12): 8-13.

Sharma, S., Chander, G., Sharma, R. and Dogra, A. (2006). Jhum cultivation - Cause of concern. Agrobios Newsletter 4 (11): 11-13.

Singh, M. J. and Kukal, S.S. (2013). Extent nature and management of soil erosion in Himalayan region of India (pp. 48-65). In Mountain Ecosystem and Man (Eds. S. Arora, B.S. Negi, S. Bhan, J.S. Bali and V.K. Bharti). Soil Conservation Society of India, New Delhi.

Suraj Bjan (2019). Significant problems and alternative strategies of watershed management and development. Souvenir. A Joint International Conference of ISCO and SCSI, New Delhi, India.

17

Importance of Micro-nutrients in Sustainable Agriculture

Abstract

Soil, which is known as "stomach of plant growth", serves as a key source of food production. As a source of food production, it plays a key role in determining the sustainable productivity of an agrosystem. Of the various functions rendered by the soil to sustain its crop productivity, one is to supply essential nutrients necessary for the growth of plants and their development which also become imperative components of human and animal food. However, it is point to mention that the functions of primary (C, H, O. N, P, and K) plant nutrients cannot be performed by those of secondary (Ca, Mg and S). Similarly, the functions of micro-nutrients such as Fe, Cu, Mn, Zn, Mo, B and Cl canot be substituted by the addition of primary and secondary nutrients in soils. Hence, the soil resource must be well supplied with micro-nutrients for the full growth of plants and synthesis of human food vis-a-vis that of animals. In this chapter of the book, thus, importance of micro-nutrients in sustainable agriculture has been detailed. It deals with the causes of micro-nutrients deficiency, extent of micro-nutrients deficiency in soils, micro-nutrients management in terms of use of micro-nutrients - Zn, Fe, Mn, Cu, B and Mo through application of inorganic and organic sources for sustenance of crop production. Apart from this, chelate, chelating agent and chelating effect has also been given. In the end conclusion, suggestions and priorities for the future, study questions and references have been presented.

Keywords: Soil; plant growth; sustainable productivity; crop productivity; essential nutrients; human food; animal food; primary nutrients; secondary nutrients; micro-nutrients; inorganic sources; organic sources; chelate; chelating agent; chelating effect

It is an established fact that there is no further scope for horizontal expansion of area under plough, as the per capita land is decreasing every year due to unabated increase in human population. Hence, the future need of food has to be met through vertical expansion i.e. more intensification of agriculture. Nearly 40 million tonnes of nutrients in the form of chemical fertilizers will have to be utilized to feed upon an estimated population of 1.5 billion by 2050 AD (Sakai, 2001). However, presently the crop productivity especially rice and wheat has stagnated not only in India but in other South Asian countries also (Mahajan and Gupta, 2009). This stagnation in crop productivity has been found due to a number of reasons including deficiency of some secondary and micro-nutrients. Hence, micro-nutrients such as Zn, Fe, Mn, Cu, B and Mo have assumed increasing importance in sustainable crop production under modern agriculture. In modern agriculture i.e. Green Revolution, in fact, increase in yield occurred through intensive cultivation with high yielding crop varieties of rice and wheat, intensive use of high analysis chemical fertilizers free from micro-nutrients and use of pesticides, limited use of organic manures and restricted recycling of crop residues, which as a consequence accelerated exhaustion of micro-nutrients from soil (Dangarwala, 2001).Global scale deficiencies of Zn and B are the most widespread. It has been reported from the global study that deficiencies of Zn and B could be suspected in almost every country (Mehra et al., 2011)

It is point to mention that at many places in spite of balanced use of NPK fertilizers, the yield of crops was not found normal. These incessant efforts for increasing food grain production from shrinking land resources will further magnify the depletion of limited micro-nutrients reserves and would cause the deficiency of other micro-nutrients besides accentuating the existing ones. If these emerging problems are not timely tackled well in time, the efficiency of used NPK fertilizers in bulk and other agricultural inputs will be remarkably reduced. Hence, there is a great need to exploit feasibility of sound technology for managing the fertilizer nutrients more judiciously and effectively in balanced proportion so as to make crop production on more sustainable basis. In this chapter, therefore, the deficiency of micro-nutrients in various soils of India and their management have been described to increase agricultural production on sustained basis.

Recent studies have indicated that of the micro-nutrients essential for plants also indispensable for human beings, Zn and Fe are the most important nutrients and nearly half of the Indian soils are deficient in Zn and 15 per cent in Fe (Shukla, 2012). It has been estimated up to two-thirds of the world's population might be at risk of deficiency of one or more essential mineral elements (White and Broadley, 2009; Stein, 2010)

Micro-nutrient deficiencies have been reported in recent years (Behera and Shukla, 2013) due to intensive cropping, loss of top soil by erosion, losses of micro-nutrients through leaching, excessive liming in acid soils, less application of organic manures and growing high demanding modern crop cultivation. Deficiency of zinc, copper, manganese and iron have been reported worldwide (Anonymous, 2008) and in Indian soils over the years (Singh and Behera, 2011). Deficiencies of Zn and B are the most common in soils of Lakhimpur district of Assam (Bhuyan et al., 2014). Copper deficiency is also emerging in low scale and cause decline in crop yield.

17.1 Causes of Micro-Nutrients Deficiency

- Soils of India are generally low in fertility as these have been consistently depleted of their finite nutrient resource due to continuous cultivation since centuries ago. As a result, the country had faced food shortages and relied upon imports till the mid-1960s (Takkar, 1996). However, a break through in food production was achieved during late nineteen hundred sixties with introduction of high yielding varieties of cereals which paved the way for "Green Revolution". The Green Revolution made the country not only self-sufficient to feed the burgeoning human population but it also brought India on the list of "Exporting Countries". This could not have been realized without the application of heavy doses of chemical fertilizers. Dr. Norman E Borlough, after acquiring Nobel Peace Prize in 1970, *stressed upon a point that if the dwarf varieties of wheat evolved by him were the vehicle, the fertilizer* was *the fuel to this vehicle which produced high yields and triggered off Green Revolution in many countries. He also stated that even the high yielding varieties or miracle seeds would not have been able to create the miracles of high yield without the fertilizer use.* In the past six decades, consumption of fertilizers ($N + P_2O_5 + K_2O$) has increased tremendously from 0.07 million tonnes to 13.5 million tonnes in 1994-95 and about 18.39 million tonnes in 2004-05 Thus, fertilizers have remained undoubtedly the kingpin of the "Green Revolution".

- The increasing use of N, P_2O_5, K_2O fertilizers has, no doubt, increased the food production substantially. However, it brought about a number of problems, especially in the areas where its indiscriminate use was made. Firstly, there was a sharp decline in the available micro-nutrients with continuous cropping and use of chemical fertilizers generally devoid of these nutrients. Secondly, use of nitrogenous fertilizers alone produced negative effects, particularly in low pH (acidic) soils. Palampur (Himachal Pradesh) and Ranchi (Jharkhand) India, provide good evidence, where plots receiving N-alone produced low crop yield even when compared with those of control (Gupta and Sharma, 2006). It is worth mentioning that in *Kandi* belt of Jammu (Jammu and Kashmir), the soils of some localities under grass cover have acidic soil reaction (pH 5.2 to 6.5). The reason for development of acidity in these soils has been attributed to the continuous use of nitrogenous fertilizer alone in the form of urea which is known to create acidity (Gupta et al., 1992).
- Thirdly, extensive use of fertilizers not only contaminated the ground water but also produced eutrophication in the water bodies and accumulation of heavy metals in the underground water (Gupta, 2016).
- Fourthly, application of nitrogenous fertilizer alone without using phosphatic and potassic fertilizers renders to form some specific types of compounds in food grains which serve as carcinogenic agents. Such compounds are named as nitrosamine (Gupta and Singh, 2006).

- Before occurrence of "Green Revolution" almost all the farmers used to utilize organic manures, but after its revolution many of the farmers neglected their use. This resulted in the deficiency of micronutrients in soils on large scale.

17.2 Extent of Micro-Nutrients Deficiency in Soils

Analysis of more than 3 lakh soii samples carried out under the aegis of All India Co-ordinated Project on micro and secondary nutrients of the ICAR, showed that about 44, 33, 13, 15, 6 and 8 per cent soils are deficient in Zn, B, Mo, Fe, Mn and Cu, respectively (Bhuyan et al., 2014).

17.2.1 Zinc (Zn)

Field scale deficiency of Zn was noticed for the first time in rice crop in Tarai soils (Nene, 1968) and in wheat on sandy soils of Punjab during 1969-70 (Randhawa and Takkar, 1975). Subsequently, the deficiency of this nutrient was observed in most of the intensively cultivated areas wherever these crops were grown. Later on, deficiency of Fe in rice, sugarcane, chickpea, and groundnut was seen on sandy soils (Takkar and Nayyar, 1979; 1981). Deficiency of Mn was reported in wheat grown on sandy soils of Punjab, where rice-wheat cropping system used to be followed (Takkar and Nayyar, 1981). The deficiency of B in chickpea and rice was noted on calcareous soils of the then Bihar (Sakai, 2001).

Now among micro-nutrients, Zn deficiency has been found widespread in many of the Indian soils. This is quite evident from the analysis of 2,51,547 soil samples from different states of India showing 48, 12, 5 and 3 per cent deficient in available Zn, Fe, Mn and Cu, respectively (Takkar, 1996; Sakai, 2001). The deficiency of Zn was the highest in soils of southern India followed by northern, eastern and western parts (Table 17.1). Singh (1999) also reported more deficiency of Zn in south and north Indian soils. Soils from arid and semi-arid regions, representing Inceptisols, Entisols, Aridisols, Alfisols and Mollisols of Punjab, Haryana and Uttar Pradesh exhibited higher magnitude of Zn deficiency to the extent of 47, 61 and 45 per cent, respectively. Vertisols, Alfisols and Inceptisols from Madhya Pradesh, Tamil Nadu and Andhra Pradesh, leached soils of Meghalaya and West Bengal, Ultisols of Orissa also showed widespread deficiency of Zn in the range of 51 to 63, 36 to 50 and 54 per cent. Magnitude of Zn deficiency in alluvial soils of north Bihar ranged from 42 to 79 per cent in old alluvial soils of South Bihar from 32 to 73 per cent and in Alfisols of plateau region (now Jharkhand State), from 23 to 37 per cent (Sakai et al., 1996-97).

According to Rattan et al. (2009), as high as 49 per cent of the soil samples tested were deficient in available Zn having less than 0.6 ppm DTPA extractable

Table 17.1: Extent of micro-nutrients deficiency in soils of different states and union territories (Adapted from Takkar [1996)

State/UnionTerritory"	Zn		Cu		Fe		Mn		B		Mo	
	No. of samples	PSD""	No. of samples	PSD	No. of samples	PSD	No. of samples	PSD	No. of samples	PSD	No. of samples	PSD
1. Andhra Pradesh	6563	51	5219	<1	5219	2	5260	20	-	-	-	-
2. Assam	12165	34	12165	Nil	12165	Nil	12165	Nil	687	17	-	-'
3. Bihar	19078	54	17802	3	17802	6	18080	2	4735	39	-	-
4. Delhi	210	20	-	-	-	--	-	-	-	-	-	
5. Gujarat	29532	24	29532	4	29532	8	29532	4	2480	25	1991	10
6. Haryana	21598	61	21598	2	21598	20	21598	4	-	-	-	-
7. Jammu & Kashmir	93	12	-	-	-	--	-	-	-	-	-	
8. Karnataka	25542	78	24411	5	25411	39	25360	19	21562	32	-	-
9. Kerala	650	34	650	31	650	<1	650	-	-	-	-	-
10. (a) Madhya	12000	63	11264	Tr.	11204	3	11259	3	1699	2	927	18
Pradesh (b)	17000	23	17000	Tr.	17000	9	17000	Tr.	-	-	-	-
11. Meghalaya	95	57	95	2	65	Nil	95	23	-	-	-	-
12. Orissa	16040	54	16040	Nil	16040	Nil	16041	Nil	882	69	-	-
13. Pondicherry"	4108	8	4117	4	4093	2	4093	3	-	-	-	-
14. Punjab	15897	47	15897	<1	15897	14	15897	2	-	-	-	-
15. Rajasthan	183	21	-	-	-	--	-	-	-	-	-	
16. Tamil Nadu	20580	53	19559	3	20173	18	20311	8	-	-	-	-
17. Uttar Pradesh	25122	45	24425	<1	25004	6	24503	3	45	24	-	-
18. West Bengal	6547	36	6547	Nil	6547	Nil	6547	3	3177	68	-	-
Over all	233003	47	246987	2	228400	13	259108	4	34385	35	2918	7

Per cent deficient soil samples; a = AICRP on Micro-nutrient; b = IBFEP + HFC

Zn content. Coarse texture, high pH, calcareousness, diminishing organic carbon and leaching often accentuated the Zn deficiency (Katyal and Rattan, 1993). There is no doubt that "Green Revolution" has greatly increased the food grain production in India. But use of high analysis of NPK fertilizers in high yielding varieties of rice and wheat catalyzed the depletion of micro-nutrients, especially Zn in irrigated areas of the country (Rattan et al., 1999).

Deficiencies of Cu, Fe and Mn are of lesser magnitude than that of Zn. The deficiency of Fe in various soils was found to range from 0 to 39 per cent (average 13%). Its deficiency was mostly in coarse textured soils and those which have free Ca CO_3, high pH and low organic carbon content. Deficiency of Mn on field scale was first observed in highly permeable coarse textured soils of Punjab in wheat and berseem grown in rotation with rice (Takkar, 1996). In Bihar soils, the deficiency of Cu, Fe, Mn ranged from 0 to 16, 0 to 22 and 0 to 11 per cent, respectively.

Deficiency of B has been found from eastern part of the country, covering States of Bihar, West Bengal, Orissa, Assam and Eastern Uttar Pradesh, to the extent of 39, 68, 69, 17 and 24 per cent, respectively (Sakai, 2001). In soils of Karnataka, the deficiency of this nutrient to the extent of 32 per cent has also been reported (Takkar, 1996).

17.2.1.1 Use of Zn-Amended Organic Manures

Application of Zn alone or in combination with organic manures significantly increased the rice yield (first crop) over control (Sakai, 2001). Grain yield response ranged from 4.8 to 19.5 q ha^{-1}. Mixing of 2.5 kg Zn either with 5 t FYM/compost or 10 t FYM/compost produced the similar yield. Grain yield response obtained due to amendment of 2.5 kg Zn with 5 t FYM/compost was equivalent to 5 kg Zn ha^{-1} application. Hence, with the use of organic manure, the optimum Zn level could be reduced to about 50 per cent. Uptake of Zn progressively increased with increasing levels of Zn and organic manure. The residual effect of Zn amended organic manure was quite distinct on second crop of barley and third crop of rice. The combined application of Zn and biogas slurry was found effective in augmenting crops yield, Zn-uptake and available Zn in post-harvest soil.

Application of 12, 5, 2.5 t ha^{-1} year^{-1} of FYM, poultry and piggery manures to maize crop in maize-wheat rotation on a Zn-deficient loamy sand soil of Punjab proved as efficient as 11 kg ha^{-1} in meeting the Zn requirement of the whole cropping system (Nayyar et al., 1990). Conjoint use of organic manures and $ZnSO_4.7H_2O$ has given encouraging results For example, application of 5t FYM enriched with 5 kg Zn ha^{-1} in black soil of Madhya Pradesh produced similar quantum of rice grain yield response as 10 kg Zn ha^{-1} applied alone.

Incorporation of 50 per cent crop residues along with a starter dose of 5-10 kg Zn ha^{-1} can sustain crop productivity (Prasad et al., 2010). Among the soil factors, organic carbon and available micro-nutrients contributed to crop yield. The combined use of crop residues and zinc can enhance the nutrient uptake by crops (Prasad, 2005).

17.2.1.2 Time Dimension of Sustainability

The longevity of soil productivity was also found to be associated with the emergence of Zn-deficiency under intensive cultivation. Sustainable productivity of wheat-maize-cowpea fodder system was studied at 50, 100, 150 per cent recommended doses of NPK (Takkar, 1996). The productivity and sustainability of the maize in the system started declining significantly after 10 annual cropping cycles of the system. It was because of depletion of available Zn pool (DTPA-extractable) from adequate level of 1.1 mg kg^{-1} to deficient level of 0.68 mg kg^{-1}. The high productivity of maize, however, got restored progressively with the application of Zn. Similarly, sustainable high productivity of rice-wheat-cowpea fodder cropping system at 100 per cent recommended rate of NPK on Hapludoll of Pantnagar started declining significantly after 12 annual cycles. This was again ascribed to depletion of available Zn status from adequate 2.54 mg kg-^{1} to low level (0.97 mg kg-^{1}). The productivity of these crops of the system got restored with application of Zn.

Of the various chemical sources evaluated for their efficiency under different soil-crop situations use of $ZnSO_4$, $7H_2O$ has proved better or equal to other sources in correcting Zn deficiency in most of the soil crop situations (Takkar, 1996). However, Zn-EDTA was far better than $ZnSO_4.7H_20$ in controlling Zn deficiency in rice on a Zn deficient loamy sand soil of Punjab but its high cost make it uneconomical.

Large number of field experiments have indicated that 2.5 to 25 kg Zn ha^{-1} as $ZnSO_4.7H_2O$ or about half of its quantity as Zn-chelates proved most effective in correcting its deficiency in many of the grown crops under diverse Zn- deficient soils. Deficiency of Zn could best be alleviated with application of 11 kg Zn ha^{-1} in wheat and rice, 5.5 kg Zn ha^{-1} in maize, soybean and sugarcane, and 2.5 kg Z ha^{-1} in groundnut, raya and ragi (Takkar, 1991). However, the rate of Zn application was higher 22 kg ha^{-1} for rice in highly sodic and flood plain soils as compared to 11 kg ha^{-1} in moderately alkali soils (pH 9.4 - 9.7) and 5.5 kg ha^{-1} in case of sandy alkaline soils as reported by Takkar (1996).

Although on the Zn-deficient Ustifluvents, application of 5 kg Zn ha^{-1} has been found optimum for rice yet for the rice-wheat system as a whole 10 kg Zn ha^{-1} exhibited superiority because of its superior residual effect on the subsequent wheat crop. However, use of Zn-frits could be an excellent source provided it is broadcast and mixed uniformly in the soil (Rattan et al., 2008).

17.2.1-3 Materials for Zn Fertilization

The most common fertilizers used to supply Zn to rice and wheat crops as well as fruit plants and other crops are given below:

Material	Formula	Zn (%)
Zinc sulphate	$ZnSO_4.7H_2O$	36
Zinc oxide	ZnO	78
Zinc carbonate	$ZnCO_3$	32
Zinc sulphide	ZnS	67
Zinc chelate	Na_2Zn EDTA	14

In addition to the above sources, FYM and poultry manure also contain some Zn along with other nutrients.

17.2.2 Boron (B)

Boron has been known to be a constituent of plants since 1857. Much attention was, however, not paid to this fact till the beginning of the twentieth century. A French scientist (H. Agulhon) discovered during 1910 that boron was essential for plant life (Nijjar, 1990). New research has evinced that among micro-nutrients, B stands next to Zn with its growing deficiency in the soils of eastern and northeast hill region of India. Optimum dose of B in calcareous and heavy textured soils ranged from 1.5 to 2.0 kg ha^{-1} and in those of light textured, it varied from 1.0 to 1.5 kg ha^{-1} (Sakai, 2001).

Boron has now been universally recognized as the most important micro-nutrient for peanut production. It is essential for all plant growth and aids in the translocation of sugars and nutrients from leaves to fruit, and increases pollination and seed development. Boron is a highly mobile element that rapidly leaches from the soil and as such its deficiency symptom mostly occur in sandy soils liable to have more leaching losses. Borax application should be made every year in such type of soils where peanuts are extensively grown, because soluble Boron present in borax gets easily leached from the root zone especially in sandy soils in high rainfall regions. Application of borax @ 10 kg ha^{-1} at the time of groundnut sowing has been found useful in increasing yield of peanut (Reddy et al,, 2014).

17.2.2.1 Response of B to Various Crops

By and large, borax ($Na_2B_4O_7.10H_2O$ - 11% B) is used as a carrier of B in soil application and boric acid ($H_2B_2O_3$ - 17% B) (0.2 to 0.4%) as foliar spray. The magnitude of yield response to soil applied B ranged from 0.10 to 11.9 q ha^{-1} (Table 17.2) in cereals, 1.2 to 15.5 q ha^{-1} in pulses, 0.14 to 5.0 q ha^{-1} in oil seeds, 28.3 to 79.5 ha^{-1} in onion and 105 q ha^{-1} in sugarcane (Sakai et al., 1996). Among pulses, chickpea was found to be highly responsive to B application. Response of chickpea to B application was found more in calcareous soils. Soils extremely deficient in available B (0.15 to 0.17 mg B kg^{-1} of soil) responded even to 3 kg B ha^{-1} application.

The relative efficacy of soil application of borax @ 20 kg ha^{-1}, foliar spray of boric acid @ 0.2 per cent + lime twice and potato tuber soaking in 0.05 per cent solution for 3 hours before planting were reported to increase the tuber yield, irrespective of the method of application. In B-deficient acidic loam soil of Chhotanapur plateau (Jharkhand), application of B produced maximum pod yield of groundnut 1704 kg ha^{-1} with application of 3 kg B ha^{-1} as band placement in the form of boric acid (Singh, 1990-93). As red and lateritic soils of Jharkhand, Assam, Orissa and West Bengal were highly deficient in available B so they showed more response to added borax to rice grain and wheat grain yields (Table 17.2). The data indicated that with application of 5, 10, 15 and 20 kg borax ha^{-1} produced rice grain yield from 1.3 to 15.0 q ha^{-1} and wheat grain yield response from 1.7 to 5.7 q ha^{-1}.

Table 17.2: Response of rice and wheat grain yield (q ha^{-1}) to boron application in soils of eastern states (Adapted from [1]Dutta et al. [1992]; [2]Sahay et al. [1992]; [3]Maharana et al. [1992j; [4]Sur and Ali [1992])

Borax levels (kg ha^{-1})	Assam[1]		Old Bihar[2]		Orissa[3]		West Bengal'	
	Rice	Wheat	Rice	Wheat	*Kharif* rice	Summer rice	Rice	Wheat
Control yield (No B)	38.0	23.1	37.1	35.8	39.0	27.3	25.2	20.2
5	1.8	2.3	2.9	3.1	4.6	7.9	1.3	1.7
10	2.8	2.5	3.9	4.2	6.6	9.9	2.0	3.0
15	3.9	2.2	4.6	4.9	4.0	12.3	2.6	3.4
20	4.6	3.2	5.0	5.7	5.7	15.0	2.7	3.2

17.2.2.2 Direct and Residual Effect of B in Cropping System

The direct and residual effect of B was evaluated in calcareous soils of north Bihar under various cropping systems e.g. sesame-chickpea, rice-wheat and maize-mustard. Results showed that application of 16 kg borax ha^{-1} to sesame increased its seed yield by 112 kg ha^{-1} and that of succeeding chickpea crop by 475 kg ha^{-1}. Residual effect of 0 kg borax ha^{-1} was negligible. Application of 16 kg borax ha^{-1} to sesame produced a residual response on chickpea which was 90 per cent of the direct response of chickpea to 8 kg borax ha^{-1}. In case of rice-wheat and maize-mustard rotations, the ideal rate and frequency of application was 16 kg borax ha^{-1} to alternate crop showing its residual effect on one crop only. Maximum average grain yield increase in rice, wheat and maize was 39, 18 and 18 per cent, respectively over control in treatment receiving 16 kg borax ha^{-1} in alternate crop (Table 17.3).

Table 17.3: Effect of rate & frequency of B application on averaae grain yield (q ha^{-1}) under different cropping systems (Adapted from Sakai [2001])

Treatments (Borax ha^{-1})	Rice-Wheat[3]		Maize-Mustard[3]		Sesame-Chickpea[b]	
	Yield	Response	Yield	Response	Yield	Response
T, Control (No borax)	31.37	-	17.66	-	10.54	-
t_2 8 kg borax (1st crop only)	33.71	2.34	18.48	0.82	10.86	0.32
T, 8 kg borax (alternate crop)	36.63	5.26	19.90	2.24	11.22	0.68
t_4 8 kg borax (each crop)	38.17	6.80	21.51	3.85	12.46	1.92
t_6 16 kg borax (1st crop only)	35.16	4.19	19.10	1.44	11.90	1.36
T_e 16 kg borax (two crops interval)	38.22	6.85	21.27	3.61	12.68	2.14
T, 16 kg borax (alternate crop)	40.C5	9.28	22.27	4.61	12.68	2.14
t_8 16 kg borax (each crop)	36.82	5.45	21.15	3.49	12.36	1.82
t_9 32 kg borax (1st crop only)	35.32	3.95	19.87	2.21	11.32	0.78
T,o 32 kg borax (two crops interval)	38.19	6.82	21.26	3.60	11.66	1.12
T,, 32 kg borax (three crops interval)	37.10	5.79	21.08	3.42	12.18	1.64
T,2 0 25% BAFS (thrice)	34.18	2.81	19.69	2.03	11.66	1.12

BAFS = Boric acid foliar spray, 32 kg borax to 1st crop + "Average of 5 crops. 8 kg borax to alternate crop, 32 kg borax to 1st crop + 8 kg borax to each crcp, Average of 10 crops.

17.2.2.3 Organic Manures and B-Use Efficiency

Application of organic manure to B-deficient soils has been found to increase the utilization and recovery of applied B. Complexation of added B and coating of the surface of Fe and Al oxides by soluble organic compounds may be the possible reason for increasing B efficiency. Complexation i.e. formation of complexes become slowly available to plants and hence enhance the recovery of added B. Application cf organic manures along with liming increased the recovery of added B in all the extractable forms. The effect was more pronounced in fine than in coarse textured soils. A study regarding the effect of B and FYM on yield of crops under maize-lentil rotation showed highest yield when 16 kg borax and 5 t FYM ha^{-1} was applied in Table 17.4 (Sakai, 2001).

17.2.3 Iron (Fe)

Extent of Fe deficiency in Indian soils is found to be less than that of Zn. However, its deficiency causes a serious problem to crop production on highly calcareous soils as well as saline sodic soils. The Fe deficiency in rice soils (sandy loam and loam texture) resulted not only because of inherent low Fe

Table 17.4: Effect of boron and FYM on yield (q ha^{-1}) of crops under maize-lentil rotation (Adapted from Sakai [2001]

T.No.	Treatments level		1st Crop		2nd Crop		1rd Crop		1th Crop	
	Borax (kg ha^{-1})	FYM (t ha^{-1})	Y	R	Y	R	Y	R	Y	R
T_1	0	0	28.3	-	10.0	-	22.0	-	7.2	-
T_2	8	0	31.0	2.7	10.4	0.4	22.4	0.4	7.2	0
T_3	16	0	32.0	3.7	11.5	1.5	23.3	1.3	7.5	0.3
T_4	32	0	31.5	3.2	12.7	2.7	24.7	2.7	8.0	0.8
T_5	64	0	29.5	1.2	13.3	3.3	24.3	2.3	8.3	1.1
T_6	0	2.5	32.3	4.0	10.8	0.8	22.7	0.7	7.3	0.1
T_7	8	2.5	33.5	5.2	12.0	2.0	23.3	1.3	7.3	0.1
T_8	16	2.5	36.0	7.7	13.3	3.3	25.0	3.0	7.8	0.6
T_9	32	2.5	34.5	6.2	14.0	4.0	25.7	3.7	8.3	1.1
T_{10}	64	2.5	33.2	4.9	14.0	4.0	25.7	3.7	8.8	1.6
T_{11}	0	5.0	33.5	5.2	12.2	2.2	24.0	2.0	7.7	0.5
T_{12}	8	5.0	35.0	6.7	13.5	3.5	25.0	3.0	7.8	0.6
T_{13}	16	5.0	36.5	8.2	14.8	4.8	26.7	4.7	8.5	1.3
T_{14}	32	5.0	35.8	7.5	14.8	4.8	27.6	5.6	9.0	1.8
T_{15}	64	5.0	35.3	7.0	14.6	4.6	26.3	4.3	9.5	2.3
T_{16}	0.1% BAFS	-	29.0	0.7	11.0	1.0	22.6	0.6 .	8.0	0.8
T_{17}	0.2% BAFS	-	30.8	2.5	12.3	2.3	22.9	0.9	8.2	1.0
T_{18}	0.4% BAFS	-	33.5	5.2	12.7	2.7	24.0	2.0	8.5	1.3
	LSD (0.05)	-	2.5	-	1.3	-	1.9	-	0.7	-

Y = Yield: R = Response; 1st and 3rd Crop = Maize; 2nd and 4m Crop = Lentil; BAFS = Boric acid foliar spray content but also due to un-favourable conditions for transformation of native Fe^{2+} into Fe^{3+} form, caused by impounding of water for a longer period. Rice, particularly in seedling stage, is highly susceptible to Fe chlorosis on calcareous soil and this was successfully prevented by raising rice seedlings in puddled beds

Iron chlorosis in crops, however, can be controlled by foliar spray of ferrous sulphate ($FeSO_4$) solution on standing crop. Soil application mode was not so effective and economical as compared to foliar feeding. Foliar sprays of 1 to 2 per cent FeS04 solution at 10 days interval prevented Fe chlorosis and significantly increased the grain yield of rice, barley, black gram and chickpea (Sakai, 2001). In rice, grown on coarse textured soils of Punjab, the foliar spray of $FeSO_4$ proved superior to soil application (Nayyar et al., 1990). Gupta et al. (1994) reported that in soils of Haryana foliar spray of 1 per cent $FeSO_4$ solution on wheat significantly increased its grain yield and was more economical than soil application.

17.2.3.1 Direct Effect of Fe Carrier and Compost/Manure on Yield and Fe Uptake by Rice

Application of 50 kg $FeSO_4$ ha^{-1} alone significantly increased the grain yield of rice by 8.7 q ha^{-1} over control (Sakai, 2001) in calcareous soil of Bihar which was at par with 100 kg $FeSO_4$ ha^{-1} dose. In its effectiveness, compost alone @ 10 t ha^{-1} was akin to 50 or 100 kg $FeSO_4$ application. The magnitude of yield response due to $FeSO_4$ or pyrite was enhanced in presence of compost. This could be attributed to the production of natural chelating agents from compost that helps in keeping Fe soluble and more available to plants. Poultry manure was found one and a half times more effective than compost to increase rice grain yield. Residual effect of Fe-amended compost/poultry manure was found in black gram where pyrite proved better than $FeSO_4$.

In another experiment, application of $FeSO_4$ ha^{-1} alone increased the grain yield of rice by 3.8 q ha^{-1} over control (Sakai. 2001) and in conjunction with 10 t FYM ha^{-1} by 6.6 q ha^{-1} over control. In barley, the residual response of 100 kg $FeSO_4$ ha^{-1} alone on grain yield was 1.6 q ha^{-1} but in combination with 10 t FYM ha^{-1}, the residual effect was 4.3 q ha^{-1}.

A marked increase in water soluble and DTPA (Diethyldiaminepenta acetic acid)-extractable soil Fe over control with the addition of *Sesbania aculaeta's* and *Dalbergia sissoo's* loppings under submerged conditions has been seen by many scientists as reported by Sakai (2001). It is because of chelation of Fe and other micro-nutrient cat:ons with organic anions produced after decomposition of *Sesbania* and *Dalbergia*'s loppings (Saravana and Indirani, 2008).

According to Takkar (1996), generally Fe-chlorosis in rice is encountered in upland soils or highly permeable coarse textured soils because of less mobilization of Fe^{2}+ as the desired degree of reduction does not occur. However, the puddling and/or manuring of such soils has been shown to markedly reduce the extent of Fe-chlorosis in rice. In a three year field experiment on an alkaline Fe-deficient loamy sand soil, combination of green manure and foliar spray of 1 per cent $FeSO_4$ solution showed the largest rice grain yield followed by green manure and/or foliar sprays. It is point to mention that green manuring assisted in mobilization of internal soil iron during its decomposition.

Iron deficiency or Fe-chlorosis has also been recorded in other upland crops (like sugarcane and chickpea) grown on wide variety of soils especially on coarse textured alkaline soils of Punjab low in organic matter (Takkar, 1996). Its deficiency has also been seen in wheat, groundnut and rice on yellow brown alluvial sandy loam soils of Gujarat, in groundnut on sorghum on Alfisols of Andhra Pradesh and Tamil Nadu, in sorghum on calcifluvent soils of Bihar, and in wheat on alkaline calcareous black clay soils of Madhya Pradesh.

17.2.3.2 Carriers of Fe

Ferrous sulphate ($FeSO_4.7H_2O$ -19 to 20.5% Fe), Fe-EDTA (ferric ethylene diamine tetra acetic acid - 9 to 12% Fe), Fe- EDDHA (ferric ethylene diamine di-[D-hydroxyphenyl acetate] - 10.0% Fe), ferrous oxide (FeO -23% Fe), ferric oxide (69% Fe), FYM (0.15% Fe), poultry and piggery manure (0.16% Fe) have been used as source of Fe to correct its deficiency in crops. However, out of these, $FeSO_4.7H_2O$, Fe-EDTA and FYM are most commonly used. Chemical sources such as pyrite (FeS_2), goethite (FeOOH), magnetite (Fe_3O_4), olivine (Fe_2SiO_4), limonite ($Fe_2O_3.3H_2O$) have proved inferior to $FeSO_4.7H_2O$ in increasing rice yield and Fe uptake in Vertisols.

17.2.3.3 Rate of Fe Application

By and large, foliar application of 10-12 kg ferrous sulphate ha^{-1} or soil application of 50-150 kg ha^{-1} ferrous sulphate alleviated Fe deficiency in most of the crops depending upon the crop, type of soil, deficiency status and environment. As the rates of soil application of Fe were high as compared to foliar application so these were uneconomical. Similarly, high cost of organic carriers (Fe-chelates) discourages farmers to use such carriers.

17.2.4 Manganese (Mn)

Like Fe, Mn also exists in soils in divalent and trivalent forms i.e. Mn^2+ and Mn^{3+}. It also occurs in tetravalent (Mn^{4+}) form. However, tetravalent Mn^4+ exists in soils as a very inert oxide.

Deficiency of Mn is not a major problem (Takkar, 1996; Sakai, 2001; Sharma et al., 2003) as has been expressed by soil and plant analysis. However, the deficiency of Mn in wheat crop has been emerged since 1979 in Punjab as a result of intensive cultivation of rice-wheat cropping system in place of maize-wheat or groundnut-wheat cropping pattern for a number of years on highly permeable coarse textured alkaline soils (Takkar, 1996). Manganese deficiency in Bihar soils has also been found but sporadically. On such soils, application of manganese sulphate ($MnSO4.H_20$) containing 32 per cent Mn has proved 1.5 and 10 times more effective than Mn-fruits and manganese oxide (MnO), respectively in increasing grain yield of wheat. Sakai (2001) reported that application of Mn @ 5 kg ha^{-1} produced wheat grain yield response of 5 q ha^{-1}. On an average, response of rice and wheat to Mn application was 4.7 and 4.1 q ha^{-1}, respectively (Sakai et al., 1996).

17.2.4.1 Efficient and Economical Use of Mn Application

Although both soil and foliar application of Mn caused a significant and marked increase in the production of wheat yet the rates of soil applied Mn (40-50 kg ha^{-1}) are found uneconomical as compared to foliar application of 0.5 to 1.0 per cent $MnSO_4$ solution (7.5 to 15 kg Mn ha^{-1}). This foliar spray was done before the first irrigation of wheat crop (Takkar, 1996).

With a view to ameliorate the Mn use efficiency, the number and time of foliar sprays of $MnSO_4$ solution were further evaluated. Results suggested that Mn deficient soils, one foliar spray of $MnSO_4$ solution to wheat before first irrigation protects the crop from latent deficiency and helps to attain the maximum yield. Four sprays of 0.5 per cent $MnSO_4$ solution (one before and three after the first irrigation) help to correct Mn deficiency.

Foliar application of $MnSO_4$ solution in the form of its three sprays (0.5% $MnSO_4.3H_2O$ solution i.e. 1 kg of $MnSO_4.3H_2O$ in 200 litres of water) is found far superior to soil application of 60 kg ha^{-1} of $MnSO_4$ for correcting Mn deficiency in wheat crop grown under Punjab soil conditions (Khurana et al., 2005).

17.2.4.2 Manganese Efficient Crop Cultivars

A large number of wheat cultivars have been screened out for their relative tolerance to Mn stress under field conditions (Takkar et al., 1989; Sakai, 2001).

Although all the wheat cultivars responded to foliar spray of $MnSO_4$ solution yet the magnitude of response varied widely among the cultivar. Generally, *Durum* cultivars of wheat were found more sensitive to Mn deficiency as compared to *Aestivum* cultivars. The cultivar HD 2329 (Mn-efficient) has been adopted on a large scale by the farmers on the Mn deficient soils of Punjab. Various sources of Mn are listed below:

Source	Chemical formula	Mn (%)
Manganese sulphate	$MnSO_4.3H_2O$	26-28
Manganese oxide	MnO	41-68
Manganese chelate	Mn-EDTA	12
Manganese dioxide	MnO_2	63
Manganese frits	-	10-15

Manganese sulphate is the most widely used source of Mn. It is water soluble and can be used in soil and foliar spray. Manganese oxide is slightly soluble and thus is not recommended for foliar spray.

Mn chelates, by and large, perform poorly as soil application. In some cases, they aggravate Mn deficiency rather than ameliorating. It is attributed to substitution of Fe for Mn in chelate source, thereby enhancing Fe availability. Mn chelates can be used for foliar spray. Fe availability in some soils interfere Mn uptake by plants.

17.2.5 Copper (Cu)

Although copper has been used for its beneficial effect on the fruit trees for a very long time yet its deficiency in cereal and other crops has been seen in few locations only. As early as Sharma and Deb (1974) reported that alluvial soils of Delhi under rice-wheat crop rotation contained adequate Cu to sustain high yields, while most of the soils supporting bajra-wheat were found to be deficient in available Cu. Sakai et al. (1996) recorded mean yield response of rice, wheat and onion to the extent of 4.4, 3.9 and 49.7 q ha^{-1}, respectively in alluvial soils of Bihar. Copper sulphate ($CuSO_4.5H_2O$) is the most common carriers used as soil and foliar spray. Soil application of Cu @ 5-10 kg ha^{-1} was found to be the most effective than foliar spray (0.25% $CuSO_4$ solution). Soil application of 10 kg $CuSO_4$ ha^{-1} has been found superior to foliar spray of 0.2 per cent $CuSO_4$ solution in increasing the soybean and wheat yields in acid hill soils of Uttar Pradesh now Uttarakhand as reported by Sakai (2001).

There are several other materials available which can be applied to the soils to overcome Cu deficiency both in crops and fruits. These are cuprous oxide (CuO_2 - 89% Cu), cupric oxide (CuO - 75% Cu), chalcopyrite ($CuFeS_2$ - 35% Cu), chalcosite (Cu_2S - 80% Cu), copper chelate (Na_2Cu-EDTA - 9 to 13% Cu)

17.2.6 Molybdenum (Mo)

Molybdenum is required by plants in very small amounts. Although its essentiality for plants was established in 1939 by Arnon and Stout (Evans, 1956) yet its role was known to be highly beneficial in the atmospheric N≡N fixation by symbiotic and non-symbiotic bacteria (Purvis, 1955). It is because Mo has been found to be a constituent of two major plant enzymes, nitrogenase and reductase. The nitrogenase enzyme contains both Fe and Mo in a ratio of about 9:1.

The role of Mo-containing enzyme, nitrate reductase in N metabolism is well understood. The research work on the assimilation of NO_3 in the plant system had shown quite early that there was an accumulation of NO_3 in the tissues of Mo-deficient plants. It was later on established that Mo is a constituent of nitrate reductase, the enzyme which catalyses the reduction of NO_3 to NO_2 in the N-assimilation and finally to amino acids and proteins.

Molybdenum is taken up by the plants in the form of molybdate (MoO_4) anion. Generally Mo content of most soils ranges from 1-10 ppm but some soils may have up to 20 or 30 ppm or more Mo. In some soils of India (Gujarat and Saurashtra) the total Mo content varied from 0.5 to 4.1 ppm having average to the tune of 1.81 ppm. The available Mo content in these sols, on the other hand, ranged from 0.012 to 0.26 ppm, most of the soils analysed being above the deficiency limit (0.05 ppm or less) and well below the toxic level (10 ppm or more). Soils of Madhya Pradesh, Uttar Pradesh, Punjab and Haryana also did not show any deficiency of Mo as reported by Nijjar (1990).

In spite of not so much deficiency occurred in Indian soils, the optimum level of Mo as soil application for some field crops appeared to be at 0.4 kg Mo ha^{-1} which produced the grain yield response in maize, groundnut and soybean to the tune of 10.2, 6.2 and 4.0 q ha^{-1}, respectively (Sakai et al., 1996). In a black sandy loam soil of Coimbatore, soil application of 2 kg sodium molybdate ha^{-1} was found superior to its foliar spray of 0.05, 0.10 and 0.20 per cent solution (Takkar et al., 1989). Singh et al. (1979) reported the grain yield response of 130 to 880 kg ha^{-1} in rice and 20 to 680 kg ha^{-1} in wheat to Mo application. Seed treatment of soybean with sodium molybdate produced yield response ranging between 2 and 4 q ha^{-1} in Alfisols of Himachal Pradesh (Sharma and Minhas, 1986).

17.3 Micro-Nutrients Management

As there is no scope to increase the net cultivated area (horizontal expansion of production) which has now stabilized at about 143 m ha with decreased land to man ratio, so the future requirement of food grain production would be through intensive agriculture (vertical growth). Nearly 40 million tonnes fertilizer nutrients are to be required for producing 380-400 million tonnes of food grains to feed an estimated population of 1.5 billions by 2050 AD (Sakai, 2001). Following of intensive cultivation would cause more deficiency of micro-nutrients. It is point to mention that stagnation in crop productivity had already been found in many parts of India. Of the many factors involved in stagnating the crop productivity, deficiency of micro-nutrients vis-a-vis secondary nutrients, is one of them. Hence, nowadays, use of micro-nutrients (Zn, Fe, Mn, Cu, B & Mo) has assumed of great importance to sustain the yield of the crops.

Micro-nutrient requirement for lowland/upland rice soils and their management practices given elsewhere (Gupta et al., 1997) indicated deficiency of at least one micro-nutrients i.e. Zn not only in India but has also been observed in other Asian countries. Besides, under upland soil conditions the deficiency of Fe has also been noticed in rice growing soils especially on Latosols due to increased uptake of Mn (Fe/Mn antagonistic effect).

Appearance of iron-rust coloured brown spots on the leaves after two or three weeks of transplanting which eventually dried up within a month. This- disorder due to Zn deficiency is known as *Khaira* dicease. Sometimes, there is bleaching of midribs while leaf tip remains green. Moreover, the plants show stunted growth i.e. they remain dwarf in size.

The symptoms of Fe-deficiency are initially a typical intervenal chlorosis of the youngest leaf develops and later on veins also become yellow. In case of severe deficiency, chlorosis of large veins followed by the loss of green color in the finer veins results in. Hence, the entire leaf and even the emerging leaves lose their green color or become bleached.

The symptoms of Mn-deficiency appear soon after the third leaf unrolls to expand symptoms appear as chlorotic mottling which later on coalesce or form large irregular areas leading to chlorotic stripings (Jaggi et al., 2001b). At 5-6 week growth, the margin of the affected leaves wither and dry.

17.3.1 Role of Organics in Micro-Nutrient Management

The use of organics helps in utilization of native micro-nutrients effectively. Apart from this, organics after decomposition also release micro-nutrients. Thus, application of organic manures like FYM, compost including vermicompost, green manure as well as use of oil seed cakes and humates, fulvates etc., increase the utilization efficiency of micro-nutrients. The beneficial effect of these organics is attributed to the formation of chelating agents, which form soluble complex with metallic micro-nutrients and increase carrying capacity of the soil solution. The lowering down of pH due to organic acids production during decomposition of organics by accelerated microbial activity and increase in cation exchange capacity leading to, thereby, greater adsorption of the nutrients. As a result, the concentration of micro-nutrients is enhanced in the soil solution which eventually increases their uptake in the growing plants and help in sustaining crop productivity.

Zinc fulvate (Prasad et al., 1995) was found superior of Zn to $ZnSO_4$. It was also better than zinc humate for rice crop (Muralidharudu, 1991). These findings provide insight into the beneficial effects of organics. Materials such as distiller spent wash, spent slurry and press mud, which are acidic greatly help in micronutrient management (Dangarwala, 2001) especially in cal careous soils.

As availability of organics is limited, hence, unconventional sources like coir dust, lignocelluloses, coir pith and other lignin rich organic wastes have been exploited to supply micro-nutrients in soils (Dangarwala, 2001). In a study conducted at Bangalore, coir dust compost enriched with micro-nutrients along with 50 per cent NPK gave the highest yield and increased the uptake of metallic nutrients (Kadalli et al., 2001). Depletion of micro-nutrients in intensive cropping system like rice-wheat takes place quickly in two cycles, particularly when high N levels are applied without organics. But when one-third of 180 kg N ha^{-1} was substituted with equivalent FYM or biogas slurry, micro-nutrient content was made up and sustained the crop production.

Bio-fertilizers along with organics further increase use efficiency of Zn fertilization. Application of Zn with organics and *Azospirillum* increased the turmeric rhizome yield in calcareous soil of Tamil Nadu (Krishnasamy and Associates 1994). Inoculation of *Rhizobium* culture increased the use efficiency of Mo as reported by Dangarwala (2001). The beneficial interaction effect on nodules per plant, fresh and dry weight of nodules and N content in grains of chickpea was also seen.

17.4 Chelate and Chelating Effect

Chelate is a term which has been derived from Greek word 'claw'. As the name indicates, the chelates have the tendency to hold certain cation tightly which are attracted to them. Actually, they tend to protect these cations, often to the benefit of growing plants.

A chelate is an organic compound which combines with and protects certain metallic cations like iron, manganese, zinc and copper. The cation-chelate combinations make complex ring structure; the metals so bound lose their usual ionic characteristics and are less apt to take part in reactions with other soil constituents, which may be advantageous. In other words, when a ring structure is formed, complex formation becomes more stable. This is known as Chelating Effect. This chelating effect may be due to the decrease in entropy as there is lesser strain on the molecule due to the formation of ring structure.

Although chelated-metals or metal complex formations are protected against soil reactions and sometimes become less available to the plants at certain pH, at which the soils are good enough to a crop. Nevertheless, it is only through chelation that the metal ions especially micro-nutrients are solubilized and assimilated fairly readily by growing plants. Thus, so long as the nutrients remain in these combinations, they are considered as being in the available form.

The following diagram can be used to explain the following points:

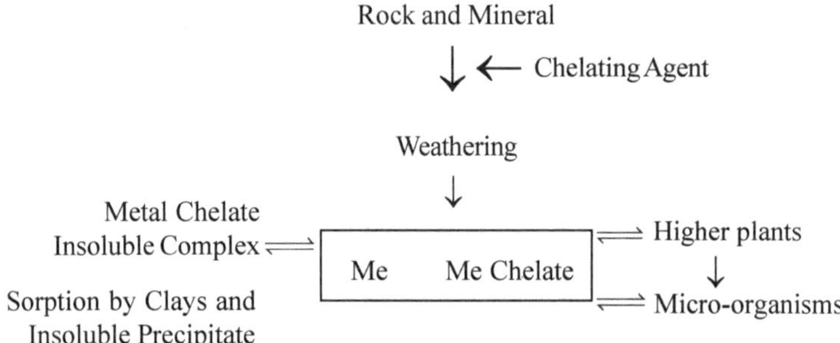

Naturally occurring organic complexes or chelating agents have been known since long. Chelation or complexing by organic molecules, particularly of Al or Fe, followed by leaching or translocation to lower position in the profile, is essential for the 'podzolization' soil forming process to occur. This phenomenon is also known as chevulation.

Since metal chelate become more stable and stability constant refers to a system in equilibrium. These, therefore, may be used to calculate the free energy ΔG and other thermodynamic functions of reactions involved by means of the following fundamental equation.

$\Delta G = - RT \log K = \Delta H - T\Delta S$

Stability constant can be explained by quoting following example:

$M + nL \rightleftarrows \Delta MnL$

Where, M = metal, n = number of lignads and L = Lignad.

By law of mass action:

$Bn^{\circ} = [MnL] / [M] [L]^{n}$

Whereas, Bn° = Thermodynamics stability constant of metal ion complex.

When we determine the negative log of Bn°, this gives the stability index of metal ion complex.

17.5 Chelating Agents

Natural chelating agents are of two types:

17.5.1 Non Humic Substances

The number of bio-organic compounds is included in this group are:

- Organic acids (citric acid, tartaric acid, malic acid, lactic acid, aromatic and dicarboxylic acids)
- Polyphenols
- Amino-acids
- Proteins
- Peptides
- Polysaccharides.

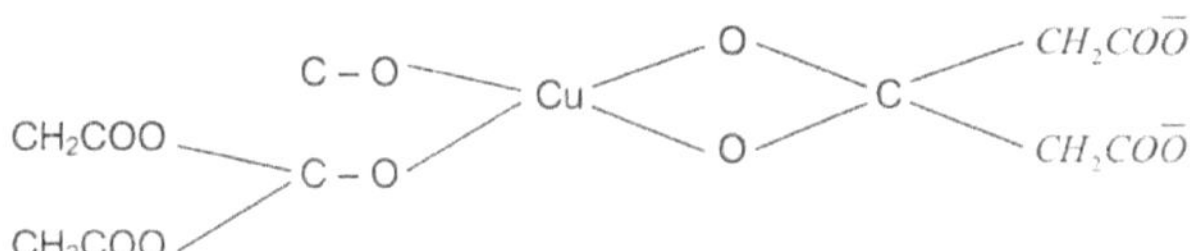

Citric Acid Chelate with Cu

O
||
O – C
NN_2
CH_2
Cu
CH – R
C – O
NH_2
||
O

Amino Acid Chelate with Cu

O = C – O
O – C = O
M
H – C – OH
OH – C – H
$(CHOH)_3$
$(CHOH)_3$
CH_2O
CH_2OH

Gluconic Acid Metal Complex

17.5.2 Humic Substances

The functional groups such as COOH, OH (alcoholic and phenolic), enolic and carbonyl present in HA (Humic Acid) and FA (Fulvic Acid), are responsible for causing stable chelate formation. The various sites of HAs and FAs which can go to form complex formation with micro-nutrient cations are:

Cu^{2+} reacts with HA in the Following Manner

COO⁻ + Cu^{2+} —(At Low pH)→ COO, O — Cu $(H_2O)_x$ +

At High pH

COO, O — Cu(OH) $[H_2O]^{+H}_{x-1}$

In case of FA

Cu^{2+} + 2 (COO, OH) —(pH 5.0)→ [COO, O — Cu — COO, O]

OH, COOH, COOH, OH, NH_2, OH, OH, O, OH, N, H-IMINO

Humic substances have the capacity to interact with metal ions, metal oxides, metal hydroxides and more complex minerals to form metal-organic associations. These metal-organic associations then differ widely in their chemical and biological stabilities and characteristics. The capacity of the humic substances to form metal-organic associations or complex formations is due to large numbers of oxygen containing functional groups (COOH, OH, C=O) which can attack and degrade soil minerals by complexing and dissolving metals which are ultimately transport within soils and waters. Not only this, organic matter is also found to be present as clay metal organic matter complexes. Interactions between humic substances and metal ions have been described as ion exchange, surface adsorption, chelation, coagulation and peptization reactions.

17.6 Conclusion

Among various micro-nutrients, the deficiency of Zn has been noticed widespread in many Indian soils followed by B especially in the eastern states of the country. The deficiencies of Fe and Mn have been confined mainly to the highly permeable light textured, calcareous and saline-sodic soils of the country. Contrary to this, the deficiency of Cu is sporadic and is found in certain local areas. Crops grown under such soil conditions, their yield becomes low. In rice and wheat cropping sequence, even application of NPK fertilizers has not proved worth in enhancing their yields. However, annual application of 5 kg Zn ha^{-1} in Zn deficient soils and 1.5 kg B ha^{-1} to alternate crop in B deficient soils has emerged effective in augmenting yield of crops. Severe Fe chlorosis in rice seedlings under calcareous soil conditions in puddle beds with nominal dose of FYM/compost followed by compounding water in beds check this disorder. Foliar spray of 1 to 2 of $FeSO_4$ solutions proved superior to soil application. The efficiency of micro-nutrients can be increased when applied in conjunction with organic manures. Recycling of crop residues in rice-wheat cropping pattern was found effective in sustaining the productivity of the system by way of increasing micro-nutrients use efficiency and their availability from the soils. In long term experiment under rice-wheat-sorghum cropping system, the yields of these crops showed a declining trend over time. But application of 10 kg Zn + 10 t FYM ha^{-1} at yearly interval resisted the declining trend with a consequence increase in yield. In wheat crop, Mn deficiency was controlled by foliar spray of 0.5 per cent solution of Manganese sulphate ($MnSO_4.H_2O$).

17.7 Future Research Strategies/Priorities

- As the magnitude of response to micro-nutrient application varied widely from crop to crop, varieties within a crop and on different soils for the same crop, so periodical monitoring of micro-nutrient status of soils and plants is an essential component to direct micro-nutrient problem well in time prior to appearing their visible symptoms to recommend remedial measures.
- Development and refinement of micro-nutrient of soil and plant test methodologies is required to be carried out. It will help in detecting heavy metal toxicities.
- Identification and detailed mapping of micronutrient deficient and toxic areas up to soil series/family level is the need of the day for taking strategic remedial measures for achieving high productivity and better soil health.
- Role of micro-nutrients in improving food quality vis-a-vis human health should be conducted on priority.

- For minimizing environmental pollution, rational agricultural use of city and industrial waste requires more emphasis.
- Studies on the dynamics of micro-nutrient transformation, availability and assessment, characterizing chemical equilibria, organic and inorganic solid phases, metal organic complexes, movement in soil and soil root interface, modeling to assist management decision in major benchmark soils, are required to be done.
- Long-term studies involving organics need to be taken-up under different soil types and cropping patterns.
- As crganics in INM mode help in utilization of native micro-nutrients effectively as well as their release through decomposition of organics so INM system requires to be advocated among the farming communities.
- There is an urgent need to enhance facilities required for the analysis of available micro-nutrients in soil from farmer's fields.

Study Questions

1. What is the ultimate goal of sustainable agriculture?
2. Name the various micro-nutrients essential for the plant growth.
3. Which type of soils show more deficiency of Zn?
4. Pick up out of the following micro-nutrients which are more deficient in soils - Zn, Fe, Mn, Cu, Mo and B.
5. Name various sources which can meet the micro-nutrients need of crops.
6. What is the optimum rate of zinc application through zinc sulphate in rice and wheat cropping system?
7. Name one micronutrient which is deficient in Indian soils after zinc.
8. What is chelate and chelating agent?

References

Alloway, B.J. (2008). Micro-nutrient deficiencies in global Crop (pp. 1-539). Crop Production. Springer, New York.

Behera, S.K. and Shukla, A.K. (2013). Depth-wise distribution of zinc, copper, manganese and iron in acid soils of India and their relationship with some soil properties. J. Indian Soc. Soil Sci. 61 (3): 244-252.

Bhuyan, N., Barua, N.G., Borah, D.K., Bhattacharyya, D. and Basumatri, A. (2014). Geo-referenced micro-nutrient status in soils of Lakhimpur district Assam. J. Indian Soc. Soil Sci. 62 (2): 102-107.

Dangarwala, R.T. (2001). Need for sustaining balanced supply of micro-nutrients in soil rather than their correction. J. Indian Soc. Soil Sci. 49 (4): 647-652.

Dutta, H., Rahman, S., Kanjilal, A. and Ali, M.H. (1992). Response of major crops to zinc and boron application in old alluvial soils of Assam (pp.180-192). In Proceedings of the workshop on micro-nutrients (January 22-23, 1992), Indo-British Fertilizer Education Project HFC Ltd., Bhubaneswar (Orissa), India

Evans, H.J. (1956). Role of molybdenum in plant nutrition. Soil Sci. 81 (3): 199-208.

Gupta, R.D. and Sharma, S.P. (2006). Soil pollution and its solution through integrated nutrient management (pp. 32-38). In Balanced Fertilization for Sustainable crop Production (Eds. S.P. Sharma, S.K. Sharma, R.P. Sharma and T.S. Verma). Proceedings National Seminar on INM for Balanced Fertilization held on November 17-18, 2006.

Gupta, R.D. and Singh, H. (2006). Indiscriminate use of fertilizer poses health hazards. Farmers' Forum 6 (9): 20-24.

Gupta, R.D., Gupta, J.P. and Singh, H. (1992). Problems of agriculture in Kashmir Himalayas (pp. 53-67). In Conserving Indian Environment (Ed. S.K. Chadha). Pointer Publishers, Jaipur (Rajasthan), India.

Gupta, R.D., Sharma, A.K. and Bhagat, G.R. (1997). Micro-nutrients management for submerged (lowland) and upland rice soils. Popular Science 5 (1): 26-29.

Gupta, R.D. 2016. Wildlife and Wetland Ecosystems of Jammu and Kashmir, Himalayas. N.R. Bookds International. Pacca Danga, Jammu, J&K State, India.

Gupta, V.K:, Gupta, S.P., Ram Kala, Potalia, B.S. and Kaushik, R.D. (1994). Twenty-five years of micro-nutrient research in soils and crops of Haryana. Technical Bull., Department of Soil Science, CCSHAU, Hisar (Haryana), India.

Jackson, M. (1968). Chemical composition of soils. In Chemistry of the Soil, Chapter 2 (Ed. F.E. Bear). Oxford and IBH Publishing Co., New Delhi.

Jaggi, R.C., Mahajan, A., Choudhary, A.K. and Dogra, R. K. (2001b). Plant nutrient deficiency symptoms are partly dependant. Indian Farmers' Digest 3 (49): 9-11.

Kadalli. G.G., Devi, S., Saddaramappa, R. and Patil, C.R. (2001). Quality and efficiency of value added coir dust based compost. J. Indian Soc. Soil Sci. 49 (1): 141-144.

Katyal, J.C. and Rattan, R.K. (1993). Distribution of zinc in Indian soils. Fert. News 38 (3): 15-26.

Khurana, M.P.S., Baddesha, H.3. and Gill, M.S. (2005). Management of manganese deficiency in wheat. Indian Farmer's Digest 38 (12): 31-32.

Krishnasamy, R. and Associates. (1994). Twenty-five Years of Micro-nutrients Research in Soils and Crops of Tamil Nadu. Technical Bull., Department of Soil Science and Agricultural Chemistry, TNAU, Coimbatore, pp. 192.

Mahajan, A. and Gupta, R. D. (2009). Integrated nutrient management (INM) in a sustainable rice-wheat cropping system (pp.1-267) Springer Science + Business Media (www. Springer.com).

Maharana, D.P., Sarengi, S.K., Singh, R.N.B. and Ali, M.H. (1992). Response of rice to application of zinc sulphate in different broad soil groups of Orissa (pp. 254-262). In Proceedings of the workshop on micro-nutrients (January 22-23, 1992), Indo-British Fertilizer Education Project HFC Ltd., Bhubaneswar (Orissa), India

Mehra, M., Kushwaha, H.S. and Kumar, A. (2011). Current status of micro-nutrients in soil and their deficiency symptoms in plants. Indian Farmer's Digest 44(1): 8-16.

Muralidharudu, Y. (1991). Studies on availability of Zn applied through chelated form in rice. J. Indian Soc. Soil Sci. 39 (1): 114-117.

Nayyar, V.K., Takkar, P.N., Bansal, R.L., Singh, S.P., Kaur, N.P. and Sadana, U.S. (1990). Micro-nutrients in soils and crops of Punjab. Res. Bull., Department of Soils, PAU, Ludhiana (Punjab), India.

Nene, Y.L. (1968). Khaira disease of paddy and its control. IRRI Newsletter 20 (1): 1-7.

Nijjar, G.S. (1990). Nutrition of Fruit Trees (pp. 273-284). Kalyani Publishers, Ludhiana, New Delhi (India).

Prasad, B,, Sarangthem, I. and Choudhary, K.C. (1995). Transformation and availability of applied zinc to maize in calcareous soil. J. Indian Soc. Soil Sci. 43 (1): 84-89.

Prasad, R.K. (2005). Studies on microbial activities and micro-nutrients availability in a long-term experiment on crop residues and zinc application under rice-wheat system. M.Sc. Thesis, R.A.U., Pusa (Bihar), India.

Prasad, R.K., Kumar, V., Prasad, B. and Singh, A.P. (2010). Long-term effect of crop residues and zinc fertilizer on crop yield, nutrient uptake and fertility build-up under rice-wheat cropping system in calciorthents. J. Indian Soc. Soil Sci 58 (2): 205-211.

Purvis, E.R. (1955). The role of molybdenum in soils and plants. J. Agri. Food Chem. 3 (8): 666-669.

Randhawa, N.S. and Takkar, P.N. (1975). Micro-nutrient research in India. Fert. News 20 (5): 11-19.

Rattan, R.K., Datta, S.P. and Katyal, J.C. (2008). Micro-nutrient management research achievements and future challenges, Indian J Fert. 4 (12): 93-100, 103-106. 109-112, 115-118.

Rattan, R.K., Patel, K.P., Manjaiah, K.M. and Datta, S.P. (2009). Micro-nutrients in soil plant, animal and human health. J. Indian Soc. Soil Sci 57 (4): 546-558.

Rattan, R.K., Saharan, N. and Datta. S.P. (1999). Micro-nutrients depletion in Indian soils - Extent, causes and remedies. Fert. News 42 (12): 75-89.

Reddy, G.K., Reddy, P.M., Reddy, P.K. and Prathima, T. (2014). Boron deficiency leads to lowest shelling percentage and pod yield in peanut. Indian Farmer's Digest 47 (2): 32-33.

Sahay, R.N., Ghosh, T.K., Verma, S.K. and Ali, M.H. (1992). Crop response to micro-nutrient application in farmers' field in project areas in Bihar (pp. 193-198). In Proceedings of workshop on micro-nutrients (January 22-23, 1992), Indo-British Fertilizer Education Project HFC Ltd., Bhubaneswar (Orissa), India

Sakai, R. (2001). Efficient management of micro-nutrients for sustainable crop production. J. Indian Soc. Soil Sci. 49 (4): 593-608.

Sakai, R., Singh, A.F., Singh, R.B. and Bhogal, N.S. (1996-97). Annual Progress Report AICRP on Micro and Secondary Nutrients and Pollutant Elements in Soils and Plants. Department of Soil Science, RAU (Pusa), Bihar.

Sakai, R., Singh, A.P., Sinha, R.B. and Bhogal, N.S. (1996). Twenty-five years of research on micro and secondary nutrients in soils and crops of Bihar. Res. Bull., Department of Soil Science, RAU, Pusa, Samastipur (Bihar), India.

Saravana, P. and Indirani, R. (2008). Role of organo-metallic complexes on plant nutrient. Agrobios Newsletter 7 (9): 46-47.

Snarma, B.M. and Deb, D.L. (1974). Copper status of soils of the Union Territory of Delhi with special reference to cropping sequence. J. Indian Soc. Soil Sci. 22 (2): 145-150.

Sharma. C.M. and Minhas, R.S. (1986). Effect of molybdenum application on the yield and uptake by soybean grain in an alfisol. J. Indian Soc. Soil Sci. 34 (2): 314-316.

Sharma. R.P., Singh, M. and Sharma, J.P. (2003). Correlation studies on micro-nutrients vis-a-vis soil properties in some soils of Nagaur district in semi-arid region of Rajasthan. J. Indian Soc. Soil Sci. 51 (4): 522-527.

Shukla, A.K. (2012). Micro-nutrients beyond soil-plant systems. J. Indian Soc Soil Sci. 60 (12): 549-559 (Supplement).

Singh, D., Leelavathi, C.R., Krishnan, K.S. and Sarup, S. (1979). Monograph on Crop Responses to Micro-nutrients (pp 1-136). I.A.S.R.I., New Delhi (India).

Singh, H.P. (1990-93). Consolidated Report of the ICAR Adhoc Scheme on Status and Response of Crops to Micro-secondary Nutrients with Special Emphasis on B, Mo and S in Soils of Chhotanagpur and Santhal Paraganas of Bihar. BAU, Ranchi (Jharkhand), India.

Singh. M.V. (1999). Micro-nutrient deficiency delineations and soil fertility mapping. In Proceedings of National Symposium on Zinc Fertilizer Industry - Whether to Lucknow (Eds. R. Singh and A. Kumar) on April 24, 1999, Lucknow (Uttar Pradesh), India. Zinc Sulphate Manufactures Association.

Singh, M.V. and Behera. S K. (2011). All India Co-ordinated Research Project of micro and secondary nutrients and pollutants elements in soils and plants - A Profile Research. Bulletin No. 10 IISS, Bhopal, pp. 1-57.

Stein, A.J. (2010). Global impacts of mineral malnutrition. Plant and Soil 335 (1-2): 133-154.

Sur, S.C. and Ali, M.H. (1992). Response of kharif rice, wheat and mustard to soil application of zinc sulphate and borax in farmers' fields of project area in West Bengal (pp 273-279). In Proceedings of workshop on micro-nutrients (January 22-23, 1992), Indo-British Fertilizer Education Project HFC Ltd., Bhubaneswar (Orissa), India

Takkar P.N. (1991).). Zinc deficiency in Indian soils and crops (pp. 55-66). In Zinc in Crop Nutrient. International Lead Zinc Research Organization. Inc. and Indian Lead Zinc Information Centre's, New Delhi (India).

Takkar, P.N. (1996). Micro-nutrient research and sustainable agricultural productivity in India J. Indian Soc. Soil Sci. 44 (4): 563-581.

Takkar, P.N. and Nayyar, V.K. (1979). Iron deficiency affects rice yield in Punjab. Indian Farming 29 (9): 9-12.

Takkar, P.N. and Nayyar, V.K. (1981). Preliminary field observation of manganese deficiency in wheat and berseem. Fert. News 26 (2): 22.

Takkar, P.N., Chibba, I.M. and Mehta, S.K. (1989). Twenty Years of Co-ordinated Research on Micro-nutrients in Soils and Plants. Indian Institute of Soil Science, Bhopal (Madhya Pradesh), india.

White, P.J. and Broadley, M. (2009). Bio-fortification of crops with seven mineral elements often lacking in human diet - iron, zinc, copper, calcium, magnesium, selenium and iodine. Nevj Phytologist, 132 (1): 49-84.

18

Agriculture Production on Sustainable Basis Through Extension System Approach

Abstract

Spreading or communicating latest agricultural technology based on scientific principles is utmost important not only to increase agricultural production but it also becomes imperative to sustain crop production. Agriculture Extension is one such principle or area through which the clients (peasants) can be persuaded and motivated for adopting the latest modern agricultural technology. There are a number of Agriculture Extension Strategies to be communicated to the peasants for increasing and sustaining the agricultural production. "Extension System Approach" is one such strategy which stands discussed in this chapter of the book. What is meant by system approach? - steps involved in training strategy, guidelines for trainings and their importance, points or tips to make training programme effective, training needs and knowledge required for farmers for sustainable agriculture - are the main areas of "Extension System Approach", which are described in details. Apart from this, conclusion and suggestions and priorities for the future are also mentioned.

Keywords: Agricultural technology; agricultural production; sustainable agriculture; Agriculture Extension; Extension System Approach; training strategy; training needs

"India lives in villages and if the villages perish, India will perish too"

- Mahatma Gandhi

Agriculture is the lifeline of India that provides livelihood to about 70 per cent of its population dwelling in about 6 lakh villages. In spite of so many projects, programmes and schemes launched from time to time after independence by Government of India through Ministry of Agriculture and Indian Council of Agricultural Research, State Agricultural Universities and Departments of Agriculture, agricultural food grains production was not realized on "sustainable basis". Even "Green Revolution" which came to fore during 1967-68 in the irrigated areas alone, could not sustain the productivity of rice and wheat. Rather, the productivity of these crops has either got declined in many cases and/or in other cases it has stagnated. This is attributed owing to inadequacies like no well defined goals, frequent shifts in emphasis and approach, sporadic attempts and unplanned extension strategies etc., which has resulted in fluctuations or elasticity in food grain production over the years. The irony of fate is that the increase in food grain production is not keeping pace with the demand of our burgeoning population which is increasing in geometrical progression. This therefore warrants an execution of a well planned and integrated extension approach known as "System Approach" to enhance agricultural production and productivity on sustainable basis (Tripathi, 2000; Gupta, 2004a). In doing so, appropriate strategies with various extension techniques/ methods have to be used with different clients to persuade and motivate them for adopting modern technology, which has been highlighted here as under:

18.1 What is Smart Agriculture?

Smart Agriculture is an innovative synchronized new path for reaping the benefits: (i) from the accurate timely information generated vertically in atmosphere using higher end tools through a regulated to perfect blend of consortium of international partners, (ii) for use along with agricultural technologies developed utilizing biological information tools as well as indigenous knowledge to mobilize a peak input efficiency horizontally through precise application without harming environment and to obtain higher benefit: cost ratio (iii) directing improvement and conservation of natural resources based on medium term scale and (iv) integrating with apatio-temporal "adaptive planning" for meeting a wide range of present and future threats, including pollution, climate change and any other pressures for long term stability and gains on a land scope scale, with four pronged simultaneous execution of all the above fronts, for rising socio-economic status to obtain a more peaceful co-existence of human beings on the mother earth. The USA, Canada and other developed countries are already getting the benefits of Smart Agriculture (Vittal, 2015).

18.2 What is Meant by System Approach?

In Extension Education, "System Approach" includes training strategy for development of skills. Broadly speaking, it may be defined as multipronged strategy or integrated use of interwoven variables, teaching tools and techniques in making trainings dynamic and promoting. Six steps which are involved in appropriate training strategy are as follows:

- Analysis of intended participants in terms of needs and interests, educational level, experience, cultural background and socio-economic status is the first and the foremost step to take into account.
- Principles of system approach, means taking all the interacting variables of system and subsystems i.e. trainer, trainee, project activity, organization's obligations and institution's expectations for consideration in an integrated way for planning and implementation of training programmes.
- Objectives should be well defined.
- Analysis of task in terms of What, How, When, Where and By Whom the work is done.

- Subject matter planning, deciding of course contents and sequencing of course topics are also important to consider.
- Design of learning, deciding about methods, selecting of media and materials, techniques of evaluation or monitoring should also be taken into account.

18.3 Guidelines for Trainings

The below mentioned guidelines must be strictly taken into account while framing the trainings for development of sustainable agriculture.

- As trainings have a vast potential in the development of skills and utilization of farm technologies for increasing agricultural production, so trainings of extension personnels and peasants must be given due weightage in all the rural development programmes.
- Extension trainings are required to be result oriented and for that matter these should aim at providing proper learning experience beneficial for solving new problems. Each learning experience must have a practical experience and should be within the ability of the trainees. Trainings should be of such nature, which have a vast practical application and to fulfill the satisfaction of the trainees. The trainers must remain within the ability of the farmers who have to be trained. Provide full experiences desired by the intended participants.

In order to achieve the above target, an appropriate training strategy is required to be worked out taking into account the following criteria:

- Analysis of the intended participants in respect of their training needs is of paramount importance.
- Principles of system approach with definite objectives vis-a-vis planning of subject-matter (as mentioned under guidelines for training) should be employed.
- Depending upon the potential of the area, farmers should be trained in an innovative market oriented cash crops like saffron cultivation in Pampore, Budgam and Kishtwar areas of Jammu and Kashmir. Similarly, peanut technology is required to be the deliberated among the farmers of Himachal and hilly areas of Jammu and Kashmir. Trainings with regard to technologies for growing of aromatic and medicinal plants as well as techniques for growing of local fruits like phalsa, jamun, karonda, girna etc are the needs of the farmers dwelling in the *Kandi* belt of Jammu (Jammu and Kashmir), Punjab, Haryana and Uttarakhand. Ladakh region of Jammu and Kashmir State now union territory, an abode of rare

medicinal plants is now mostly devoid of its wealth (Gupta, 2012c). Despite this, no well-defined parameters still exist for the threatening status assessment of the medicinal plants found in Ladakh region. Some of these threatened medicinal plants are: Buma (*Aconitum violoceum, Aconitum heterophyllum),* Demok *(Amebia euchroma),* Khamchu (*Artemisia brevifolio),* Lacchu (*Rheum webbianum, Rheum speciformae),* kuth (*Saussurea lappa)* etc. hence special training programmes are required to be framed to save these medicinal plants for the coming generations. For this purposs a latest book on medicinal plants written by Gupta et al (2016) should be consulted.

18.4 Significance or Importance of Trainings

Organizing of training programmes of various disciplines in agriculture and allied subjects are of much importance. It is because they not only enhance the competence of extension personnels but also improve the skills of the farmers for the success of rural development programmes like Integrated Rural Development Programme, Broad Based Agriculture (Trainings and Visit System) and Prime Minister's Development Programme. Infact, the need for training was realized with the starting of Community Development Programme during 1952 and since then a large number of trainings pertaining to agriculture including animal husbandry/sheep husbandry have been organized under National Demonstration Programme, Laboratory to Land Programme, etc. Frontline demonstrations are still being organized for updating the knowledge of the farmers and improving their skills as well as to improve the morals of the staff and finally the efficiency of production and productivity through Krishi Vigyan Kendras.

In order to make the extension personnels more effective for delivering the real goods under different rural development programmes, they are required to undergo pre-jobs and on-job trainings. The pre-job trainings are mostly for formal education while on-job trainings are usually of orientation and apprenticeship type of trainings as well as refresher and advance trainings. It becomes the duty of the master trainer to create a desire to congenial training atmosphere. The farmers are given general trainings in crop production including horticulture/ olericulture production or specialized trainings in mushroom growing, poultry and dairy farming, bee-keeping or apiculture and sericulture, farm machinery etc.

These days trainings are also imparted in cultivation of flowers and medicinal plants (Gupta et. al. 2016 and Gupta, 2019) to enhance income of the farmers

Briefly, it is summarized or concluded that training is akin to education in its intention to bring changes in behavior of the participants. Systematic training demands efforts together data on the participants, their source of learning, interests, preferences and reaction about the past training experience. Trainings to be effective must suit the convenience of participants and stimulate learning. It is further added that nowadays training in agriculture is assuming much importance. It is because:

- Farmers, extension personnels and scientists constitute the most precious human resource for speedier agricultural development (Chahal et al., 1999).
- Although some of our villages may be rich in other resources yet unless the people engaged in agriculture are first developed, they will not be able to develop and harness other resources of the village and contribute their to its growth and development.
- Many training programmes remain ineffective due to lack of activity, variety and direction even when intentions are sincere.

18.5 Points/Tips to Make Training Programme More Effective

The following points or tips, which are basic to the every training, can make a training programme more effective:

18.5.1 Identifying Trainings of Trainees

The training programmes based on the accurate training needs of the trainees are not only always sound but also cost effective. Hence, identification of training needs requires to be followed strictly.

18.5.2 Keep Moderate Level of Course Contents

In the training programmes we should not try to cover every thing about a given subject, but its solid points are required to be discussed. This will not only enable the trainees to grasp the subject more vividly and clearly but the trainers also have lot of time to design activities more effectively which have to introduce, present and reflect upon before the trainees.

18.5.3 Balance between Attitude, Knowledge and Skill Learning

The basic objective of every training is to foster the attitude to promote understanding of concepts and models, and finally to develop practising skills. Thus, while imparting trainings to the farmers or other trainees all the aforesaid areas must be focused and should not be excluded any one of them from schedule of training programme.

18.5.4 Creation of Effective Learning Environment

As the communication learning is a two-way process so it demands openness on the part of both the trainers and the trainees. A trainer who thinks that he knows each and everything will not be able to train the trainees effectively. A relaxed calm atmosphere always contributes to effective trainer-trainee relationship. Similarly, credibility of the trainer has a positive and indelible impact on the trainees. Gentle repetition, strategic reminders and informality have always been proved helpful in creating effective learning environment during training.

18.5.5 Use Variety of Training Methods

As adults learn in different ways, so use of various methods of trainings keep their interest alive. Time allocation, group formats and the physical settings should also be varied to heighten the training experience. The use of lecture method without slides should always be discouraged.

18.5.6 Encourage Learning by Sharing and Doing

All the trainees must be permitted to participate actively in all the sessions as it will lead to a learning by "sharing and doing". By "sharing and doing", the trainees feel free to express themselves and become sensitive to each other's views and finally they will serve a active partner with the trainer. The "sharing and doing" can be achieved using a number of training methods like business games, brain storming, buzz groups and various other forms of group discussions, assignments and exercises.

18.5.7 Utilization of Trainee's Expertise

Sometimes, the trainees themselves have lot of expertise and experience and, therefore, such experiences must be utilized into trainer's training design. This would enable to learn from each other's experience.

18.5.8 Start the Process Well in Advance for Selecting Trainees and Trainers

Training process requires to be started well in advance by selecting needy and highly motivated trainees. The selection of the right trainers is also equally important. The trainers must have good knowledge of communication technologies and training technology. Hence, make a plan ahead and select those trainers who are well versed with latest agricultural technology and pedagogy.

18.5.9 Problem Solving Trainings

Sometimes application of learned attitude, knowledge and skill is not just something to solve real life problem after acquiring training. Hence, it must be a major focus during training. For this purpose, opportunities must be provided to the trainees so that they can utilize course content to address and help in solving actual problems which they are commonly experiencing in the farmers fields.

18.5.10 Evaluation of Training Outcome

Evaluation of training outcomes i.e. knowledge and skill gained and attitude developed in trainees, is essential because it helps to assess as to how far the objectives of the training programme have been fulfilled.

18.5.11New Models of Agricultural Extension

Now we have multiple models of agricultural extension in operations', public sector, private sector, NGOs led programmes, bank led models, etc. we have more than 600 Krishi Vigyan Kendras (KVKs), Krishi Darshan Programmes in Door Darshan Programmes. We have Kisan Call Centre's across India. But each one of them has their sets of problems (kumar, 2009).

Agriculture Technology Management Agency (ATMA) appears to be the new hope amongst the public sector as it is demand led model. ATMA is a society under the chairmanship of the District Collector with participation of the Line Departments, Commodity Interest Groups. Farmers, NGO's etc. ATMA is now present in all the districts and is expected to play role in the Twelfth Five Year Plan and Thirteenth Five year Plan.

18.5.12 Timely Follow-up

The timely follow-up has distinct advantages. Firstly, it serves as a feedback to the trainees, which in turn helps to improve upon the training in future. Secondly, guidance can be given to the trainees by the trainer to tackle the specific problems and difficulties, which the trainees are facing in their working situations.

18.6 Training Needs

Various training programmes for the farmers in different areas must be based upon their needs. This calls for identification of farmer's training needs which serve as a challenge and constitutes a prime task before an extension worker in planning the training courses.

Training needs may vary from place to place. As for example, about 91 per cent of the peasants of hilly areas of Sirmour in Himachal Pradesh were reported to require trainings in crop husbandry (Gupta and Kanwar, 1986 Mahajan and Gupta, 2009). It is due to the fact that their livelihood depends mainly on growing of crops. Soil and moisture conservation was the second important training need of these farmers (Table 18.1). Soil and water conservation has also been found to be the main training needs of the peasants dwelling in the sub-mountainous areas of Jammu (Jammu and Kashmir) Siwaliks as well as other similar areas of Himachal Pradesh, Punjab and Haryana (Arora and Gupta, 2019). The next relative training need was animal husbandry, followed by horticulture, olericulture, home science, poultry farming and bee-keeping.

Table 18.1: Frequency of farmers responding to different types of training needs for hilly areas in Sirmcur (Adapted from Gupta and Kanwar [1986))

Sr.No.	Training needs	Frequency	Percentage
1.	Crop Husbandry	127	90.71
2.	Soil and Moisture Conservation	94	67.14
3	Animal Husbandry	69	49.29
4.	Horticulture	55	33.29
5.	Olericulture	35	25.00
6.	Home Science	31	22.14
7.	Poultry Farming	16	11.43
8.	Bee-keeping	9	6.43

According to Gupta (2004b) for the success of any training programme, it must be based on the needs of the people, who have to be trained. He has suggested that the "participatory rural appraisal" should be used to realistically identify the training needs of the people. Besides the training need as perceived by field functionaries at different levels in Jammu Province ascertained through discussion and thrust areas for training so identified in agriculture, horticulture, animal husbandry and sericulture are as follows:

18.6.1 Agriculture - Crop Production

- Maize production technology in hilly areas.
- Wheat production technology under water stress conditions.
- Pulse production technology.
- Integrated Pest Management for crop production.
- Crop production under rain-fed conditions.
- Paddy production technology.
- Wheat production technology.
- Fodder production technology around the year.

18.6.2 Horticulture/Olericulture

- Post-harvest technology for fruits.
- Fruit production technology of indigenous fruits like jamun (*Syzygium cumini*), phalsa (*Grewia asiatica*), amla (*Phyllanthus emblica*), girna (*Carissa spinarum*), karonda (*Carissa carandes*) and lassora (*Cardia dichotoma*) and their medicinal values (Gupta et al., 2016) is one of the training programme which is reguired to give more emphasis or stress.
- Vegetable production technology.
- Nursery propagation technology.

18.6.3 Animal Husbandry

- Technology for increasing milk production.
- Health and disease care in animals.

18.6.4 Sericulture

- Propagation of mulberry.
- Selection of cocoons for breeding operations.
- Rearing of silkworm.

Apart from the above, grass and fodder production technology, agro-forestry, cultivation of ber (*Zizyphus mauritiana*) and use of fertilizers are the other areas of training needs, especially for the women of Uttar Pradesh (Singh et al., 2000) or technology blending (Kumari, 2000) i.e. improved practices versus farmer's practices.

18.7 Knowledge Requirement for Farmers for Sustainable Agriculture

Looking into our past and future requirements of knowledge and skills of farming community, the Government sponsored extension machinery will have to play a vital role and equip itself with various new skills in developing farmers as agri-entrepreneurs. Knowledge requirement for farmers in this regard is as follows:

18.7.1 Identification of Sustainable Crop Varieties

Sustainable crop varieties suitable for growing in particular agro-climatic zone based on the productivity of land and climatic conditions must be identified.

18.7.2 Weather Conditions

Interpretation of weather conditions suitable for particular crop is of very important to sustain the yield of a crop, especially under the changed climate due to Global Warming. The change in climate is not only going to set back development and food production but around 10 million people world wide have also been driven out of their houses by falling rains, incre asing desertification, rising sea level and other climatic factors (Gupta, 2013).

18.7.3 Skill and Knowledge Requirement

Various kinds of skills and knowledge in assessing the input and their management like use of seeds, pesticides, systems of irrigation, agricultural implements etc. are required to be advocated to the farming community.

18.7.4 Skill Requirement

Technological skills related to production, business and marketing skills in analysis of cost and benefit as well as skills for value addition to agriculture produce should be imparted to the farmers.

18.7.5 Diversified Farming

As these days, the yield of rice and wheat has either become stagnated or decreased, so the need of the hour is to grow pulses, oil seed crops, vegetables and fruits, flowers and medicinal plants. Growing of such crops will not only bring higher revenue for the peasants but also serves as cushion against crop failure and provide more balanced nutritious food. This is known as diversification of crops. The detailed accounts of diversified farming has recently been highlighted (Gupta, 2019, Gupta, 2009b; Sharma et al., 2009). Diversified farming has proved beneficial to the changed climatic situation especially the Kandi belt of Jammu, Himachal Pradesh, Uttarakhand, Haryana and Punjab. Thus, the farmers of these areas are required to get well familiarized with the latest technologies with regard to diversified farming. For this purpose, training programmes have to be arranged from time to time.

After a period of prosperity, the Green Revolution today either stands for stagnation or in decreasing trend and environmental degradation (Kang, 2013). To intensify and diversify agriculture is the answer and an requires new policy, technology and markets.

Agricultural diversification is traditionally associated with subsistence farming and is an extremely important aspect of food security at the household level. It has been generally agreed upon that agricultural diversification helps humans to obtain food security and improved, healthful foods (Kang, 2013). Apart from this, it helps to create jobs in rural areas. It has favourable impact on soil fertility and enhances the capacity of the land to produce more, besides controlling destructive crops pests. Agricultural diversification is an important mechanism for economic development, especially for the rural areas as reported by Gupta (2019).

In Punjab, agricultural diversification has been receiving lip service for the past nearly one decade All this while the minimum support price for paddy kept rising, and new tube wells started getting installed and sunk deeper. It was announced by the Punjab government that two alternatives viz., agroforestry and dairy will be promoted (Anonymous, 2013c) as well as growing of medicinal plants (Gupta et al. 2016; Kumar and Namgyal, 2019).

In order to achieve sustainable agricultural development to meet the future challenges, the farmers are required to empower so that they have say in policy matters of Government in Agriculture Sector and support from the State-run extension services in the aforesaid points.

18.8 Conclusion

Hitherto, after introduction of Green Revolution in many Asian and Latin American countries, food production was increasing rapidly than population growth (Vandenban and Hawkins, 1996). However, nowadays, the yields, especially of rice and wheat in Asia particularly in South Asian countries are no longer increasing (Mahajan and Gupta, 2009). In India, their yields has either stagnated or declined. Food production per capita is also decreasing in many African countries. Previously much of their increase in food production occurred from an increased area under cultivation. But this is now no longer possible because much of the land has decreased its productivity due to soil erosion (Ramanathan, 2000) and/or other land degradation causes like unchecked commercial establishment, population explosion, salt affected and waterlogged soils etc. (Gupta, 2005c). It is worthwhile to mention that soil loss at global level through soil erosion is 26 billion tonnes (t) year^{-1} and out of this soil loss, India contributes about 12 billion t year^{-1} (Bangarwas, 2001). Not only the soil erosion, the stagnated and decreased yield of rice and wheat in India and other South Asian countries after introduction of Green Revolution is attributed to number of other problems like indiscriminate use of chemical fertilizers and pesticides (Gupta and Singh, 2006; Gupta and Singh, 2008). Hence, it becomes an imperative to build up such a sound extension approach which not only helps to render a farming vocation viable but also sustainable one.

In this chapter, therefore, agricultural production on "sustainable basis" through extension education, in terms of "Integrated Extension Approach" known as System Approach, has been highlighted.

18.9 Future Research Strategies/Priorities

- There should be enough opportunities for field staff to make use of communication media on different projects. Sufficient exhibition materials must be provided to the extension personnels for demonstrations done at the time of field days and village level training camps arranged for the farmers.

- More number of community training centers are required to be set-up for a cluster of villages. It is because the success of any agricultural development programme depends mainly upon the "preparation and participation" of the target population which is implemented through trainings.
- These days, due to change in climate, the farmers are required to be made aware of the importance of the diversified farming on scientific lines. Diversified farming will not only meet their diversified needs but will also help to safeguard against erratic behavior of weather.
- In order to modernize the agriculture in hilly areas of the country like Jammu and Kashmir, Himachal Pradesh, Uttarakhand and North East Hilly States, the strategies employed should focus on land resource development, rain-fed farming ecosystem and diversified agriculture.
- Land resource development requires soil and water conservation on watershed basis. Drainage of water-logged soils is absolutely necessary.
- All kinds of soil and water conservation measures viz. agronomical practices, engineering and biological control measures should be adopted to control the soil erosion in eroded and denuded areas.
- To impart trainings to the farming communities with regard to the impact of the modern agriculture, especially of "Green Revolution" in terms of the effects of indiscriminate use of chemical fertilizers and pesticides on environment - land, air, water, animals and plants - is the need of the day.
- Organize meetings to finalize the training courses where extension personnels (agents), teachers and researchers can discuss research findings and experiences with farmers.
- A robust extension system requires being involved in terms of technology development through "on-farm trials" and "adaptive research experiments" to be more appropriate and viable. After conducting at least three years research on these lines, the so developed technologies must be transferred to the farmer fields to increase production of agricultural crops (Gupta, 2008).
- Since nowadays, in field of agriculture like other sectors, the progressive use of new scientific and technological equipments has changed the spectrum of its functioning, so quick dissemination of technological information from agricultural research system to the end users and vice-

versa is one of the critical inputs to increase the farm productivity and production on sustainable basis.

- The potential areas for increasing the women's participations in agricultural development should be studied. Some of the key areas have been suggested in Table 18.2 for women's involvement and participation and inter-institutional support and linkages that would be needed to activate the process (Lakshmi et al., 2000).

  Some latest ICT (The Information and Communication Technolog) is paying a key role in agricultu ral growth and development in the country by providing timely and useful information in a demand driven mode (Bharti et al. 2019) so it must be followed urgently.

Table 18.2: Key areas for women's participation (Adapted from Lakshmi et al. [2000])

Practice	Pre-knowledge Status	Post-knowledge Status
A. Agriculture		
	Inadequate	Adequate
Crop rotation and multiple cropping	- do -	- do -
High yielding crop varieties of rice, wheat, sunflower, urd, mung, etc.	- do -	- do -
Early and late varieties	- do -	- do -
Soil testing	- do -	- do -
Soil sample collection	- do -	- do -
Correct sowing treatment	- do -	- do -
Irrigation schedules	- do -	- do -
Seed treatment	- do -	- do -
Use of *Rhizobium* culture	- do -	- do -
Use of Azotobacter	- do -	- do -
Weedicide application	- do -	- do -
Insecticide application	- do -	- do -
Plant protection measures	- do -	- do -
Green manuring	- do -	- do -
Spraying technique	- do -	- do -
Use and dosage of fertilizers as Urea and DAP	- do -	- do -
Use and dosage as Zn and MOP	- do -	- do -
Scientific grain storage	- do -	- do -
Agro forestry	- do -	- do-
Kitchen gardening	- do -	- do -
B. Animal Husbandry		
	Inadequate	Adequate
Improved breeds	- do-	- do -
Care of animals	- do -	- do -
Preparation of balanced feed	- do -	- do -
Correct of animals	- do -	- do -

Contd.

Preparation of balanced feed	- do -	-do -
Correct milking method	-do -	- do -
Control of parasites and common diseases	- do -	- do -
Feeding green fodder	- do -	-do -

Source: Lakshmi et al. 2000

- There is need to use *new mantra* laying emphasis on "Teaching by doing" and "Learning by doing" has been felt more effective by the proverb, "Tell me, I will forget" but when you "Show me, I will always remember". This has proved useful in teaching the farmers the latest agricultural technology by the Scientists of Krishi Vigyan Kendras and eventually its use in sustained crop production.
- Presently, there is about 2.5 lakh (one-fourth of a million) extension personnels in our country to serve about 130 million families (Kumar, 2009). Thus, the ratio of extension personnel's to farm families works out to be 1:520, which is very large gap. Hence, there is an urgent need to recruit more extension personnels in each and every state of the country.
- Wherever it is possible, mobile based Agro-Advisory System should be used. It is the system where- under the farmers become able to consult experts to improve farm produce and profits by sitting in his field in any part of the country. It is just as patients can take customized advice from a doctor using telemedicine's services.
- As sustainable development of agriculture is an essential ingredient for achievement of prosperity of the farmers and the nation, so it becomes imperative to make familiarize the peasants with the good quality of seeds and other inputs. For this purpose, the *Kisan Melas* must be organized by way of displaying the seeds and other inputs as exhibits.
- Diversification of agriculture is one of the ways to stabilize the economy and growth of the farming community. In this regard, the paramount need of the hour is to educate the farmers about the new technologies in respect of diversified agriculture, motivate them to adopt and secure their effective participation in the training programmes organized in the various fields of agriculture from time to time.

Study Questions

1. How we can define system approach in agricultural extension education?
2. What are the main guidelines for trainings in extension education?
3. What are the main points which make the training programme more effective?

4. Write a paragraph about the importance of trainings in extension.
5. Write a note on Smart Agriculture.
6. Write the main criteria upon which the training programmes depend or vary.

References

Arora, Sanjay and Gupta, R.D. (2019). Soil and Water resources conservation and management for sustainable for rural livelihoods in foot hills of north west Himalayans India. Souvenir. A joint International Conference of ISCO and SCSI Organised by SCSI, New Delhi, India

Anonymous. (2013c). Punjab: The leader of agriculture sector. Agriculture Today 16 (1): 40-43.

Bharati, V.K., Bhan, S., Meetalp, D. and Bhart, S. 2019. e-Resources for effective management of agricultural knowledge. Journal of Soil and Water Conservation. 18(1):85-92

Bangarwas, K.S. (2001). Stop the march of deserts. Intensive Agriculture 39 (5-6): 15-16.

Chahal, V.P., Lata, V. and Waldia, R.S. (1999). Making training effective. Agricultural Extension Review 11 (5): 13-15.

Gupta, M.P. (2004a). New strategy for agriculture development. The Kashmir Times 61 (328): 11 [November 23, Tuesday].

Gupta, M.P. (2004b). Trainings for development. Director Extension Education, SKUAST-J, Directorate of Extension Education, Railway Head, Jammu (J&K State), India .

Gupta, M.P. (2008). Green revolution, 2nd. The Kashmir Times 65 (275): 13 [October 2, Thursday].

Gupta, R.D. (2005c). Environmental Degradation of Jammu and Kashmir Himalayas and Their Control. Associated Publishing Company, Karol Bagh, New Delhi (India).

Gupta, R.D. (2009b). Diversified agriculture in Kandi belts. The Daily Excelsior 45 (320): 6 [November 18, Wednesday].

Gupta, R.D. (2012c). Needed a policy for medicinal plants. The Daily Excelsior 48 (134): 6 [May 15, Tuesday].

Gupta, R.D. and Kanwar, B.S. (1986). Constraints and training needs of farmers in a hilly tract of Sirmour District of Himachal Pradesh Cooperative Bulletin 29 (9): 3-7.

Gupta, R.D. and Singh, H. (2006). Indiscriminate use of fertilizers poses health hazards. Farmers' Forum 6 (9): 20-24

Gupta, R.D. and Singh, H. (2008). Indiscriminate use of pesticides in agriculture: Public health issues and their control Indian Farmers' Digest 41 (1): 8-13.

Gupta, R.D. Gupta, S.K. and Bhardwaj, S. D. (2016). Agrotechniques and Uses of Medicinal Plants. Associated Publishign Company A Division of Astral International Pvt. Ltd. New Delhi - 110002, India.

Gupta, R.D. 2013. Climate change and rendening refuges. The State Times. 18(185):6.

Gupta, R.D. 2019, Diversification for enhancing income sucess story form J&K. Soil and Water Conservatin Today. 14(2)3-4

Kang, M.S. (2013). Achieving crop diversification in Punjab. The Tribune 133 (62): 9 [March 4, Monday].

Kumar, P and Namygyal, D. 2019. Medicinal diversity in flora of cold arid Ladakh. The State Times. 24(209):6.

Kumar, S. (2009). Agricultural technology transfer. Agrobios Newsletter 7 (9): 57.

Kumari, K. (2000). Technology blending. Agricultural Extension Review 12 (1): 26-28.

Lakshmi, K.S., Ravikumar, K.N. and Rani, K.U. (2000). Women in agriculture - for a bright future. Agricultural Extension Review 12 (2): 14- 18.

Mahajan, A. and Gupta, R. D. (2G09). Integrated nutrient management (INM) in a sustainable rice-wheat cropping system (pp.1-267) Springer Science + Business Media (www. Springer.com).

Ramanathan, K.M. (2000).-New approaches in wasteland development and watershed management, J. Indian Soc. Soil Sci. 48 (4): 697- 703.

Sharma, A.K., Gupta, R.D. and Rai, P.K. (2009). Crop diversification is the need of the hour for conserving soil and water in Kandi belt of Siwaliks (pp. 69-71). In Soil and Water Conservation Training Manual (Eds. B.S. Hansra, P. Vijaykumar, R. Nanda and P.K Rai) organized by School of Agriculture, IGNOU, New Delhi and KVK, SKUAST of Jammu, R.S. Pura Jammu (Jammu and Kashmir State), India.

Singh, A., Kumar, B. and Kumar, A. (2000). Strategy for training of farm women. Agricultural Extension Review 12 (1): 3-6.

Sood, S.K. 2019. Agroforestry for boosting farmers income. The State Time. 24(210):6.

Tripathi, B.D.(2000). Human resource development of agricultural extension personnel in new millennium: A prospective view. Agricultural Extension Review 12 (2): 3-13.

Vandenban, A.W. and Hawkins, H.S. (1996). Agricultural Extension. Second edition, Blackwell Science Ltd. Editorial Office: Osney Meac, Oxford OX20EL 25 John Street, London WCIN 2BL.

Vittal, K.P.R. (2015). Smart agriculture for developing countries. Agriculture Today 18 (3): 54-56.

Appendix A

List of Indian Magazines Related to this Book

Name of the Magazines	Periodicals	Publishers
Agriculture Today	Monthly	Centre for Agriculture and Rural Development 502, Rohit House Telstoy Road, New Delhi
Agrobios Newsletter	Monthly	Agrobios (India), Behind Nasrani Cinema, Chopasani Road, Jodhpur
Command Areas Development Subzar	Quarterly	Command Areas Development, Talab Tilo, Jammu, J&K
Co-operative Bulletin	Monthly	General Secretary, J&K Co-operative Union Vir Marg, Jammu, J&K
Down to Earth	Monthly	Centre for Science and Environment, New Delhi
Dharti	Annual	Soil Conseration Training School, Miran Sahib, Jammu, J&K
Environment and People	Monthly	Society for Environment and Education, Hyderabad-500 073
Employment News	Weekly	General Manager and Chief Editor, East Block IV, Level-7, R.K. Puram, New Delhi - 110 066
Farmers' Forum	Monthly	Bharat Krishak Samaj, New Delhi
Farmers and Parliament	Monthly	Executive Committee of the Farmers Parliamentary Forum, New Delhi
Farm Progress	Monthly	Associate Publisher, Madras Ltd., 856 Annasaloi, 6002
Forest Newsletter	Quartely	Conservation of Forest, 73 Bhagwati Nagar, Jammu
Gram Vikas Jyoti	Quarterly	Council for Development of Rural Areas, 2, Vigyan Lok, Vikas Marg Extension, Delhi
Indian Farmers' Digest	Monthly	Director, Communication Centre, G.B. Pant University of Agriculture & Technology, Pantnagar
Indian Farming	Monthly	Indian Council of Agricultural Research, Krishi Anusandhan Bhavan, New Delhi
Intensive Agriculture	Monthly	Krishi Vistar Bhavan, Directorate of Extension, Ministry of Agriculture, IASRI Campus, Pusa, New Delhi
Kurukshetra	Monthly	Nirman Bhavan, Ministry of Rural Development, New Delhi
Progressive Farming	Monthly	PAU, Ludhiana
Yojana	Monthly	Ministry of Information and Broadcasting, Yojana Bhavan, Sansad Marg, New Delhi

Appendix B

List of Indian Newspapers Related to this Book

Name of the Newspaper	Periodicals	Publishers
The Tribune	Daily	Printed and published (Jammu and Kashmir Edition) by Arun Joshi for the Tribune Trust at Danik Jagran Printing Press, Bari Brahmna, Samba, J&K
Tne Daily Excelsior	Daily	Printed and published by Kamal Rohmetra and owned by Daily Excelsior Executive Editor Neeraj Rohmetra, Printed at Daily Excelsior and Excelsior Printer Pvt. Ltd., Janipur, Jammu - 180 007, J&K
The State Times	Daily	Printed and published at State Times publications, Plot No. 110, 111. 112, Phase-ll, Gangyal, Jammu, J&K
The Kashmir Times	Daily	Printed and published by Pramodh Jamwal, for and on behalf of Kashmir Times Publication, Printed at KT Press Pvt. Ltd. at Diglana and published from Residency Road Jammu, J&K
Employment News	Weekly	Printed and published by General Manager cum Chief Editor East block Iv, Level-7, New Delhi-110 066

Appendix C

Nutrient Composition in Various Fertilizer Materials

Sr.No.	Fertilizer	Nutrient content (%)		
		N	P_2O_5	K_2O
1.	Nitrate fertilizers			
	I. Sodium nitrate	16.0		
	II. Calcium nitrate	15.5	-	*
2.	Ammonium fertilizers			
	I. Ammonium sulphate			
	II. Ammonium phosphate	2020 or 16	20	-
	III. Ammonium chloride	24-26		.
	IV. Anhydrous Ammonia	82		.
	V. Ammonium solution	20-25	-	
3.	Nitrate and ammonium fertilizers			
	I. Ammonium nitrate	33-34		
	II. Calcium ammonium nitrate	26	.	.
	III. Ammonium sulphate nitrate	26	-	-
4.	Amide fertilizers			
	I. Urea	46		
	II. Calcium cynamide	21		-
5.	Phosphatic fertilizers containing water soluble phosphoric acid or monocalcium phosphate			
	I. Single superphosphate		16-18	
	II. Double superphosphate	-	32	
	III. Triple superphosphate	-	46-48	
	IV. Ammonium phosphate		20	16
6.	Phosphatic fertilizers containing citric soluble phosphoric acid or dicalcium phosphate			
	I. Basic slag		14-18	
	II. Dicalcium phosphate	-	34-3S	-
	III. Rhenania phosphate	-	23-26	
7.	Phosphatic fertilizers containing phosphorus which is not soluble in water or citric acid or tricalcium phosphate			
	I. Rock phosphate		20-40	
	II. Raw bonemeal	3-4	20-25	.
	III. Steamed bonemeal		22	
8.	Potassic fertilizers			
	I. Potassium chloride			60
	II. Potassium sulphate	-	-	48-52
	III. Potassium nitrate	13	-	44
	IV. Sulphate of potash-magnesia	-	-	25-30

Appendix D

Nutrients, Their Functions, Deficiency and Control of Deficiency

Plant nutrients	Function of nutrients	Deficiency symptoms	Control of deficiency
Nitrogen (N)	• N-being building blocks of aminoacids are essential in the formation of proteins. • N is also an integral part of chlorophyll which is the primary absorber of light energy needed for photosynthesis. • It imparts dark green colour to plants. • When N is used properly in conjunction with other needed soil fertility inputs especially P. it can speed up the maturity of various crops.	• N deficiency is characterized by yellowish and pale green colour of older leaves, but necrotic and chlorotic spots appear at an early stage. • N deficiency results in the collapse of chloroplasts and also in a disturbance of chloroplast development. • The plants remain small, stems have a spindly appearance, and leaves become small and premature which later on begin to fall.	• Use of nitrogenous fertilizers such as urea, CAN, ammonium phosphate etc. as soil application. • Foliar spray of urea. It can be done in the morning and evening hours on the fully grown up leaves of various crops. Urea can be used up to 3 per cent. 7.5 kg of urea in 250 litres of water is sufficient for a hectare of crop.
Phosphorous (P)	• The unique function of P in the metabolism is the function of adenosine mono, di and triphosphates (AMP, ADP and ATP) bond which allow energy transfer. • Since P is readily mobile in plants, it can be translocated both in upward and downward direction. • Phytic acid are organic P compounds which occurs in seed and regarded as P reserve. It is also an integral part of phospholipids. • In plants, P increased root growth and early maturity of crops.	• Deficiency appears on the older leaves which are characterized by purple/ violet or darkish green colour. • P deficiency accumulates low molecular weight of N compound. • There is reduction of RNA synthesis owing to deficiency of P in soils. • Maturity of the crops gets delayed. • There is poor seed formation which eventually results in poor growth rate. • In cereals, tillering is affected and in fruit trees, growth of new shoots is reduced	• Application of phosphatic fertilizer in the soil, e.g., super phosphate or diammonium phosphate (DAP).

Contd.

	• It makes plants more tolerant to drought, cold, insects and diseases attack	due to its deficiency.	
Potassium (K)	• K has high active uptake rate among all the essential mineral cation species. K is the only nutrient which can be transported against electrochemical gradients into plant cells. • K in plants is readily mobile, so its main transport action is towards the meristematic tissues. Often K from older plant organs is redistributed to younger tissues. • K enhances the translocation of assimilates. It also promotes rate of CO_2 fixation. This fact is probably via ATP synthesis. • The main function of K is its activation of various enzyme system and over 80 plant enzymes require K for their activation. • Activator of enzymes involved in protein and carbohydrate metabolism, stomatal opening, membrane permeability, and pH control.	• K deficiency does not result immediately in visible symptoms. At first, there is only a reduction in growth rate which is called as *Hidden Hunger* and later on necrosis and chlorosis occur. • Symptoms generally appear on the older leaves. • Its deficiency results in collapsing of chloroplast and mitochondria. • There is decrease in turgour pressure and plants become flaccid. • There is poor formation of xylem and phloem tissues due to its deficiency in soil. • Lignification of vesicular bundles is impaired. • K deficiency leads to synthesis of toxic amine such as Putrescine and Agnethine chemicals as well. • Due to deficiency of K, the plant becomes susceptible to lodging *vis-a-vis* insect and pest attack.	• Use of potassic fertilizer in the soil, e.g., muriate of potash (MOP).
Calcium (Ca)	• The most important function of Ca in plants is the formation of middle lamella layer of their cell wall in the form of calcium pectate. • It is required for cell elongation and cell division. • It plays an essential role in biological membranes.	• Its deficiency is characterized by the reduction in growth of meristematic tissues. • The deficiency can be first observed in the growing tips and youngest leaves. • These became deformed and chlorotic and at a more advanced stage, necrosis occurs at the leaf margins. • The affected tissues become soft due to	· Use of Ca carbonate or hydroxide in the soil.

Contd.

	• It is required for maintaining optimum membrane permeability. Ca is required in plants io detoxify the divalent metal cations. • It plays an important role in abscission and delays leaf senescence. • Keeps the soil neutral in reaction.	the dissolution of the cell walls. • In apple, Ca deficiency causes a disease called as bitter pit while in tomato and pepper, it causes blossom end rot. • A deficiency of Ca manifests itself in the failure of the terminal buds of plants and apical tips of roots to develop. • Inadequate supply of ca to celery causes black heart disease.	
Magnesium (Mg)	• Mg acts as a cofactor in almost all enzymes activating phosphorylation processes. It forms a bridge between the pyrophosphate structure of ATP or ADP and the enzyme molecule. Like N, Mg plays a key role as the centre and the only mineral of the chlorophyll molecule. So its deficiency in soils immediately affects the chlorophyll formation which ultimately will affect the photosynthesis phenomenon influencing, thereby, the crop growth. • Since the fundamental process of energy transfer occur in photosynthesis, glycolysis, tricarboxylic acid cycle and respiration, so Mg has importance throughout the plant metabolism. • Mg is required both in photophosphorylation as well as in phosphorylation reaction which otherwise limits the regeneration of ribulose diphosphate in the calvin cycle.	• Since Mg is mobile in plants, so deficiency always begins in the older leaves and then moves to the younger leaves. • Interveinal yellowing and chlorosis occurs and in extreme cases, the area becomes necrotic. The plants give a withered appearance. • Mg deficient leaves often fall prematurely. Individual leaves suffering from its deficiency, however, are stiff and brittle and the intercoastal veins are twisted. • Inadequately supply of Mg often delays of reproductive phase in plants. Its deficiency can inhibit CO_2 assimilation and protein synthesis.	• Foliar application of Mg sulphate.

Contd.

	• It is involved in nitrogen metabolism. • Mg with S increases oil content of many oilseed crops.		
Sulphur (S)	• S helps in chlorophyll formation and encourages vegetation growth. • It promotes nodule formation on roots of legumes. • It is an essential constituent of proteins and enzymes • S is one of the constituents of amino acids Cystine, cysteine and methionine are constituent of protein. Hence S promotes protein formation. • S is also needed in oil formation. It also constitutes allyle sulphide and vinyl sulphide found in onions and garlic. • Plant growth is stimulated and extra leaf yields of alfalfa, clover and tobacco is obtained. • Secures crops against deficiency diseases. • Promotes nodular growth of N-fixing bacteria.	• General deficiency symptoms are similar to N deficiency but persist even when N supply is adequate. • The whole leaf in plant has light green colour. • Chlorosis occurs in young leaves. • Stiffing of leaves is also common. • Flooding aggravates S deficiency by converting soluble sulphates to insoluble sulphides	• S deficiency can be corrected by application of S- containing fertilizers such as ammonium sulphate, super phosphate, potassium sulphate, ammonium sulphate, nitrate, S containing complex fertilizers and gypsum.
Iron (Fe)	• Fe plays an important role in synthesis of chlorophyll and carbohydrate synthesis. • It is essential for the synthesis of proteins present in the chloroplasts. • As a constituent of enzyme system, it regulates photosynthesis, respiration and reduction of nitrates and sulphates • Fe is a constituent of leghaemoglobin (like haemoglobin of blood) found in the nodules of leguminous plants.	• Young ieaves become chlorotic but principal veins remain green. • This has been noticed in groundnut grown caleareous soils. • In severe deficiency, leaves became pale white due to loss of chlorophyll. • Death of young growing tissues can occur and rossetting appearance can be seen.	• Spraying of 0.5 per cent ferrous sulphate on groundnut foliage.

Contd.

	• Fe is constituent of porphyrin protein complex such as cytochromes and catalase. It is present as a constituent of catalase and peroxidase enzyme. • It is involved in the biological nitrogen fixation mechanisms. In fruit trees, chlorosis is cured by the supply of Fe. Chlorophyll does not fully develop without it.		
Manganese (Mn)	• It is essential for the splitting of water molecule during photosynthesis i.e., *Hill reaction* • It helps in chlorophyll formation. It also helps a key role in nitrate assimilation, in the conversion of soluble nitrogen compounds to protein and in the splitting of peptide bonds. • It helps carbon assimilation and chlorophyll synthesis. • It acts as a catalyst in oxidation and reduction reactions within the plant tissues • A good supply of Mn sometimes helps in counteracting the bad effect of poor aeration • It prevents the disease known as grey speck on oats and sugar beet	• Chlorotic patches between the veins are first younger leaves where Mg deficiency is usually observed on older leaves. • Spots of death tissues scattered over young leaves are seen.	• Soil or foliar application of visible on Mn sulphate
Boron (B)	• It helps in uptake of Ca and efficient utilization of Ca by plants. • It tends to keep calcium soluble and increases its mobility in plants. • It is necessary for the translocation of sugars in plants.	• Its deficiency affects the growing points roots, shoots of and young leaves • Its deficiency leads to cracking and cork formation in stocks stems and fruits and shortened internodes. • Leaves become thicken and margins roll	• Foliar spray of boric acid or borax. • Use of boron in soil

Contd.

	• B can help to reduce the incidence of club root which is responsible for fungal diseases in brassica. • It promotes the growth of nodule bacteria in leguminous crops. • It saves the root crops from rot diseases • Indirectly, it affects the intake of other elements from the soil by the plants.	upward. • The leaf tip and margin of older leaves die prematurely.	
Zinc (Zn)	• As a constituent of enzyme systems, it regulates various metabolic reactions in the plant • It is associated with water uptake and water relations in the plant. • It helps in the formation of growth hormones like IAA in plants. Many of the enzymes active in hydrogen transfer are activated by Zn, and it is constituent of the protein which forms the enzyme carbonic anhydrase This enzyme catalyses the reaction of carbonic acid with water vice- versa. • It acts as a catalyst in chlorophyll formation. • Mottle leaf of citrus and little leaf of peaches are the peculiar deficiency diseases or disorders of Zn.	• Development of light green yellowish bleached spots in the leaves. • Chlorotic bands on either side of the midrib of the monocot leaves. • Poor tillering. Stunted growth. In maize, leaves show white patches which turn to bluish red. In wheat, white to yellow patches turn to brown. Ear formation and maturity are delayed. In paddy, light yellow spots appear which turn to deep brown commonly known as *'Khaira'* disease of rice.	• Soil application of Zn sulphate @ 25-50 kg ha^{-1}. • Foliar application of 5 kg Zn sulphate +25 kg lime ha^{-1} in 1000 litres of water.
Copper (Cu)	• Cu forms many compounds with amino acids and proteins in plants. • It acts as an electron carrier in enzymes which bring about oxidation-reduction reaction in plants. • It helps in utilization of Fe in chlorophyll	• Young leaves are permanently wilted without spotting. Moreover, tip of the leaves may turn white. • Yellowing and chlorosis of normally green leaves. • At maturity, panicles and pods are	• Foliar and soil application of Cu.

Contd.

	synthesis. • It helps in preventing dieback disease in citrus. • Cu has been shown to facilitate the movement of Ca in barley and wheat.	poorly filled, sometimes even empty if deficiency is severe	
Molybdenum (Mo)	• Mo helps in the function of atmospheric nitrogen - symbiotic as well as non- symbiotic. • It helps in protein synthesis. Actually it is essential in the reduction of NO_3 after its up taken i.e., conversion of NO_3 into amino acids, which are building blocks of proteins. Lack of Mo, growth of plants is limited by the inability to use NO_3; thus NO_3 may accumulate in the plant.	• Marginal scratching, rolling and cupping of young leaves are the main symptoms. Mo deficiency is due to N deficiency and its active role in biological N fixation.	• Soil and foliar application of sodium molybdate or ammonium molybdate
Chlorine (Cl)	• Cl is associated with the production of oxygen during photosynthesis for raising cell osmotic pressure and maintaining tissue hydration. Encourages growth in crop plants.	• Chlorosis in younger leaves and overall wilting due to possible effects on transpiration.	• Potassium chloride application in the soil.

Colour Plates

Chapter 7: Integrated Nutrient Management (INM) System and Sustainable Agriculture

Plate 7.2: Farmyard manure

Plate 7.3: Green manures incorporated in the field

Plate 7.4: Crop residues of rice placed in the field

Plate 7.5: Wheat residues incorporated in the field

Plate 7.6: Compost placed in the field

Chapter 8: Use of Bio-Fertilizers in Sustainable Agriculture

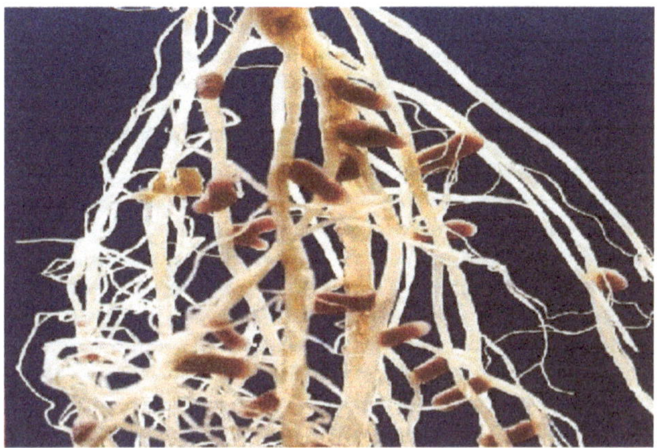

Plate 8.1: *Rhizobium*

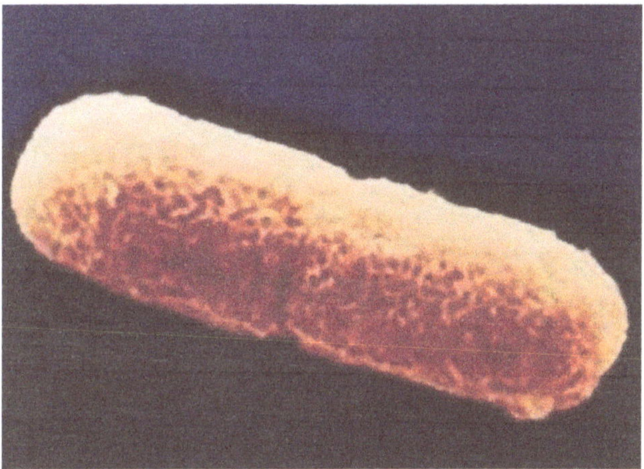

Plate 8.2: *Azotobacter*

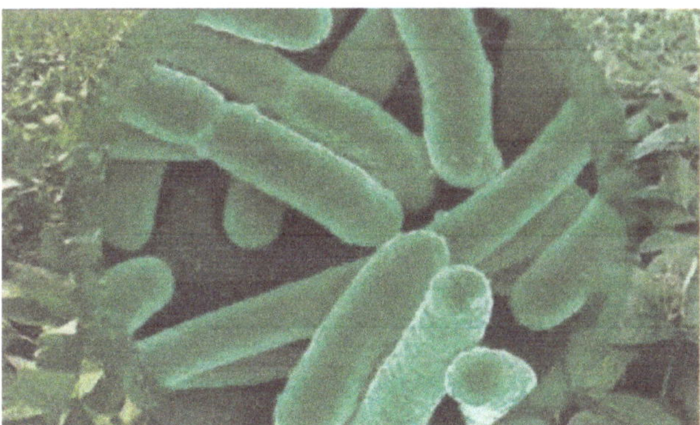

Plate 8.3: *Azospirillum*

Plate 8.4: *Azolla*

Plate 8.5: Blue green alge

Index

A

B

C

D

E

F

G

H

I

J

K

L

M

N

O

P

Phosphate solubilizing micro-organisms 142, 163, 175, 213, 224

Q

R

S

T

U

V

W

Z